PHYSIOLOGY OF INSECTS

PHYSIOLOGY OF INSECTS

By
MANJU YADAV
Lecturer
Department of Zoology
M.M.H. College
Ghaziabad (U.P.)
(India)

DISCOVERY PUBLISHING HOUSE
NEW DELHI-110002

First Published-2003

ISBN 81-7141-743-4

Published by

DISCOVERY PUBLISHING HOUSE
4831/24, Ansari Road, Prahlad Street,
Darya Ganj, New Delhi-110002 (India)
Phone: 23279245 • Fax: 91-11-23253475
E-mail:dphtemp@indiatimes.com

Printed at:

Tarun Offset Printers, Delhi-53

Preface

The fundamentals of Physiology of Insects are presented within the framework of scientific discovery. Researches in Entomology have been made almost increadible strides in the past few decades. Consequently, existing concepts of Insect biology have been expanded. These has been a revolution indeed in this direction.

The text integrates the descriptive, experimental and biochemical approaches into a conceptual framework. All important points are illustrated diagramatically. The title is not intended to be comprehensive nor could it be at length, but it concentrates as putting across the basic principles of the subject as briefly and lucidly as possible. It does this with the aid of careful selected examples–some recent and other classic of the field and with numerous illustrations. The aim is to enthuse the reader with this active and exciting area of research and to lay a solid foundation on which further study of its various facets may be based.

The author has freely consulted various articles, discussion notes, reviews and extracts from the various scientific journals in the preparation of the present book, in able to make it comprehensive and upto date.

Though every care has been taken by the printer, publisher and me, it is quite likely that some errors might have found their way into the book but I hope these are of very insignificant nature. However, sugestions to improve book and pointing out of errors and mistakes, if any, will gratefully be appreciated.

The author expresses grateful to her friends and colleagues whose constant inspiration have initiated her in bringing out this book.

Special thanks are given to Mr. Vasan and staff of M/s Discovery publishers for their whole hearted co-operation in the publication of this book.

Author

CONTENTS

1

FOOD REQUIREMENT

Food requirement of an organism is called the *nutrition.* These studies may include the constituents of the normal diet, or the studies may be expanded to include substances that produce energy in the most efficient way. In the basic terminology of the nutritionist, food constituents are classified as *essential* or *nonessential.* The first term designates those components that must be included in the diet because they cannot be synthesized by either the metabolic system of the animal or the normal or the normal compliment of symbionts. The second term indicates food materials that may have to be consumed to produce energy and that can be converted into a form in which they can be utilized through metabolic processes. Vitamins, amino acids, and certain mineral salts are common examples of essentials. Although carhohydrates, proteins, and fats are individually nonessential, the animal must eat at least one of these "nonessentials" to maintain life. Many animals have the ability, which may be born of necessity or may be the result of evolutionary development, to convert one type of food to another through metabolic processes. The nutritionist directs his investigations in various ways, depending upon the type of animal he studies.

Human nutritionists work on the diets that bring the best possible health to humans and study the effects of deficiencies and their correction. The classic example of this work was the introduction of vitaman C to prevent and cure scurvy in the British navy by feeding the sailors lime juice. Animal husbandrymen who study nutrition emphasize the importance of the ability of animals

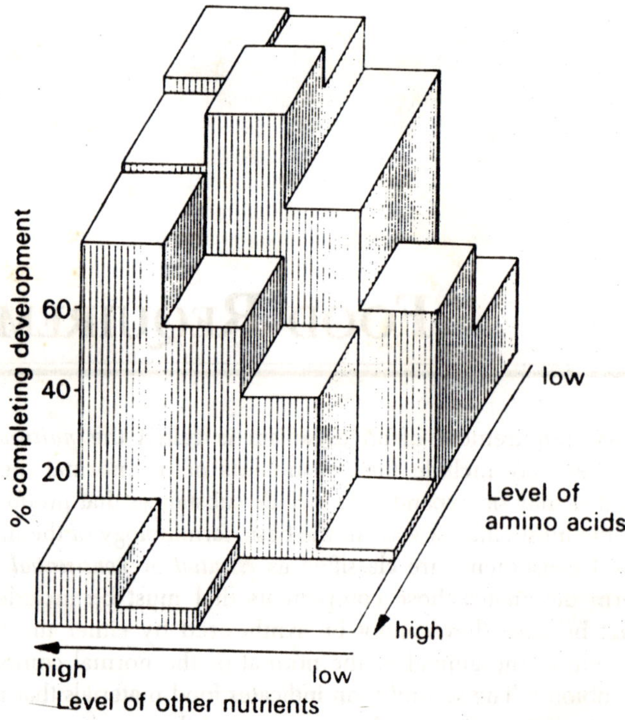

Fig. 1.1. Percentages of Agria larvae completing development in four days with amino acids present in different amounts and constituting different proportions of the diet.

to utilize foods efficiently to increase fecundity and develop larger animals. The conversion of carbohydrate to protein and the processes that control this conversion are of great economic importance. Insect nutritionists have worked in at least three directions. The first is the study of the natural food of insects. In a group so highly diverse as insects, it is expected and (experienced) that for almost everything comestible there is an insect that will eat it. Brues has discussed this in his *Insect Dietary*. The second direction is the study and search for food combinations that can maintain normal populations for laboratory study or produce large numbers of insects for sale as bird food, fish food, or fish bait. These food combinations include growing plants, dehydrated or extracted plant tissue, commercial feeds such as dog food, and combinations of ingredients that can be described by specific chemical analysis.

A large part of the research in insect nutrition falls into this category. The third direction is the study of essential food components with diets derived from chemically defined ingredients, often in bacteriologically sterile formulation, and with insects reared from externally sterlized eggs. This study is an approach to the determination of essential food components, particularly if the contributions of symbiotic fauna-flora and food reserves that pass through the eggs are taken into account.

Values of Food

The monograph of Uvarov provided the table of contents that could serve well as a guide for nutritional research, and his introduction clearly. Uvarov described the importance of food requirements in his monograph. He began his treatise as follows:

It is well-known that the enormous losses regularly caused by insect pests to cultivated plants, domestic animals, and human life are directly or indirectly connected with their feeding habits. On the other hand, the products of a few useful insects, like the honey bee, silkworm, lac insect, etc., are from the physiological point of view, substances of definite importance in the metabolism of insects and their quality and output depends on the character of the food taken. It would seem, therefore, that the problem of the nutrition and the metabolism of insects should be regarded as a key both to the successful control of injurious insects and to the progress of the industries dependent upon the products of useful insects.

It would be natural to expect that the attention of entomologists always has been and still is concentrated on its solution. If this were so, we might possess a through knowledge of the nutritional physiiology of, at least, all the most important insects, and be in a position to apply it to practical purpose. Regarding his compilation of data, Uvarov continues:

The first glance at the summary, and, particularly, at the voluminous bibliography, including nearly six hundred titles, may give the impression that a very large amount of work has been done on the problems of insect nutrition and metabolism, but when all the data available on each particular subject are put together it becomes clear that very few points have been touched and that the results achieved are but a drop compared with the ocean of unknown phenomena. Thus, to begin with, existing knowledge of the food of various insects from the chemical point of view extremely meagre. It is, of course, known that many insects feed on green plants, but the question which substances are utilized by phytophagous insects and which are not, is doubtful and has been very little studied. The problem has, however, an enormous practical importance, since the immunity of certain varieties, or species, of plants against particular insects must be closely connected with it. With regard to another group of insects of great economic importance, viz., those boring in timber, destroying forests and buildings, only one fact has so far been established that either they feed not on wood, but on fungi in the burrows, or that the assimilation of their food depends on the presence of special micro-organisms in their bodies; thus the whole problem of their nutrition is highly peculiar and still unstudied from a physiological point of view.

In the case also of the blood sucking insects which cause enormous losses in human and animal life in the tropic and elsewhere, there is practically no information on the chemical side of their nutrition, although there is a voluminous literature on the anatomy of the digestive apparatus, on the mechaniscs of its action, on the selection of hosts for feeding purposes, etc... it is, perhaps, worthwhile pointing out that there is still no information available regarding the actual food of the larva of the commonest insect, viz., the housefly; it is known that the larvae breed in various decomposing substances and that some bacteria seem to be necessary for their development, but what the larvae actually feed on has never been properly investigated. As a matter of fact, there are very few insects on the food of which there is a reasonable amount of information. Curiously enough, the insect best studied in this respect is the banana fly (*Drosophila*) which is largely used in experiments on genetics and is therefore breed in numbers in laboratories all over the world; unfortunately, the data obtained on this insect are of very little practical value.

Another well studied insect is the cockroach, again a common laboratory species, and again one presenting but little interest to the general entomologist. The same may be said with regard to flesh flies also very favoured in physiological laboratories, where the objects for the study are naturally enough chosen as a rule without regard to their interest for entomologists. It is the duty of all entomologists to draw the attention of physiologists and biochemists to the existence of many other insects which can be raised in the laboratory with equal ease, and which are equally interesting from physiological and chemical points of view, but are at the same time very important from other points of view.

The ideas expressed by Uvarov will remain worthy of serious consideration for many years.

The possibility is worth considering that nutrition may have an effect upon resistance, either by improving the general physiological status of insect populations or by providing a particular factor in natural food that imports the resistance characteristic. Actually, all experimental studies should begin with a thorough understanding of the nutrition of laboratory species and, in some instances, the laboratory strain. This understanding will become possible only after more careful work has been completed in the study of insect nutrition.

Nutritional Requirements

According to House the nutritional requirements of insects as follows: The present understanding is based upon research that ranges from investigations of the natural foods to the studies, most of which have been conducted during the last 10 years, of the effects of single components of chemical defined diets. Thus far, the data are limited almost entirely to the needs of insects during part of the development of one generation, and in only a few cases more than one. Most data have been derived from the study of a few representatives of the orders Coleoptera, Diptera, Lepidoptera and Orthoptera. The factor of symbiosis which occur particularly in species that feed on plant juice, blood–and stored products, and also includes cockroaches and some other ominoverous forms–has made it difficult to reach finite conclusions.

Other factors that must be considered include: the nutrient reserves that are passed along in the egg or accumulate in the immature stages to be used in later stages of development, the possibility of the inclusion of small quantities of harmful substances that may stimulate a metabolic system and result in imporved nutrition, and the effects of the presence or absence of phagostimulants that can render an otherwise suitable diet unsatisfactory.

Nutritional requirements of young insects may vary with sex and developmental state, and may depend on the nutritional state of the parent; also data indicate that nutritional deficiencies may have the greatest effect on the smallest individuals. The requirements for specific factors (e.g., amino acids) tends to be more or less uniform during larval life by may vary widely in adults. This variation may be especially true in adult females, which require changes in nutrition during egg formation and maturation. Some adult forms do not feed but rely entirely on nutrients carried over from the immature stage. Others feed on carbohydrates, etc. Despite these variable, the accumulated research indicates that insects require, or can utilize, carbohydrate, protein, mineral salts, sterol, and most water-soluble (vitamin B) vitamins.

Fats are utilized by many insects, but relatively few species have been shown to have a definite requirement. The fat-soluble vitamins (A, D, K and E) are considered unnecessary. Insects grow readily on vitamin A-free diets, and vitamin A has been detected in extracts of large numbers of insects. However, no data are available to prove or disprove that a deficiency of vitamin A affects visual acuity in insects. There is a visual pigment in insects that responds to light in a manner similar to that of vertebrates, but this pigment probably is not chemically derived from pigment vitamin A.

The insect exoskeleton does not have the calcium-phosphorus complex found in vertebrate bone, which may explain why vitamins A and D are not essential. Vitamin K is required for normal blood coagulation in vertebrates, but blood coagulation in insects is accomplished by another means. The case of vitamin E, sometimes called the fecundity vitamin, seems to be more difficult to rationalize, but its essentiality is doubtful. Vitamin C (ascorbic acid) apparently is snythesized by many insects in sufficient quantity, but this is not so in the migratory locust (*Schistocerca gregaria*), which requires ascorbic acid.

Application of the Study of Insect Nutrition

Commercial animal feeds have been recommended for many of the ominverous species, and some of these feeds are satisfactory. It is important that the diet used for a laboratory colony is nutritionally sufficient in all respects. The presence of deficient insects is a colony very likely alters the results of other experimental work. So far, no single diet has been developed that can be used for all insect species, although the requirements of insects are qualitatively similar and basically the same as for verebrates, with the exception of vitamins. With the deverse feeding habits of the different insect groups, quantitative differences in diet are important, and the factor of phagostimulation can be decisive. Commerical dog food, nonmedicated chick mash, fish food, and rabbit food supply all the essentials for many insects, colonies can flourish for generations on any of these. The necessary nutritional and chemical constituents are present in all of these prepared diets, but the physical properties may require some adjustment.

The problem of determining essentials for insect growth and development has lead to a considerably more complicated experimental process, and through necessity, a specialized terminology. The development of this specialized terminology is an attempt to define both procedures and diets in terms that have absolute meanings and are free of connotative meanings, which can lead to confusion. The practice of implying connotative meanings, which the author calls "willful misunderstanding," is a real hazard in the communication of ideas. Such is the case with the terminology that has been used in nutrition; to correct these semantic difficulties, new words derived from classical languages have been introduced. In this vein, the word *defined* (or *chemically defined*) has not always been applied to components that are characterized with chemical precision (e.g., a diet containing dextrin, agar, or a racemic amino acid would not be defined).

The new word that indicates a medium composed of components that have a precisely known chemical structure before compcunding is *holidic*. This definition ignores the possible changes through interaction during compounding or by association with the organism, but these possibilities must be considered in actual practice. A diet with a holidic base, to which is added at least one substance of unknown structure or uncertain purity, is described as *meridic;* a diet comprised of crude organic material (e.g., dog food) is *oligidic*. In the description of colonies or cultures of organisms, a similar situation exists. Those who object to the word *pure,* as it applies to a culture, on the basis that purity is relative, prefer to describe a culture that contains just one species as *axenic*. When the species present in the culture are known, the system is *gnotobiotic*. An axenic culture has a single known species growing gnotobiotically. If more than one species is present and if they are positively indentified, the culture is *synxenic*.

The opposite of a gnotobiotic culture is an *agnotobiotic* or *xenic* system, in which the number of species (or their identification) is not certain. This is the case with insects reared in association with their normal complement of symbionts (either inter or intracellular) or in a normally infected or infested environment. These terms are being adopted by most writers in insect nutrition. The practical impossibility of satisfying some of the more rigid definitions will undoubtedly lead to the use of comparative terms, which will negate the value of the definitions.

Importance of Suitable Diet

The most popular method for assaying nutritional factors is growth. This method may be described in terms of the increase in size or weight of individuals, or of whole colonies. Much of the data has been drawn from observations on one generation, or even on one or two stages of development. Gordon suggested a series of improvements that might be added to experimental procedure. The first improvement concerns weight as a criterion and relates this to the duration of culture under the conditions of the experiment.

As an illustration of this method, House indicated that arginine is essential for the first generation growth of the German roach, because arginine-free diets produced individuals that grew at a rate that was 80 percent that of the control insects. These results might also be interpreted as indicating that while arginine accelerates growth, it must be present in the reserves carried over in the egg or synthesized by the insect, in which case it would not be essential. To avoid such uncertainty, the term *essential nutrient* should be reserved for substances tht must be present in the diet to maintain growth and reproduction *indefinitely.* The second improvement suggested by Gordon is the elimination of all nonessential elements to attain a minimal diet. Conventional experimental procedure begins with a basic diet, and the obvious procedure is to compound each new diet as the basic, minus one component. This assumes that nutrients are not interconvertible, which is not necessarily true. The third improvement suggests the investigation to alternate nutrients whenever possible.

These alternate substances fall into three classes–precursors that have no activity of their own but are converted into chemially active sustances in the animal, comprising nutrients that have chemical activity but that are not readily formed from or converted to other nutrient materials, and precursor nutrients that exhibit chemical activity but are also convertible into other nutrients. The last category can be further subdivided, depending upon the extent to which conversion takes place and the degree of activity of the converted nutrient. The problems of assaying nutrients for essentially are complex, and strict adherence to definitions proposed by

Dougherty is at the present stage impossible. Adherence to these definitions must be considered a desirable goal toward which to work, with the realization that they may be approached but not attained. That essential nutrients must be considered both qualitatively and quantitatively, makes the number of possible combinations of nutrients almost infinite. The number of assays necessary may be limited by the use of an advanced statistical design in the experiments, and some of the initial problems can be resolved by intrpreting the data for essentially as a destinctly positive or negative reaction.

In any case some compromises are necessary, and Gordon made several good suggestions, based on his extensive study of the nutrition of the German roach, reared in colonies under xenic conditions on a meridic diet. His criteria are expressed numerically. The first criterion is the survival fraction, defined as the ratio of the final number of adults to the initial number of nymphs. This cannot be used if adults are not produced. The second criterion is the growth index. This index is defined as the ratio of the product of the average length times the number of nymphs at the end of the nymphal growth period divided by the product of the average length times the number of nymphs at the beginning of the experiments (zero days). The third criterion is the maturation period–the interval in days between the appearance of the first and the last adult.

The maturation period is a rough indicator of growth rate but not necessarily a criterion of adequacy. The fourth criterion is the reproduction index–the number of nymphs produced per egg capsule (as in the German roach) in the first hatch divided by the number of days from the start of the experiments. This has significance not only in that it proves that reproduction has occurred, but also because it gives a record of the duration of the life cycle.

Nutritional Values of Diet

A diet must contain a source of energy, usually in the form of a carbohydrate; a source of amino nitrogen, a protein, a source of sterol, often supplied as a minor component of a fat or oil; vitamins, particularly the B series; and mineral salts. The addition, the physical and chemical properties of the diet must make it acceptable to the species to be fed. An inert carrier is often necessary to provide bulk and texture for synthetic diets.

Carbohydrates

Carbohydrates play a fundamental role in the life of animals and plants. Plants synthesize carbohydrates by a process of photosynthesis and thus utilise solar energy. Plant starch is the principal source of carbohydrate, but various types of sugar may be ingested in the normal diet from plant sap, nectar and tissue. The utilization of the carbohydrate depends upon the ability of the insect species to conert the complex polysaccharides and oligosaccharides into assimilable simple sugars. The carbohydrates in the order of utilization by insects generally include: dextrin, fructose, glucose, lactose, maltose, mannitol, raffinose, sorbitol, starch, sucrose, and trehalose. Glucose and fructose are usually utilized well: sorbose and glactose not so well; and in general, pentoses are not used at all. Results from the study of the requirements of *Schistocerca gregaria* indicate the quantitative requirements, for carbohydrate.

These insects grew poorly and failed to complete development when no digestible carbohydrate was incorporated into the diet. Normal growth was obtained when glucose and sucrose formed 13 percent of the diet, but the rate of development was not maintained beyond the third instar. Satisfactory growth and development occurred with 26 percent carbohydrate, and 39 percent gave normal development but poor growth. That a quantitative optimum for carbohydrate in the diet exists is indicated by the statement of Lipke and Fraenkel that carbohydrates may be nutritionally inert; may be satisfactory, but unacceptable from a physical or chemical standpoint; or may be toxic.

Proteins and Amino Acids

Proteins are the most characteristic chemical compounds found in the living cell. Proteins are complex nitrogenous compounds composed of chains of a-amino acids. They have a high molecular weight depending upon the number of amino acids forming them. They are the principal component of animal tissue. From a nutritional standpoint, the requirement for protein is a requirement for the individual amino acids that comprise them. Twenty-one amino acids make up the list currently accepted as occurring in nature. This means that these 21 amino acids are present in natural plant or animal proteins. It should be immediately apparent that asymmetric carbons are present in all but glucine, and that isoleucine and threonine have more than one asymmetric carbon each. This configuration suggests special isomersm, which does exist and is indicated by "D" or "L" used as a prefix. (*d* and *l* in the older literature). In addition to the D and L forms, isoleucine and threonine have an *allo* form. It is generally that only the L amino acids are utilized by animals, and there is always the possibility that the allo amino acids are physiologically different. At least one D amino acid (D-alanine) was found in the blood of the milkweed bug (*Oncopeltus fasciatus*), and it is apparently synthesized by the insect. Amino acids incorporated into diets may come from proteins, from protein hydrolysates, or from purified compounds isolated from hydrolysates. Some synthetic preparations are also all present.

Table 1.1. The Amino Acid Composition of Common Proteins.

Amino Acid	*Protein (percent amino acid on the basis of 16 grams of nitrogen)*			
	Casein	*Galatin*	*Zein*	*Albumin (Egg)*
Arginine	4.2	8.2	1.8	6.1
Histidine	3.2	0.9	1.7	2.4
Lysine	8.5	5.0	0.0	6.5
Tyrosine	6.4	0.5	5.2	4.2
Tryptophan	1.3	0.0	0.1	1.5
Phenylalanine	6.3	2.3	6.4	7.5
Cystine	0.4	0.1	1.0	2.4
Methionine	3.5	0.8	2.3	5.5
Serine	6.8	3.5	7.7	8.5
Threonine	4.5	1.9	3.0	4.2
Leucine	10.0	3.5	23.7	9.4
Isoleucine	7.5	1.7	7.3	7.5
Valine	7.7	2.8	3.0	6.4
Glutamic acid	23.0	11.0	26.6	16.0
Aspartic acid	7.0	6.2	5.6	9.0
Glycine	2.1	23.6	0.0	3.6
Alanine	3.3	8.2	11.4	7.4
Proline	13.1	15.3	10.4	8.1
Hydroxyproline	0.0	13.0	–	–

The most common proteins available in a reasonably pure state include albumin, casein, gelatin and zein. As ingredients in commercial feeds these proteins come from meat scrap, grass and leaves (e.g., alfalfa meal), yeast, whole wheat flour and bran. The calcualted amino acid content of each of these proteins is shown in Tables 1.1 and 1.2. The values from these tables should be accepted with some caution because of probable losses of amino acid during hydrolysis. The tubular figures are percentages calcualted on the basis of 16 grams of amino acid per gram of nitrogen on a water-and ash-free basis. To convert the figures to actual amounts, it is necessary to find the percent nitrogen content of the sample by analysis and to correct the figures by a factor obtained by dividing the percent nitrogen by 16.

Table 1.2. The Amino Acid Composition of Crude Feed Components.

Amino Acid	*Protein (percent amino acid on the basis of 16 grams of nitrogen)*				
	Meat Scrap	*Grass and Leaves*	*Yeasts*	*Whole Wheat*	*Bran*
Arginine	7.0	7.0	4.0-5.0	4.3	7.5
Histidine	3.5	2.0	3.0	2.1	1.7
Lysine	5.6	5.5	7.0-8.0	2.7	3.9
Tyrosine	3.2	5.0	3.6	4.0	–
Tryptophan	0.7	2.2	1.3	1.2	1.3
Phenylalanine	5.1	5.0-6.0	4.5	5.1	3.0
Cystine	1.2	2.0	1.1	1.8	1.5
Methionine	2.0	2.5	2.0	2.5	1.3
Serine	3.7	5.0	–	4.3	5.3
Threonine	3.9	5.4	5.5	3.3	2.5
Leucine	8.0	10.0	7.5	7.0	6.5
Isoleucine	3.4	5.0	5.8	4.3	4.1
Valine	6.1	5.0	5.8	4.3	4.1
Glutamic acid	–	11.5	14.7	29.0	–
Aspartic acid	–	5.3	–	–	–
Glycine	–	–	–	–	–
Alanine	–	–	–	–	–
Proline	–	3.0	–	–	–
Hydroxyproline	–	–	–	–	–

The proteins used most often in insect diets (albumin, casein, etc.) lack certain amino acids, making it necessary to fortify the proteins. Casein needs additional tryptophan, histidine, cystine, glycine, and possibly hydroxyproline. Gelation should be fortified with tyrosine, tryptophan, and probably cystine or methionine. Zein requires the addition of argining, histidine, lysine, tryptophan, cystine, valine, glycine, and hydroxyproline. Thus far, the amounts of each amino acid remain pragmatic, but like carbohydrate, there are quantitative as well as qualitative requirements. Some amino acids are toxic if used in excess. When racemic mixtures are used, it is the general practice to assume equal amounts of the D and L forms. Amino acid mixes that approximate synthetic proteins have been prepared for the nutrition of several insect species and are compared in Table 1.3. Some of these amino acids are sold commercially as hydrochlorides. The hydrochloride form adds stability to the compounds, but their incorporation into diets may make the food unacceptable to insects.

Table 1.3. Amino Acid Mixes for Insect Diets.

Amino Acid	*l-from (mg/ml)* German Cockroach (B. germanica)	*Onion Maggot (H. antiqua)*	*Pink Bollworm (P. gossypiella)*	*Blowfly (P. regina)*
Allanine	5.5. mg/ml	1.09 mg/ml	0.38 mg/ml	1.29 mg/ml
Arginine	4.5	0.8	7.6	1.06
Aspartic acid	6.5	1.22	0.82	1.53
Cystine	1.5	–	1.66	0.36
Cysteine	–	0.48	–	–
Glutaminc acid	23.0	4.42	2.18	5.48
Glycine	3.0	1.75	2.50	0.70
Histidine	3.0 (HCl)	0.48	3.8	0.70
Hydroxyproline	2.0	0.38	–	0.48
Isoleucine	6.5	1.26	1.02	1.53
Leucine	12.0	2.35	2.96	2.80
Lysine	7.5	1.34 (HCl)	1.18 (HCl)	1.76
Methionine	4.0	0.34	0.60	0.94
Phenylalanine	6.0	1.01	0.78	1.40
Proline	8.5	1.68	1.68	2.0
Serine	7.5	0.88	0.08	–
Threonine	4.0	0.38	0.92	0.94
Tryptophan	2.0	1.75	0.24	0.47
Tyrosine	7.0	1.24	0.52	1.64
Valine	7.0	1.36	1.04	1.5

More is know about the amino acid requirement of insects in a qualitative way than is known any other nutritional group. The results of the many investigations are difficult to interpret and compare because of the differences in the sources and in insect species, and in a number of cases because of the incorporation of amino acids into the diet from unsuspected source. The order or protein requirement for the confused flour beetle compares closely with that of the rat. Arginine, histidine, isoleucine, leucine, lysine, methionine, phenylalanine, threonine, tryptophan, and valine are essential for growth and development of the onion maggot, the pink bollworm,and many other species.

House and Hilchey showed that alanine, valine, leucine, isoleucine, serine, lysine, histidine, proline, tryptophan, and probably arginine are required for growth and development of the German roach under the conditions of their experiments. These conditions included the rearing of isolated insects from externally sterilized eggs on autoclaved diets that were nearly holidic (white dextrin–a nitrogen-free, but chemically undefined, carbohydrate source–was included). Some differences in the requirements of each sex were also noted in this work. In a comparable assay under xenic conditions, a minimal amino acid mix was shown to require arginine, lysine, leucine, histidine, threonine, tryptophan, and glutamic acid.

Faster growth resulted if valine was included in the mix and isoleucine and tryptophan were found to be partly replaceable by phenylalanine. Addition of amino acids classed as nonessential (serine, glycine, or proline) also appeared to accelerate growth. When the criterion of essentiality is *growth in terms of increase in weight,* the qualitative requirements derived from the nearly axenic culture and the xenic culture are remarkably similar. Alanine can replace glutamic acid, which is considered to be a major source of nitrogen which may explain the unusual alanine requirement of the German roach. These data have been cited to show that the differences in protein requirements which exist are subtle. For the present, many of these differences can be explained on the basis of technology, including those that have iodine in the molecule, improve growth rates. Methionine and some other amino acids are toxic in moderately high concentrations. Ordinarily, the sulfur-bearing compounds are necessary for integument formation and normal molting, but it has been shown clearly that the German roach can synthesize these compounds by incorporating inorganic sulfur into an amino acid molecule.

Fats and Sterols

Fats constitute a major class of tissue components and a major food stuff. By definition, fats are esters of one or more fatty acids and glycerol, a trihydroxy alcohol. Like carbohydrates, fats are also composed of the elements carbon, hydrogen and oxygen, the former two is larger quantity than oxygen. Enzymatic hydrolysis in the digestive tract splits the fatty acids from the glycerol, and each is metabolized separately. Sterols are compounds, with a complex ring structure, that always contain hydroxy groups. Sterols are associated with fats, but they are not esters, they will not saponify, and they are not split by lipases or esterases. Through the activity of the hydroxy groups, it is possible for sterols to react with fatty acids to form esters. A number of esters are found in animals, and most of these compounds have physiological activity.

The sterols are closely related to bile acids, and they are a continuent of many hormones. Ergosterol, which in its irradiated form in vitamin D, is a sterol. So far as is known, all insects have a requirement for sterol, but only a few have been shown to have a clear-cut fat requirement. This does not preclude the possibility that fats of fatty acids can be utilized by insects; in fact, it is well established that many insects digests, assimilate, and metabolize fats. The sterol requirement for insects can be satisfied with cholesterol, 7-dehydrocholesterol, and stigmasterol. Plant feeders usually have a broader tolerance for sterols other than cholesterol, than do the flesh feeders and omnivorous species. In the purified state, cholesterol is highly insoluble in water. This posses a serious problem when it must be incorporated into a liquid or gel for a diet. Friend solved this problem by dissolving the required amount in hot 95 percent ethyl alcohol, precipitating the cholesterol by adding water, and stablizing the suspension with Tween 80. The alcohol was removed under vacuum.

Gordon impregnated his insect diet with an ether solution and evaporated the solvent, and Beck and co-workers used cholesterol acetate, which is relatively soluble in water. Lipids that are associated with sterols are often added to insect diets as purified vegetable oils. Corn oil linseed oil, soya bean oil,and wheat germ oil are the common choices. Corn oil is approximately 45 percent oleic acid, 40 percent linoleic acid, 7 percent palmitic acid, and 3 percent steric acid. Linseed oil has 35 percent linoleic acid, 45 percent linolenic acid, 9 percent oleic acid, 6 percent palmitic acid, 4 percent stearic acid, with traces of myristic, archidic, and isolinoleic acids. Soya bean oil contains 30 percent oleic acids, 55 percent linoleic acid, 9 percent palmitic acid, 4 percent steric acid, with traces of myristic and lignoceric acids.

Wheat germ oil is composed of 44 percent linoleic acid, 30 percent oleic acid, and 10 percent linoleic acid. All of these oils have an unsaponifiable residue composed of sterols and high-carbon alcohols. The total residue varies from about 3 to 5 percent, a small part of which is made up of tocopherols, sitosterols, dehydrositosterols and stigmasterol. From these data, it would seem that the plant oils are not an effective source of metabolizable sterols, but insects grow and develop on diets in which plant oils are the only source of sterols. Only trace quantities of these sterols are required in insect diets. From this it seems obvious that some conversion of the plant sterols takes place in insects, either through the intervention of the plant sterols takes place in insects, either through the intervention of symbionts or through a more fundamental metabolic process. Cholesterol is required by insects for larval growth and oogenesis, and its function is similar to that of the juvenile hormone, which is an unsaponifiable constituent of insect lipids. Cholesterol may act as a precursor to the various steroid hormones.

A sterol deficiency deprives insects of some of their natural immunity toward bacterial infections. Most Lepidoptera have a requirement for an exogenous fatty acid, as well as a sterol requirement. Certain moths require linoleic acid for larval growth and development, the German roach shows deficiency symptoms in the progeny of deprived parents, and a species of locust requires linoleic acid for normal wing formation. As the criteria for sufficiency become more critical, it is possible that the number of insect species with a fatty acid requirement will increase.

Nucleic Acids

Nucleic acids are organic polymers and found in the nucleus as well as in the cytoplasm. In the nucleus they are associated with chromosomes and are important in transmitting information from nucleus to the cytoplasm. In the cytoplasm they are most closely concerned with the synthesis of proteins. These compounds, present in the nuclei of all cells as ribonucleic acid (RNA) or deoxyribonucleic acid (DNA), exhibit an even increasing number of metabolic functions. There is no reason to believe that nucleic acids do not have equal importance in insects as in vertebrates. Thus far, it has been demonstrated that the addition of RNA increases the growth rate in various dipterous insects. Requirements for other insect groups have not been demonstrated but might well be anticipated.

Vitamins

Vitamins are the accessory, indispensable food factors, organic in nature (organic acids, amines, amino acid, esters, alcohols, steroides, etc.), and are required in minute amounts for the growth and normal functioning of cells. They play an important part in metabolic processes, in which they function principally as part of enzymes. The vitamin requirements of insects have been the subject of many studied, but little clear-cut information is available. It is generally conceded that the water-soluble group (B series), including thiamin, riboflavin, niacin, pyridoxine, pantothenic acid, biotin, and cyanocobalamin, is required by most insects. Ascorbic acid (vitamin C) is essential for the migratory locust. A major source of confusion in the literature of vitamin nutrition comes from the use of common nomenclature. The alphabetic system by which they were originally classified has largely given way to the use of a chemical name. Common names of still exist and have rather broad synonomy.

Thiamin

The first of the series is thiamin. It is also known as B_1, thiamine hydrochloride, betabion, heat labile factor and aneurine. It is readily soluble, crystalline substance and is insoluble in ether and chloroform. Thiamin occurs in plant and in animal tissues, grain cereals, yeasts, liver, eggs, milk and in green leaves. Most of the commercial product is synthetic. Since growth is the principal criterion by which thiamin deficiency is measured in insects, a thiamin deficiency results in poor growth. From vertebrate nutrition it is known that thiamin functions as a coenzyme for pyruvate metabolism, and it is a rasonable assumption that this is its role in insects. Thiamin can be produced by the symbionts normally present in the insect body so that the deficiencies are most pronounced in axenic culture.

Riboflavin

It is also known as vitamin G and ovoflavin. Riboflavin is a slightly soluble pigment that occurs in trace amounts in all plant and animal cells. Good sources are milk, eggs, malted barely, liver and kidney tissue and yeast. Riboflavin is available in a purified commercial preparation. In insects this vitamin is concentrated in the Malpighian tubes, in the free form in the lumen, and in a phosphorylated the form in the cell walls. Riboflavin is never excreted as an intact molecule. The principal known function of riboflavin is as a prosthetic group for the flavoprotein enzymes. Insects may be raise din xenic culture on a riboflavin-free diet, which indicates that the symbionts can furnish a part of the requirement. The mortality under these conditions usually is high.

Niacin

Niacin (nicotinic acid amide or vitamin B_5) is soluble in water and alcohol but sparingly in fat solvents. It is a stable compound resistant both to heat and to acids and alkalies. It occurs in reasonably high concentrations in liver, yeast, milk, white meat, alfalfa, legumes, and whole grain cereals. Whole wheat flour averages about 60 micrograms of niacin per gram. Like the other vitamins, niacin is available commercially in a purified form. In vertebrates it functions as part of dehydrogenase co-enzyme.

Pyeidoxine

Also known as vitamin B_6. These are water soluble and sensitive to ultraviolet irradiation. Pyridoxine occurs in yeast, liver, and cereals and is available in a pure form. There is no requirement established for vertebrates, but pyridoxine takes a part in the process of amino acid conversion in the role of a coenzyme. It is considered essential for most insects.

Pantothenic Acid

Also known as vitamin B_3, the yeast factor and the chick antipellagra factor. Pantothenic acid was discovered as an essential for chicks. It is found in all living tissue, is commercially available in purified form, and is found in liver. One of the most potent natural sources of panthothenic acid is the royal jelly of the honeybee. The commercial product is synthetic. Pantothenic acid is part of coenzyme A (CoA) and as such probably is required in a number of metabolic processes. It is considered essential for insects.

Biotin

Also known as vitamin B_7, co-enzyme R or vitamin H. It is sparingly soluble in cold water and dilute acids but freely soluble in hot water and dilute alkalies. Biotin, a vitamin found in liver kidney, yeast, and milk in low concentrations, is either essential or beneficial to the nutriment of all animals. It is available in pure form, and its incorporation into insect diets improves growth. Biotin's cellular function is as a coenzyme in the fixation of carbon dioxide in 4-carbon acids.

Cyanocobalamin

Cyanocobalamin, or cobalamin is often referred to as vitamin B_{12}. It was first isolated from liver in 1948, is commercially available and is used empirically in insect diets. In cellular processes cyanocobalamin functions as a coenzyme for an enzyme that is active in the methyl transfer and nucleic and metabolism.

Other B Vitamins

Other B series vitamins have been studied in insect nutrition, and some have been found to improve growth or development. Among these is vitamin B_T (carnitine), which is the only one of the group isolated from studied with insects. Carnitine has been found imporant to the development of Tenebrionidae and *Aedes aegypti.* The requirements of carnitine are variable with different species, and the qualitative requirements, as they are known, are tabulated.

Mineral Salts

Minerals are the elements which are present in small quantities in the body to regulate the different metabolic activities of the animals. They act as general regulator of body processes. Various mineral salts are nutritionally essential for insects, but this field has received relatively little attention. The obvious approach to the problem is to ash insects and determine the salt content of the residue. The results have not been too enlightening because they show such a wide spectrum of elements, many of the elements are of questionable importance, because the analyses include the unassimilated food in the digestive tract.

There is a general agreement that the ash of insects contains potassium, sodium, magnesium, calcium, zinc, copper, aluminium, silicon, phosphorus, arsenic, sulfur, chlorine, manganese, and iron. Some of these are found in trace quantities and if they are utilized at all, it is in very minute quantities. The most common practice in diet formulation is to use one of the commercially prepared nutritional salt mixes for vertebrate work. These are designated as USP formulations, or under the initials of the name of the person who described the original preparation (e.g., salt W is Wesson's salt mix). All of the mixes contain the same elements in approximately the same concentrations; the principal difference comes from the method of formulation. Because of the lack of better information, salts are added to insect diets in empirical amounts by a trial and correction method. There is no reason to assume that the salt balance satisfactory for the mineral requirements of vertebrates should be the same or even close to that of insects. From the available literature on the subject, there seems to be very little agreement on mineral salt requirements of the various species except that they require phosphorus and potassium. This is probably due more to the lack of critical studies than to the lack of actual requirements.

Potassium, sodium, calcium, and magnesium are always present, and they undoubtedly function to maintain the ion balance necessary for the control of permeability of tissue membranes. Magnesium and potassium have been found in comparatively high levels of concentration in insects, particularly in the phytophagous forms. Magnesium may partially replace calcium as a divalent ion in insects. That the magnesium ion does not exert the anesthetic effect on insects that it has for many other invertebrates indicates an intrinsic difference in the permeability relations of the insect nerve sheaths. Zine in trace quantities is an activator for some enzyme reactions and except for this its function is unknown.

Copper occurs in fairly high concentrations in some insects, and it has been speculated that it may be a part of an unidentified respiratory pigment. No such pigment has been isolated from insects. Phosphorus in the form of a phosphate radical is the medium of exchange of energy, and it must be available for the synthesis of the phosphogen reserve. Chlorine is present as a chloride in salts of various anions, particularly potassium, sodium, calcium and magnesium. The amount of chlorine found in insects is not sufficient to combine with the anions, so that these must be present in other than chloride salts. Iron, the last of the trace elements, is a part of the cytochrome system.

Commercially Produced

The survey of literature reveals clearly that research in insect nutrition may be simple or complictated, depending upon the end result desired. When insect production is the principal requirement, the best procedure is to find a commercially available product that can grow the most individuals with the greatest size and vigor, probably in the shortest period, and with no diminution of the reproductive potential. Most of this work is entirely pragmatic. The results of a typical experiment of this type are shown. This series of curve compares the rate of growth with weight increase of nymphs of the gray house cricket (*Acheta domesticus*) reared under similar conditions of crowding temperature, and humidity and provided with a series of commercially available animal feeds. Here, only positive results are definitive.

The reasons for poor growth are obscured by the variables incorporated in a completely nondefined diet, and from experience, it is apparent that an examination of the manufacturer's specifications does nto give a dependable basis for prediction. That growth rates on one diet are low at the outset and become rapid in the later stages, or *vice versa,* may indicate that basic requirements change during growth of the nymph. This experiment demonstrates, when combined with observations of wing formation and ability to produce successive generations, that several commercial feeds are approximately equal in their requirements for this insect. Very little more can be gained from this experiment. The obvious refinement of this experiment is to prepare diets that are nearly holidic (achieving a strictly holidic diet is possible, but it may be impractical), and to rear the insects as individuals in axenic culture, taking into consideration the possiblity of intervention of intracellular symbionts.

Symbiotic Nutrition

The intestinal fauna-flora in insects that act symbiotically can be removed by external egg sterlization and rearing under aseptic culture techniques. This is not true for the intracellular symbionts. These *bacteroids,* which are found in specialized cells usually at the edges of the segmented fat body adjacent to the midgut or deep in the fat body, have been observed frequently. The cells that contain bacteriods are described as *mycetocytes* or *mycetomes.* Intracellular symbionts have never been cultured nor completely identified. They have been described as Rickettsia-like, bacteroid, or like fungi or yeasts.

One of the most complete studies of bacteroids demonstrated that it is possible to prevent the transovartal transmission of these bodies in the German cockroach by subjecting the parents to high temperatures (approximately 37°C), or by feeding the parent aureomycin (chortetracycline) or sulfathiazole. A concentration of 0.1 percent aureomycin, fed to the parents during their entire lie, assures aposymbiotic nymphs, and the lack of symbionts can be demonstrated by histological technique. In tests conducted by Brooks and Richards, aposymbiotic nymphs were virtually incapable of growth on a natural diet that was adequate for symbiotic nymphs. These observation bring up several problems that complicate nutritional study, even when conducted under aseptic conditions.

The transmission of bacteroids through the egg makes external sterlization ineffective, and it seems apparent that these poorly defined bacteroids function in supplying essential nutritional factors. Two courses are open. Either the investigator must assume that since bacteroids are intracellular and since the symbiosis that appears to exist is of great antiquity in the evolutionary origin of the symbionts, they are an intrinsic part of the animal; or an attempt must be made to remove the symbionts by one of the methods described, and essential nutrients must be added to supplement the diet. The latter idea, although attractive from an academic point of view, offers a serious obstacle to the progress of nutritional study.

A great deal more needs to be known about intracellular symbionts and their functions before their possible activity can be compensated for. Evidence indicates the procedures necessary to eliminate symbionts could confuse the result because of side reactions. Auroemycin-treated chick mash, for example, with the drug level as low as 10 grams per ton has significantly inhibited the growth and development of the gray house cricket, confirming earlier observations with the American cockroach (*Periplaneta americana*). In the case of the cricket, the addition of powdered liver increased both the growth and development rates. From these data, it may be argued that as essential growth factor is removed by the drug, and the powdered liver furnishes these essentials, or that the pathogenic, as well as the symbiotic flora, are destroyed, and the elimination of the pathogenic flora increases the growth and development rate when nutrition is adequate. It is also of possible significance that cockroaches reared at high temperatures (above 36°C) show a sharp break in their excretory efficiency at about 37°C. This indicates the inhibition of chain of enzyme reactions. There is no doubt that the intracellular symbionts of insects exist, but their presence as symbiotic organisms that are not an intrinsic part of the insect is debatable.

Selective Diet

Preparing either meridic or holidic diets that are physically and chemically attractive and acceptable to the various insect species is a technical obstacle in nutritional studies. Some insects are selective in their feeding, and observation of the residue left in the feeding dishes of a colony of cockroaches or crickets shows that these insects have selected the smaller particles and left the larger ones. Some plant feeders require an edge upon which to chew, and seed-feeding insects may refuse food until it is formed into a hard pellet. Methods of correcting these difficulties include such expedients as grinding the food (commercial animal feeds, for example) to a fine uniform particle size, formulating the food in dried gelatin or agar sheets, and forming the food into hard pellets in a pharmacist's pill press. No doubt other physical subterfuges have been used to induce feeding by altering properties of the food. Chemical factors also induce feeding, and in some case when the attractant is present, insects feed readily on a totally inadequate diet. The silkworm, for example, will feed on agar if one of several hexanols plus mulberry leaf is included. The larvae of the varied carpet beetle (*Anthrenus verbasci*) were observed to eat nylon in preference to wool after the nylon had been treated with certain finishing agents. Newly hatched larvae of nymphs are the most sensitive to phago-stimulation; after the feeding pattern is established, phagostimulation may become less important. The acceptance of the diet by the insect species is of great importance in nutritional work, and many attempts have failed because of refusal of the insects to feed.

2

Feeding and Digestion

The animals take organic compounds as the source of energy for building material, as food. Insects feed on a very wide variety of animal, vegetable and dead organic materials. Some are virtually ominovorous, but the majority are more specific, being restricted to a particular category of food or even to a particular plant or animal. Food preference may be based on nutritional or non-nutritional factors and appears to be of significance in that survival is better of fecundity greater with the preferred food. The finding and recognition of such food involves various mechanisms depending on the insect and its particular mode of life, but vision and olfaction are of widespread importance. Feeding and ingestion involve modifications of the mouthparts and physiological adaptations.

Fluid feeders often inject enzymes into the food and in blood-sucking insects an anticoagulant may be injected. Predaceous insects restrain their prey by force or by means of a venom injected with the saliva or, in Hymenoptera, via the sting. A few insects grow fungi as food and social insects store food. Amongst the social insects feeding of one insect by another commonly occurs and this behaviour may provide a basis for the whole social system.

Different Types of Feeding

As stated already that insects feed on a wide variety of animals, vegetable and dead materials. Any classification of feeding habits is arbitrary, but Brues (1946) recognises four fairly comprehensive categories–plant feeders, predators, scavengers and parasites. Nearly half of the species of insects feed on plants and these may be further subdivided into those feeding on green plants (*phytophagous*)

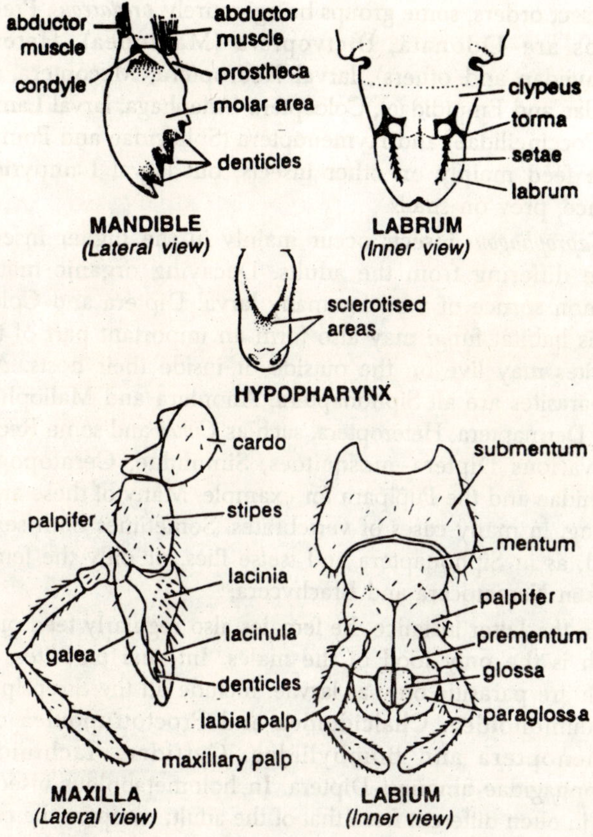

Fig. 2.1. P. americana. Mouth parts.

and those feeding on fungi (*mycetophagous*). Predominantly phytophagous groups of insects are–orthroptera, Lepidoptera, Homoptera, Thysanoptera, Phasmida, Isoptera, Coleoptera (families Cerambycidae, Chrysomelidae and Curculionidae), Hymenoptera (Symphyta) and some Diptera. Most of these feed on higher plants, but, for instance, the aquatic larvae of Ephemeroptera and some Plecoptera and Trichoptera feed on algae.

Fungus-feeding larvae are frequent amongst Diptera (particularly Mycetophilidae) and the habit also occurs in various Coleoptera. In many other insects fungi form at least a part of the diet. This is true in many dung-feeding insects and others, such as some termites, which cultivate their own fungi. Some predators occur in most of the insect orders, some groups being entirely *predaceous*. Predaceous groups are–Odonata, Dictyoptera (Mantodea), Heteroptera (Raduviidae and others), larval Neuroptera, Mecoptera, Diptera (Asilidae and Empididae), Coleoptera (Adephaga, larval Lampyridae and Coccinellidae) and Hymenoptera (Sphecidae and Pompilidae). These feed mainly on other insects, but larval Lampyridae, for instance, prey on snails.

Saprophagous insects occur mainly in the higher insects with larvae differing from the adults. Decaying organic matter is a common soruce of food for many larval Diptera and Coleoptera. In this habitat fungi may also form an important part of the diet. Parasites may live on the outside or inside their hosts. Amongst ectoparasites are all Siphonaptera, Anoplura and Mallophaga and some Dermaptera, Heteroptera, such as *Cimex* and some Reduviidae, and various Diptera–mosquitoes, Simulidae, Ceratopogonidae, Tabanidae and the Pupipara for example. Many of these are blood-sucking, in many cases of vertebrates. Sometimes both sexes suck blood, as in Siphonaptera and tsetse flies, or only the females do so, as in Nematocera and Brachycera.

In the latter instance the females also regularly feed on nectar, which is the only food of the males. Internal parasites, most of which are parasitic only as larvae, include all the Strepsiptera, the Ichneumonoidea, Chalcidoidea and Proctotrupoidea amongst Hymenoptera and Bombyliidae, Cyrtidae, Tachinidae and Sarcophagidae amongst Diptera. In holometabolous insects larval food in often different from that of the adult; compare the caterpillar and the butterfly.

Methods of Obtaining Food

There are many insect species which are surrounded by an abundance of food from the time of hatching. So the problem of finding food does not arise for them. This usually results from the oviposition habits of parent as in many *phytophagous forms* in which the female oviposits on the plant which forms the larval food. Simialrly amongst *saprophagous* and *endoparasitic* insects the female lays her eggs in suitable detritus or in the appropriate host. The larvae of social insects are presented with food by the workers and are incapable of searching for it, but, with these exceptions, food-finding mut play an important part in the lives of many insects. The initial attraction of insects to their food from a distance is ofen not very specific. Final recognition of the food usually occurs at closer quarters and involves different stimuli from those involved in attraction from a distance. The problem of finding food has been studied in a variety of insects as given under the following heads:

Plant Eating Insects

As described earlier that the phytophagous insects are those feeding on green plants. Attraction to a food plant from a distance can occur visually of by olfaction, probably varying very much with the species and the situation. *Schistocera* (Orthoptera), for instance, is attracted by the sight of any solid object of an appropriate size and especially by a pattern of vertical stripes, but such attraction can usually only be a very general because of the variability in form of most plants. Colour may also play a part in recognition. Under experimental conditions aphids are attracted by yellow and this may be related to the fact that they prefer young or senscent leaves, which are often yellowish, to mature, green ones. Form and colour are important to bees in findings flowers and colour is probably important in many species feeding on the flowers themselves. Because of the general lack of specificity of visual factors.

In Insects the olfaction plays an important part in arriving at and recognising the food. Because of the general lack of specificity of visual factors. Thorsteinson (1960) suggests that initial effect of odours is to inhibit locomotion so that once a plant is found, presumably largely as a result of random searching, there is little tendency for the insect to move away again. Such odours he calls 'aggregants' and they are only effective over short distances. There are, however, instances of attraction to a more distant source of smell. Larval *Schistocerca,* for instance, respond to the smell of food over a meter away by moving upwind towards it, but only if they are starved before hand. At close quarters chemoreception–olfaction, contact chemoreception and gustation–are important in food plant recognition. Bees, having made a visually directed approach to a flower, are guided by the flower smells and by their own colony smells previously left on the flower. Contact chemoreceptors commonly occur on the tarsi, and stimulation of the tarsi of blowflies and butterflies with sugar leads to extension of hte proboscis.

Similar receptors on the proboscis promote feeding and others within the mouthparts lead to continued feeding provided the food

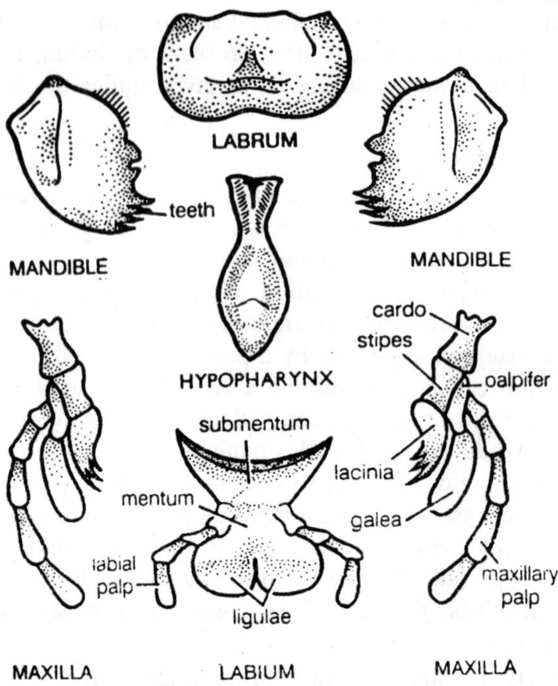

Fig. 2.2. Grasshopper. Mouth parts.

is suitable. In nectar-sucking insects such as the bee, the strength of these responses increases with the concentration of sugars so that the insect always takes the most concentrated odd available. Host plant selection in aphids occurs mainly after alighting when the insect probes the plant with its proboscis, testing the physical and chemical properties of sap. Kennedy and Booth (1951) suggest that selection is based on two sets of stimuli. Some stimulating factors are non-nutritious, but are specific chemicals relating to the taxonomic position of the plant as a result of which the aphid tends to prefer particular plant species.

Other factors are nutritional and indecative of the physiological condition of plant. These act via the plant sap or physiologically related factors. The two sets of factors may oppose each other since ofen a taxonomically suitable host is not in a suitable physiological condition and *vice versa.* The choice made by the aphid is a balance between the two. In biting and chewing insects like the locust, once contact is made with potential food the insect rapidly vibrates its palps so that these touch the surface of the food and are stimulated chemotactically. Biting, which follows, is a non-specific response and if the insect is sufficiently hungry it will bite substances which are normally rejected although these may subsequently be rejected as a result of gustatory stimuli. Continued feeding results from the presence of attractive substances, including nutritive ones, in the food and in *Camnula* (Orthoptera) feeding is induced by the sugars and amino acids normally present in wheat leaves on which the insect feeds.

In the larva of *Bombyx*, Hamamura *et. al.* (1962) have analysed the factors promoting feeding and have isolated separate factors which lead to attraction, biting and continued swallowing. The chemical b-sitosterol and morin have been isolated as biting factors, and celluloe is a swallowing factor whose effect is enhanced by a number of co-factors such as sucrose, inositol, inorganic phosphate and silica. In general it appears that suitable gustatory stimuli are important for continued feeding.

Predaceous Insects

There are several species of predaceous insects catching their prey either by sitting and waiting for it to come their way or by actively pursuing it. *Mantis,* for example, sits and waits for its prey and, as it has a very mobile head, the movements of the prey can be followed without the whole mantis moving. The eyes are large and wide apart enabling the mantis to judge its distance from the prey accurately. When striking at the prey, *Mantis* can compensate for the position of the head relative to the thorax, which is indicated by proprioceptive hairs at the front of the prothorax, and for

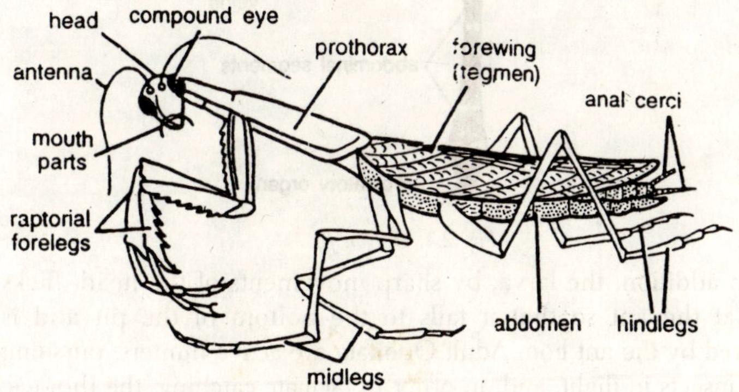

Fig. 2.3. Mantis religiosa (Praying mantis).

deviation of the prey from the optical axis between the two eyes. The front legs of the mantis are raptorial and armed with spines so that when the prey is within range it is caught and held by a rapid movement of the forelegs (extension only takes 30-60 msec) and then brought back to the mouth.

In a comparable way some dragon fly larvae wait for their prey, lying concealed in the mud at the bottom of a pool and seizing the prey with the labial mask. This is a modification of the labium in which pre- and post-mentum are elongated and the palps are modified to form grasping organs. The mask can be extended in front of the head by an increase in blood pressure and the palps are used to catch the prey. By withdrawl of the mask the prey is carried back to the mandibles. A few insects make traps in which they catch their prey. Ant lions, for instance, dig pits one to two inches in diameter with sloping sides in dry sand and then bury themselves at the bottom with only the head exposed. If an ant walks over the edge of such a pit it has difficulty in regaining the top because of the instability of the sides.

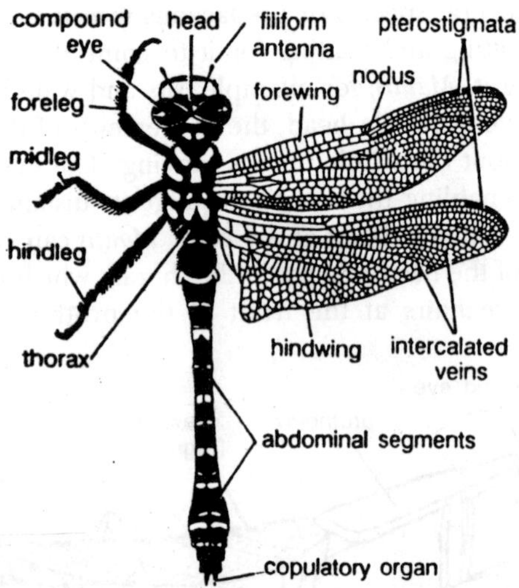

Fig. 2.4. Dragon fly.

In addition, the larva, by sharp movements of the head, flicks sand at the ant so that it falls to the bottom of the pit and is captured by the ant lion. Adult Odonata are active hunters, pursuing other insects in flight, and, in order to faciliate catching, the thoracic segments are rotated forwards so as to bring the legs into an anterior position. Tiger beetles hunts on the ground and have long legs, which increase their speed, and prognathous mouthparts with large mandibles.

The majority of hunters have well-developed eyes since only vision can give a sufficiently rapid directed response to moving prey. This reaction is usually not specific and the predator will pursue any moving object of suitable size. Thus dragonflies turn towards small stones thrown into the air and the wasp *Philanthus* orientates to a variety of moving insects of appropriate size, although it only catches bees. This subsequent recognition involves other senses and only if the insect has the smell of a bee does *Philanthus* attempt to capture it. Provided it has the appropriate smell the wasp will attack any insect of the right size, but, having caught it, stinging does not follow unless the insect really is a bee, other insects experimentally given the smell of a bee are released. This final recognition is presumably tactile. Comparable behaviour is probably common amongst other parasites and predators.

In predaceous larval forms with poorly developed eyes and often with subterranean habits, such as a tabanid larvae, the finding of prey must be largely olfactory. *Dytiscus* (Coleoptera) also responds to chemical stimuli in the water, rather than the sight of prey. Mechanical stimulation is sometimes important in finding prey and some dragonfly larvae depend on mechanoreceptors on the antennae or tarsi for this. *Notonecta* is able to locate prey trapped in the air-water interface as a result of the ripples which radiate from it, perceiving the vibrations with sensory hairs on the swimming legs.

Amongst terrestrial insects the response of ant lions to their prey depends on mechanical stimulation from the falling sand as well as on vision. Coccinellid (Coleoptera) larvae preying on asphids only respond to the prey on contact. They move across a leaf, searching to either side as they go, but after finding and eating an aphid–they tend to remain in the same area by making numerous small turning movements. Since aphids usually occur in large numbers this is clearly an advantage and this type of behaviour is only suited to prey which is more or less sedentary and locally abundant.

Blood-sucking Insects

Blood sucking insects are called *sanguivorous.* Perception of a host at a distance may arise from visual, olfactory or mechanical stimulation, depending on the species and the situation. *Glossina*

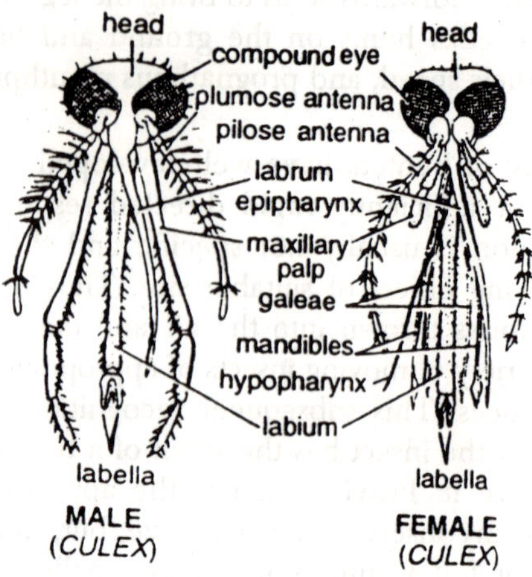

Fig. 2.5. Culex. Head and mouthparts.

swynnertoni, a tsetse fly inhabiting relatively open savannah, can see cattle moving 450 ft. away, but *G. medicorum,* from dense forest and thicket, only reacts to a moving screen at distances under 25 ft. Movement of a potential host increases the likelihood of the insect reacting to it. The smell of the host is also important and where vision is limited, as in *G. medicorum,* may be particularly important. if an insect responds to the smell of a host by taking off and then orientates upwind it will fly into the vicinity of the source of smell.

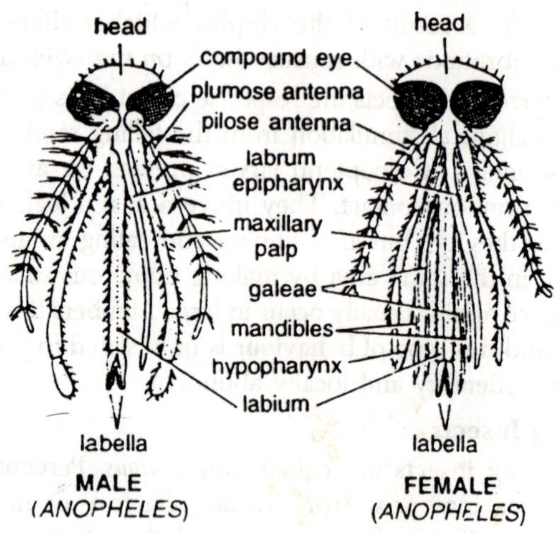

Fig. 2.6. Anopheles. Head and mouthparts.

At closer quarters vision may become more important again. Mosquitoes react to hosts at a distance in similar ways. At closer quarters other factors also play a part in attraction. In addition to smell, moisture and warmth are important to mosquitoes and settling depends on the nature of the surface. Mosquitoes settle moe repidly rough than on smooth surfaces and often on dark rather than light ones. After settling, probing the host tissues with the proboscis is induced by olfactory stimuli and also by the warmth of the host.

Parasitic Mode of Nutrition

In the majority of internal parasites the parent insect oviposits in a suitable host. Smell, and possibly also contact chemoreception, are involved in this. In some cases, however, the parent does not seek out the larval host, but oviposits or larviposits in places frequented by the host so that the larvae make their own way on to the host when the occasion presents itself. For instance, the human warble fly, *Cordylobia anthro-pophaga,* oviposits in sand fouled with urine. The larvae hatch in a day or two and then remain inactive until the area is visited by a potential host, man or some other mammal. They are activated by the vibrations and warmth of the host and bore is through the skin. In a very few insects the larvae act as the dispersive phase and reinfect new hosts. This is the case in Strepsiptera and the meloid beetles, both of which produce vast numbers of larvae known as triungulins.

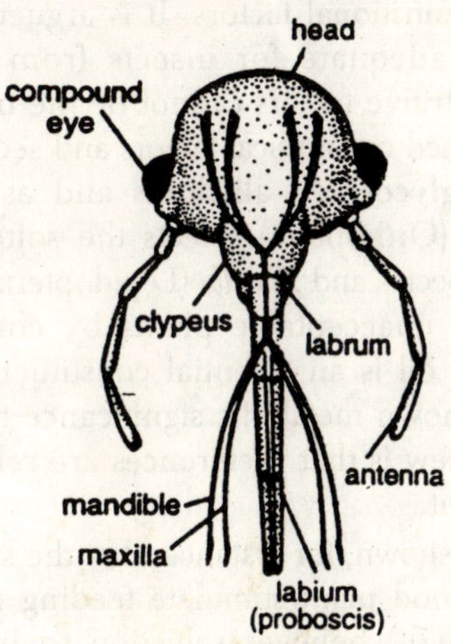

Fig. 2.7. Cimex. Head and mouthparts.

Strepsipteran triungulins escape from the female, which is an internal parasite, when the host is visiting a flower. They remain in the flower until another insect arrives and then jump on to it. If this is the correct host they remain clinging to it and, depending on the species, parasites it or its offspring; if it is not the appropriate host they jump off again. Some meloids find their hosts in essentially similar ways, while in others, parasitic in grasshoppers' eggs, the triungulins actively seek out the eggs.

Plant Eating Insects

The degree of specificity of insects to particular plants varies considerably. Some species, particularly amongst Homoptera and sawflies, are restricted to one particular plant species and are regarded as monophagous. *Coccus fagi,* for instance, only feeds on beech, and the larvae of the sawfly, *Xyela julii,* only on *Pinus sylvestris.* Some such species may feed on other plants if they are forced to do so, but others will die in the absence of the correct plant. Other insects are less restricted, but nevertheless feed on only a limited range of plants. They are called oligophagous. *Pieris rapae* (Lepidoptera), for instance, feeds only on Cruciferae and other plants which contain mustard oils.

Finally, there are polyphagous insects which feed on a very wide range of plants, but even these show preferences for particular foods. *Schistocerca* is polyphagous. There are two points of view with regard to the basis of food preference. One suggests that choice is governed by non-nutritional factors. It is argued that the leaves of most plants are adequate for insects from the nutritional standpoint so that nutritive factors cannot be the basis of selection. Rather, selection is based on physical factors and secondary chemical substances such as glycosides, alkaloids and essential oils. For instance, *Nomadacris* (Orthoptera) selects the softest and moistest grass irrespective of species and *Plutella* (Lepidoptera) can be induced to feed on normally unacceptable plants by coating them with mustard oil. Mustard oil is an essential constituent of the normal food plant with no known metabolic significance to the plant. The alternative point of view is that preferences are related to nutritive substances in the plant.

Thorsteinson has shown, for instance, that the sugars and amino acids of the normal food plant stimulate feeding in *Camnula* and, whereas Dethier (1947b) believes selection to be based on the variation in attraction by different chemicals, Thorsteinson considers

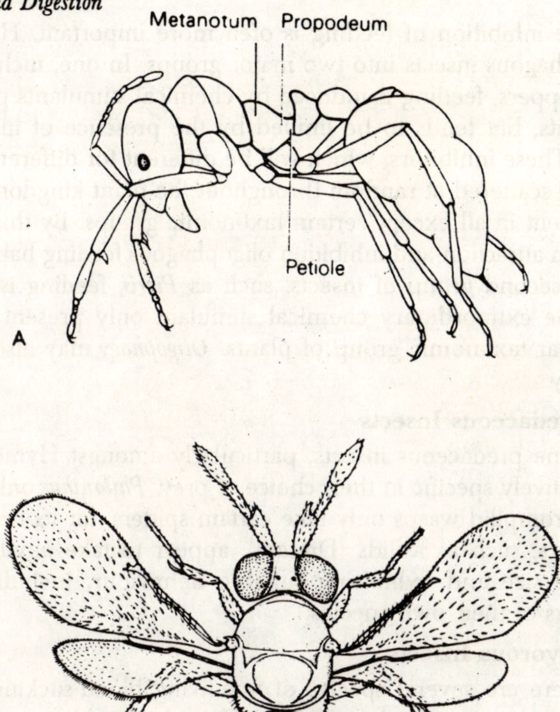

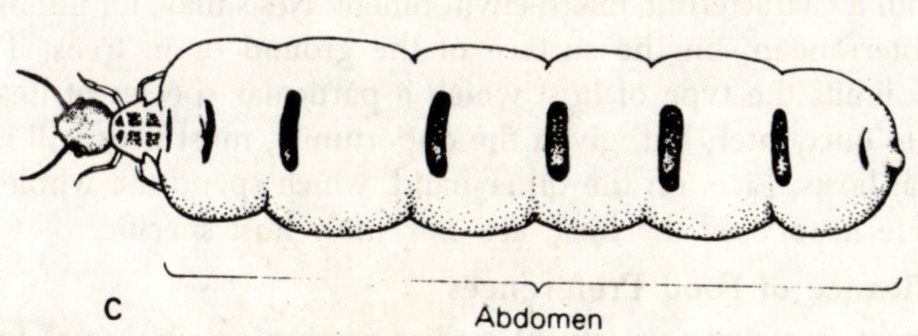

Fig. 2.8. Modification of the abdomen. A–Fire ant. B–Parasitic wasp, Eusemion sp. C–Gravid termite queen.

that the inhibition of feeding is ofen more important. He divides phytophagous insects into two major groups. In one, including the grasshoppers, feeding is induced by chemical stimulants present in all plants, but tends to be limited by the presence of inibitors in some. These inhibitors, which will be different for different insects, may be scattered at random throughout the plant kingdom or may be present in all except certain taxonomic groups. By this balance between attraction and inhibition oligophagous feeding habits result. In the second group of insects, such as *Pieris,* feeding is induced by some extraordinary chemical stimulant only present in some particular taxonomic group of plants. *Oligophagy* may also arise in this way.

The Predaceous Insects

Some predaceous insects, particularly amongst Hymenoptera, are relatively specific in their choice of prey. *Philanthus* only catches bees, prompilid wasps only take certain spiders, for its nest. Other predators, such as asilids (Diptera), appear to take anything of a suitable size and exhibiting suitable behaviour, including even numbers of their own species.

Sanguivorous Insects

There are several species of free-living blood-sucking insects, such as mosquitoes and tsetse flies. These insects bite a wide range of hosts, although most are restricted to mammals and birds. Within this range tsetse flies, at least, show distinct preferences. *Glossina morsitans* feeds primarily on ungulates, which are essential for its continued existence, while man and reptiles are particularly important of *G. palpalis.* The less mobile blood-suckers are more specific. This is true to some extent of fleas, although here the specificity is largely ecological.

Development of fleas usually occurs in the nest of the host and particular species tend to be restricted to a particular type of nest with a characteristic micro-environment. Nests may, for instance, be subterranean, on the surface of the ground or in trees. This clearly limits the type of host which a particular species of flea is likely to encounter, but, given the opportunity, most fleas will bite unusual hosts. Lice, on the other hand, which spend the whole of their life history on the host, are extremely host specific.

Significance of Food Preferences

There is a wide variety of insects rearing on abnormal food substances, food preferences are often of considerable significance.

Melanoplus, for instance, fails to survive on some food plants, but on its preferred food, hedge mustard, development and survival are good. Even so development takes longer and fewer eggs are laid than with a diet containing a variety of plants. Similarly, mosquitoes can survive solely on nectar, but many species require a blood meal if they are to produce eggs. Such species are called *anautogenous;* others, which can produce eggs on a diet of nectar alone, are called autogenous.

Forms of Food

There is some evidence that insects can become conditioned to particular food. *Carausius* (Phasmida) normally feeds on privet and will reject ivy, but if it is forced to feed on ivy it ultimately tends to prefer this to privet and, comparably, locust larvae reared on an artificial diet can only be induced to eat grass, the normal food, with difficulty. In the case of *Carausius* the altered preference is passed on to the next generation. There is some suggestion that the food first eaten by a phytophagous insects, conditions its subsequent behaviour so that it shows a preference for this food.

Ingestion

Once the insect has recognised its food as suitable it starts to feed. The processes by which food is ingested vary considerably.

In Plant-eating Insects

Phytophagous insects are typically plant-feeding insects with biting mouthparts. They bite off fragments of food and pass them back to the mouth with the aid of the maxillae while grasshoppers also help to guide the food into the mouth by holding it between their forelegs. Often such insects feed at the edge of a leaf, moving on towards the centre, and usually the more woody parts are avoided. Some insects may obtain their food from the cell sap or directly from distributive vessels and are called the fluid feeders.

Aphids, for instance, usually tap the phloem. When an aphid lands it inserts its mouthparts into the plant tissue using the protractor and retractor muscles of the stylets in the head, probably aided by a clasping action of the labium. In the course of penetration through the leaf epidermis and parenchyma it encounters mechanical resistance and this to some extent affects selection of the feeding site, but it is probable that this resistance is partially overcome by the saliva dissolving the middle lamellae between the plant cells. As the stylets penetrate the tissues saliva flows from the tip and

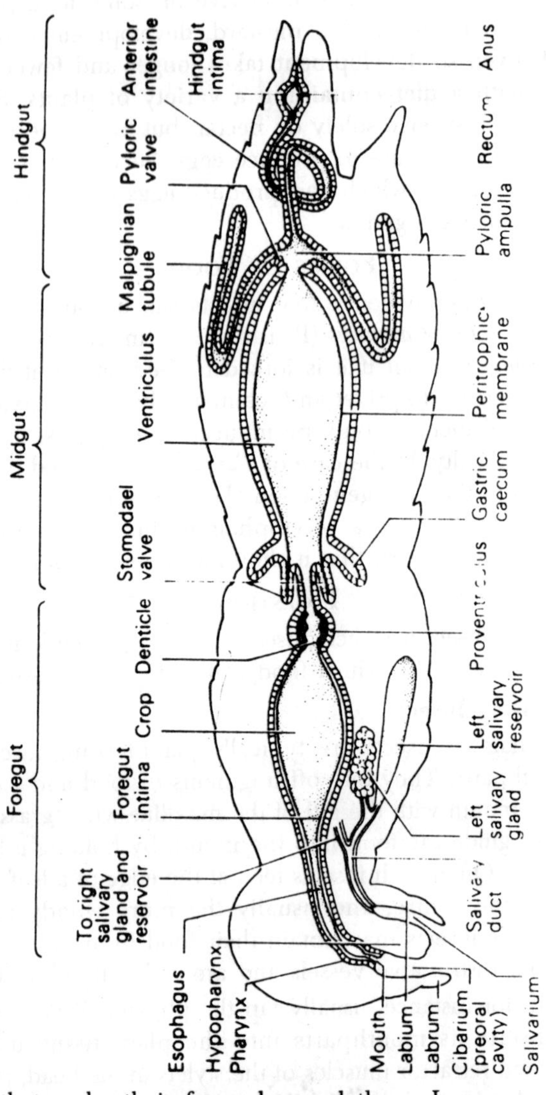

Fig. 2.9. Generalized insect alimentary system.

gels so that a sheath is formed round them. In some species, at least, no feeding takes place until the stylets reach the phloem and in *Aphis* penetrating to this tissue takes about an hour. The subsequent intake of sap is largely passive due to the pressure within the plant forcing sap into the stylets, but the rate of flow is controlled by the aphid since if it is starved or is attended by ants the rate of feeding is increased.

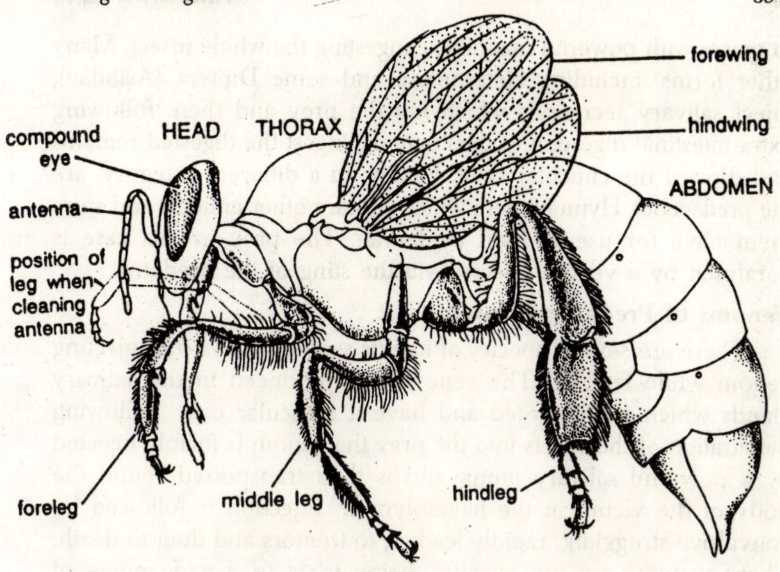

Fig. 2.10. Worker honeybee (Lateral view).

In Predaceous Insects

The predaceous insects such as mantis capture their prey, restrain the victim by sheer mechanical strength and then tear it

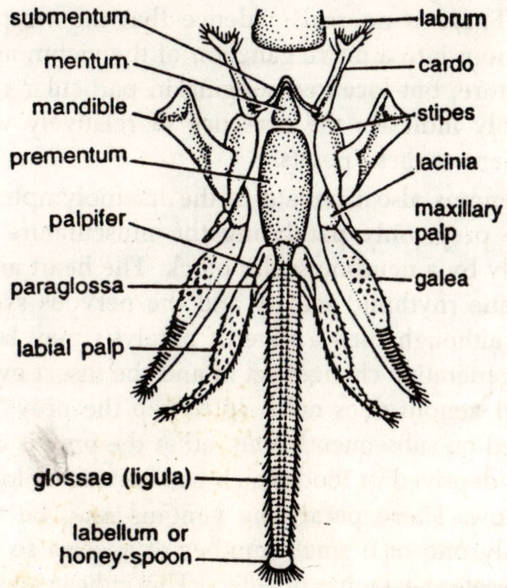

Fig. 2.11. Honeybee. Mouthparts.

to pieces with powerful mandibles, ingesting the whole insect. Many other forms, including Heteroptera and some Diptera (Asilidae), inject salivary secretions which kill the prey and then, following extra-intestinal digestion of the contents, ingest the digested remains and discard the cuticular shell. Finally, in a different category, are the predaceous Hymenoptera which capture other animals and store them alive for use as food by larvae. The prey in this case is paralysed by a venom injected via the sting of the predator.

Venoms of Predaceous Insects

There are several species of insects such as *Platymeris,* injecting venom while feeding. The venoms are produced in the salivary glands which are enlarged and have a muscular coat. Following penetration of the stylets into the prey the venom is focibly injected by a powerful salivary pump and is then transported round the body of the victim in the haemolymph. Injection is followed by convulsive struggling, rapidly leading to tremors and then to death. These venoms are non-specific, being toxic to a wide range of insects and their action is to cause a general lysis of the tissues so that nervous activity rapidly stops. Similar venoms are probably used by larval Neuroptera and by Asilidae and Empididae (Diptera) and Odonata. The member of group hymenoptera paralyse their prey by injecting the venom via the sting, which is modified ovipositor. There is no real evidence that the wasp attempts to inject its venom into a nerve ganglion of the victim as is suggested in the literature, but localised stinging in particular regions of the prey probably indicates the presence of relatively weak spots in the integument at these points.

Thse venoms also circulate in the haemolymph, but they do not kill the prey, only paralysing the musculature of the body wall, possibly by a neuromuscular block. The heart and alimentary canal continue rhythmic activity and the nervous system remains active, but, although such a state of paralysis may last for several months, degenerative changes set in and the insect eventually dies. The injected venom does not help to kep the prey 'fresh' for the larvae to feed on subsequently, but rather the prey is in the position of an insect deprived of food which can survive as long as its food reserves allow. These paralysing venoms may be very specific, affecting only one or a small number of species so that injection an inappropriate species has no effect. This indicates very specialised and specific chemical configuration for the venoms.

Sanguivorous Insects

After a tsetse fly lands on its host is spreads its legs apart, grips the skin with its claws and braces itself so as to be able to exert a downward pressure. Then, it lowers the haustellum and pierces the skin by the rasping action of the labella. Feeding never follows the initial probe and the haustellum is partly withdrawn, the head moved and a new thrust made in a different direction. In this way the fly causes a local haemorrhage in the tissues and it drinks from the pool of blood which is produced. The whole process of feeding to repletion takes about two minutes in *G. morsitans.* In mosquitoes the stylets bend inside the host tissues and often penetrate a blood capillary from which blood is then-drawn directly. This also occur in *Rhodnius* where, as in mosquitoes, the stylets penetrate the tissues as a result of the action of their protractor and retractor muscles following an initial thrust from the leg musculature. In sanguivorous insect the saliva contains an anticoagulant. These insects inject saliva into the wound. This prevents the blood from clotting in the wound and in the proboscis of the insect, but the saliva of some such insects, notably *Aedes aegypti,* contains no anticoagulant and the blood clots in the stomach within 15 minutes of feeding. The absence of the anticogulant, however, does not impair feeding.

Blood is sucked up by the action of the cibarial pump and feeding continues until the insect is enormously distended; *Rhodnius* may increase its weight by more than six times in a 15-minute feed. The size of the meal in *Rhodnius* is limited by the extent to which the abdomen will enlarge, which is determined by the epicuticle. The abdominal cuticle consists only of undifferentiated endocuticle and strongly folded epicuticle and expansion can continue only until the epicuticle is smooth. Expansion is facilitated by plasticisation of the endocuticle so that it becomes less rigid at an early stage, during the meal. This plasticisation is probably brought about by a neurosecretion which reaches the body wall through the abdominal nerves. It is only a temporary effect and decreases after feeding.

Insects Growing Fungus

Some insects grow fungi on specially prepared substrates as in termites and ants. All the Macrotermitinae are fungus growers and they produce a 'comb' of chewed wood on which the fungal hyphae grow and produce conidia. Some of the fungi, such as *Xylaria,* an

Ascomycete, are not confined to termite nests, but the Termitomyces, which are Basidiomycetes, occur only in termitaria. *Xylaria* only produces fruiting bodies when the nest is deserted by the termites. The fungus is eaten in small amounts by the workers and is fed to some of the larvae. The wood boring insects also have constant associations with particular fungi although they do not prepare specific substrates.

Amongst these are the ambrosia beetles, various species of Scolytidae and Platypodidae and all Lymexilidae. These beetles make their tunnels in and under the bark of trees, but the bulk of their food is derived from fungi growing in the tunnels. The fungi are mostly rather specialised ones some of which may be connected with a variety of different beetles while others are constantly associated with one or a few species. They are transmitted by the female beetle in depression of the body surface in which secretions of oil from associated glands accumulate. When the beetle starts actively boring the output of oil is increased and the ideal cells of the fungus are washed out. These then germinate on the tunnel walls of produce a new growth of ambrosia.

Feeding Variations

In insects the feeding is affected by the summation of a series of internal and external stimuli. Pupae do not feed at all and in Ephemeroptera, Lepidoptera and Oestridae (Diptera) there are some species which do not feed as adults, often having reduced mouthparts. Feeding does not occur in newly emerged insects and is reduced or non-existent during diapause and at the time of moulting. Female mosquitoes do not feed while they are producing eggs. The state of feeding is also important and the tendency to feed is greatly reduced after a meal and may even be physically impossible as in *Rhodnius,* which only takes a single large meal in each larval instar. Tsetse flies are not attracted to their hosts for two or three days after feeding and phytophagous insects have a period of postprandial quiescence during which feeding activity, as well as other activities, is reduced. There is often also some diurnal variation in feeding, directly related to the existing environmental conditions, such as light and temperature, which may be limiting, but, as with other activities, changes in these conditions may be important in stimulating feeding.

Nomadacaris, for instance, feeds mainly in the morning and evening when conditions are changing rapidly. Many mosquitoes are crepuscular in their biting habits and, although biting is influenced by climatic factors, the timing appears to result partly from an endogenous rhythm of activity, changes of light intensity merely acting as time cues. Biting activity also varies with the habit. In forest regions *Aedes Africans* bites by day at ground level, but is crepuscular in the canopy, and *Mansonia fuscopennata* also bites at different times in different situations.

Insects Storing Food

In a great variety of insects, temporary internal stores of food are built up in their fat bodies or in crop. However, external storage of food has been observed in the social and subsocial insects. Many solitary Hymenoptera (Sphecoidea and Pompiloidea) build cells and provision them for the use of their larvae. Some, such as *Ammophila,* exhibit progressive provisioning of the nest. *Ammophila* excavates a hole in the ground, provisions it with a caterpillar and then lays an egg and closes the cell. The wasp then starts a new nest. Each morning on her first flight the female visits each nest, of which there may be three at any one time, and examines its state of provisions. If there is an ample supply she closes the nest and leaves it until the following day, but if there is not much food she brings a fresh supply.

The female continues to bring food to the nest whenever it is required until the larva is well grown when she puts in a final store and seals the nest for the last time. Other solitary wasps, like *Eumenes,* and solitary bees put a large stock of provisions into the nest at the time the egg is laid and this suffices for the whole of larval development, for the cell is never visited again. This type of provisioning is known as *mass provisioning.* In insects like *Myrmecocystus* the honey ants, the storage of nectar and other sweet substances is done in the crops of certain workers known as *repletes.* These are fed sugars until their gasters become enormously distended so that their movement is greatly restricted and they remain hanging from the roof in special chambers in the nest. The sugar is regurgitated to other workers as it is required.

The food storage becomes most significant in some social insects like honey bee–*Apis,* living in perennial colonies in temperate regions where there is a long period during which fresh food is unobtainable. *Apis* feeds on honey and pollen. Honey is derived from nectar, which has a variable composition, but may consists of some 60% water and a very high proportion of sugars of which

40-50% is sucrose. Only traces of protein are present. When nectar is collected the enzyme invertase is secreted on ot it so that sucrose is broken down to glucose and fructose. On reaching the nest the forager gives its nectar to a house bee who has the task of reducing its water content to about 20%. To achieve this the worker regurgitates a small drop of nectar from the honey stomach, manipulates it with the mandibles and then swallows it again, repeating this process some 80 to 90 times in 20 minutes. By this process water is evaporated and the nectar concentrated. When it has reached a suitable concentration the honey is put into a cell and used for food or is sealed up for use later. Honey provides the carbohydrates and water for adults and larvae; protein is derived from pollen collecting is facilitated by the pectinate hairs characteristics of Apoidea since the pollen grains tend to become caught up in these hairs.

The bee may actively collect pollen, biting anthers so as to increase the amount released, or it may simply become dusted with pollen while in search of nectar. Pollen collected on the head region is brushed off with the forelegs and moistened with a little

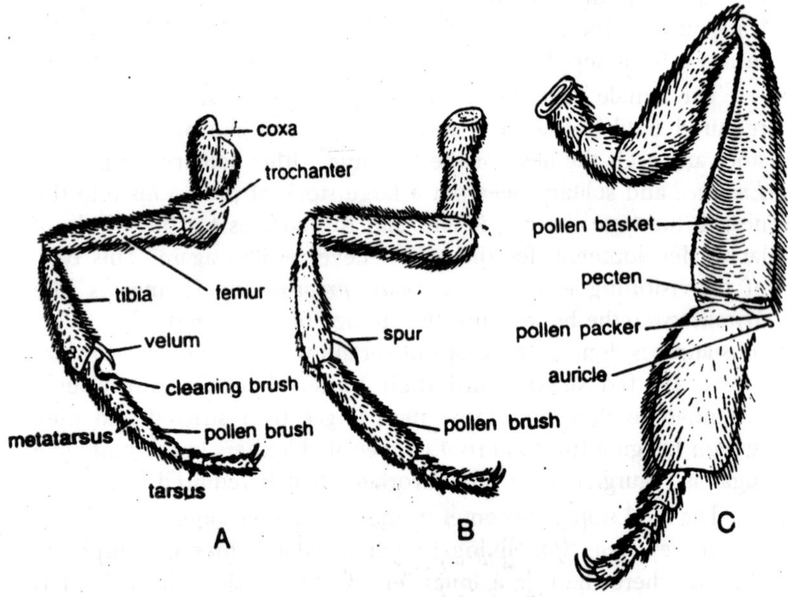

Fig. 2.12. Honeybee legs. A–Prothoracic leg; B–Mesothoracic leg; C–Metathoracic leg..

regurgitated nectar or honey before being passed back to the hind legs which also collect pollen from the abdomen using the combs on the inside of the legs. The pollen on the combs of one side is then removed by the rake of the opposite hind leg and collects in the pollen press between the tibia and metatarsus. By closure of the press pollen is forced outwards and upwards on to the outside of the tibia and is then held in place by the hairs and spines of the pollen basket. On returning to the nest the pollen is kicked off by the middle legs into an empty cell, house bee break up the masses of pollen and pack it down and the cell is then capped or left open for current use.

Social Feeding

Sometimes feeding of one insect by another occurs in non-social insects during courtship as in the presentation of food to the female empid (Diptera) by the male but in general such behaviour occurs only in the social insects. Here, there are often stages–larvae, soldiers or reproductive forms which are incapable of feeding themselves and must be fed by the workers. In wasps the larvae are fed on the masticated remains of other insects. Bees use honey and pollen, but also brood food, a secretion of the hypopharyngeal and mandibular gland of the workers containing protein derived from pollen. Brood food is fed to all larvae for the first three days after hatching and subsequently probably forms an important part of the diet of larvae destined to become queens. The quantity of food eaten also plays some part in queen determination.

In a similar way queen determination in the ant *Myrmica* is related to feeding, larvae destined to become queens apparently having more protein in their diet. This varies with the physiological condition of the workers tending the larvae and possibly again involves glandular secretions by the workers. Termites practise social feeding and in Kalotermitidae proctodael feeding occurs. These insects produce two types of excrement, solid faeces and a liquid containing fragments of wood and intestinal flagellates. Production of this second type of excrement is stimulated by other individuals placing their antennae on the dorsal or perianal region of the worker. Apart from any direct nutritive value this behaviour is important in renewing the intestinal fauna of newly moulted individuals because this fauna is lost each time a termite moults. This behaviour does not occur in Termitidae with no comparable intestinal fanua, although faeces may be eaten.

Trophallaxis

Trophallaxis is a peculiar type of breeding behaviour often seen in social insects where the organisms mutually exchange food. When, for instance, an ant feeds a larva it receives from the larva a drop of salivary fluid, which may be so attractive to the worker that it solicits saliva from the larva without giving anything in return. This mutual exchange of food has been regarded as the basis of social systems in insects, but in the wasp *Vespula sylvestris* it has been shown that the larval saliva, although taken by the workers, is not especially attractive to them. It is probable that the secretion of saliva is a means whereby the larvae eliminate excees water, and its removal by the workers prevents the nest from becoming fouled. Thus in this insect, at least, typical trophallaxis does not occur, and if such a relationship ever existed it has become modified in the course of evolution.

Selection of Food

Visual, chemical, and mechanical stimuli may act as cues which enable an insect to locate and/or select material suitable for ingestion. As food is ingested, it may be tasted, and this determines whether or not feeding continues. The extent of stimulation required to initiate feeding varies with the state of an insect. Starved insects, for example, may become highly sensitive to odors or tastes asssociated with their normal food. In extreme cases, they may become quite indiscriminate in terms of what they ingest. Feeding is terminated either by mechanical stimuli, for example, stretching of the crop wall, or by chemical means, for example, by adaptation of chemoreceptors so that the feeding reflex is not longer initiated. In some plant-feeding (*phytophagous*) species, visual stimuli such as particular patterns (especially stripes) or colours may serve to initially attract an insect to a potential food source. Usually, however, the initial orientation, however this occurs, is dependent on olfactory stimuli.

In many larval forms there appear to be no specific orienting stimuli because, under normal circumstances, larvae remain on the food palnt selected by the mother prior to oviposition. When feeding begins, food is tasted by receptors located either on the mouthparts, especially the palps, or within the pharynx. Tasting determines whether or not intake continues and depends on the presence in the food of specific stimulating or inhibitory substances. These substances may have nutritional value to the insect or may be nutritionally unimportant ("*token stimuli*"). Nutritional factors are almost always stimulating in effect. Sugars, especially sucrose, are important phagostimulants for most phytophagous insects. Amino acids, in contrast, are generally by themselves weakly stimulating or nonstimulating, though may act synergistically with certain sugars or token sitmuli. For example, Heron (1965) showed in the spruce budworm (*Choristoneura fumiferana*) that, whereas sucrose and L-proline in low concentration were individually only weak phagostimulants, a mixture of the two substances was highly stimulating.

In addition to sugars and amino acids, other specific nutrients may stimulate feeding in a given species. Such nutrients include vitamins, phospholipids and steroids. Token stimuli may either stimulate or inhibit feeding. Thus, derivatives of mustard oil, produced by cruciferous plants, including cabbage and its relatives, are important phagostimulants for a variety of insects which normally feed on these plants, for example, larvae of the diamondback moth (*Plutella maculipennis*), the cabbage aphid (*Brevicoryne brassicae*), and the mustard beetle (*Phaedon cochleariae*). Indeed, *Plutella* will feed naturally only on plants which contain mustard oil compounds. Such species, whose choice of food is limited, are said to be *oligophagous.* In exteme cases, an insect may be restricted to feeding on a single plant species and is described as *monophagous.* Species which may feed on a wide variety of plants are *polyphagous,* though it must be noted that even these exhibit selectivity when given a choice. In many *predaceous insects,* especially those which actively pursue prey, vision is of primary importance in localating and capturing food. carnivorous species, especially larval forms, whose visual sense is less well-developed, depend on chemical or tactile stimuli to find prey.

For example, many beetle larvae which live on or in the ground locate prey by their scent; ladybird beetle larvae, on the other hand, must make physical contact with prey in order to determine its presence. Species parasitic on other animals usually locate a host by its scent, though tsetse flies may initially orient by visual means to a potential host. For many species which feed on the blood of birds and mammals, temperature and/or humidity gradients are important in determining the precise location at which an insect alights on a host and begins to feed. The extent of food specificity for carnivorous insects is variable. Many insects are quite nonspecific and will attempt to capture and eat any organism which falls within a given size range (even to the extent of being cannibalisticl!). Others are selective, for example, spider wasps (Pompilidae), as their name indicates, capture only spiders for provisioning their nest. Parasitic insects, too. Exhibit various degrees of host specificity.

Thus, certain sarcophagid flies parasitize a range of grasshopper species; the common cattle grab (*Hypoderma lineata*) is typically found on cattle or bison, rarely on horses and man; lice are extremely host-specific, as would be expected of sedentary species. Apart from the specific cues outlined above that facilitate location and selection of food, there are other factors that influence feeding activity. Typically insects do not feed shortly before and after a molt or when there are mature eggs in the abdomen. In addition, a diurnal rhythm of feeding activity may occur, in response to a specific light, temperature, or humidity stimulus.

For example, the red locust (*Nomadacris septemfasciata*) feeds in the morning and evening, and many mosquitoes feed during the early evening (though this may change in different habitats). Pupae, most insects when in diapause, and some adult Ephemeroptera. Lepidoptera, and Diptera, do not feed.

Digestive System

The alimentary canal or gut is a tube like organ. The gut and its associated glands triturate, lubricatem, store, digest, and absorb food material and expel the undigested remains. Structural differences throughout the system reflect regional specialization for performance of these functions and are correlated also with feeding habits and the nature of normal food material. The structure of the system may vary at different stages of the life history because of the different feeding habits of the larva and adult of a species. The gut normally occurs as a continuous tube between

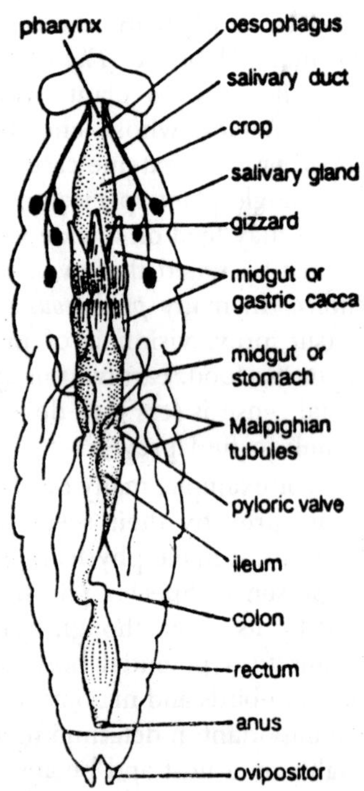

Fig. 2.13. Grasshopper. Alimentary canal in dorsal view.

the mouth and anus, and its length is broadly correlated with feeding habits, being short in carnivorous forms where digestion and absorption occur relatively rapidly, and longer (often convoluted) in phytophagous forms.

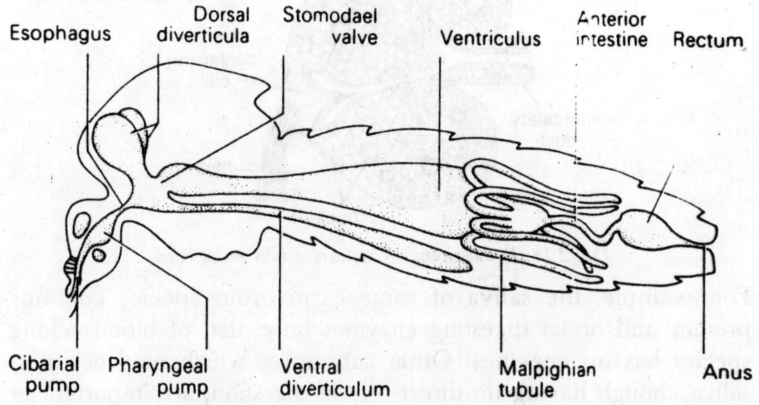

Fig. 2.14. Alimentary canal of a mosquito.

In a few species which feed on fluids, such as larvae of Neuroptera and Hymenoptera Apocrita and some adult Heteroptera, there is little or no solid waste in the food, and the junction between the midgut and hindgut is occulated. As fig. indicates, food first enters the buccal cavity, which is enclosed by the mouthparts and is not strictly part of the gut. It is into the buccal cavity that the salivary glands release their products. The gut proper comprises three main regions: the foregut, in which the food may be stored, filtered, and partially digested; the midgut, which is the primary site for digestion and absorption of food; and the hindgut, where some absorption and feces formation occur.

Salivary Glands

Salivary glands are present in most insects, though their forms and function are extremely varied. Frequently they are known by other names according to either the site at which their duct enters the buccal cavity, for example, labial glands and mandibular glands, or their function, for example, silk glands and venom glands. Typically, saliva is a watery, enzyme-containing fluid which serves to lubricate the food and initiate its digestion. Like that of man, the saliva generally contains only carbohydrate-digesting enzymes (*amylase* and *invertase*), though there are exceptions to this statement.

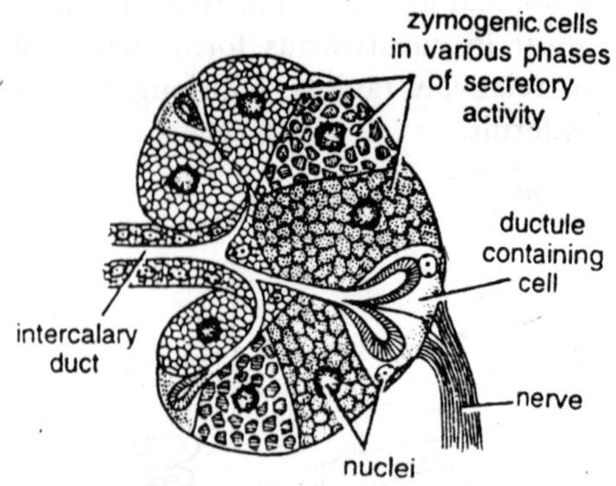

Fig. 2.15. P. americana. A salivary acinus in secretion.

For example, the saliva of some carnivorous species contains protein–and/or fat digesting enzymes only: that of bloodsucking species has no enzymes. Other substances which may occur in saliva, though having no direct role in digestion, are important in food acquisition.

The saliva of aphids contains a *pectinase* that facilitates penetration of the stylets through the intercellular spaces of plant tissues. *Hyaluronidase,* an enzyme which breaks down connective tissue, is secreted by some insects which suck animal tissue fluids. *Anticoagulants* are present in the saliva of bloodsucking species such as mosquitoes. Toxins (venoms), which paralyze or kill the prey, occur in the saliva of some assassin bugs (Pentatomidae) and robber flies (Asilidae). It is also reported that substances which induce gall formation by stimulating cell division and elongation ae present in the saliva of some gall-inhabiting species. In some species the glands have taken on functions quite unrelated to digestion. For example, production of silk by the labial glands of caterpillars and caddis fly larvae, and pheromone production by the mandibular glands of the queen honeybee.

Foregut

It is the anterior part of alimentary canal formed during embryogenesis by invagination of the integument, is lined with cuticle (the intima) which is shed at each molt. Surroudning the intima, which may be folded to enable the gut to stretch when filled, is a thin epidermis, small bundles of longitudinal muscle, a thick layer of circular muscle, and a layer of connective tissue through which run nerves and tracheae. The foregut is generally

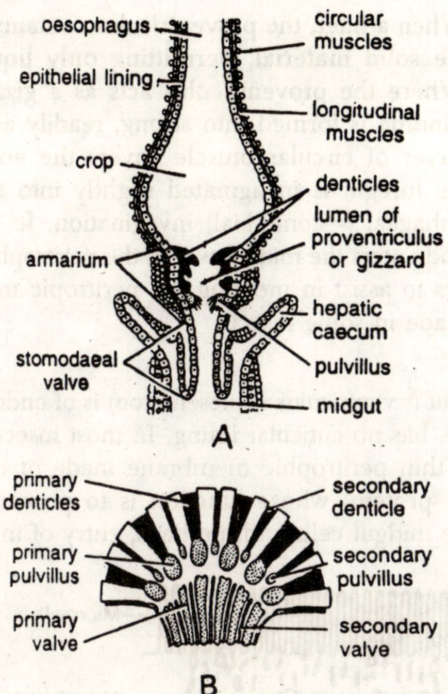

Fig. 2.16. P. americana. A–Crop and gizzard in L.S. B–Proventriculus split longitudinally and laid open.

differentiated into pharynx, esophagus, crop, and proventriculus. Attached to the pharyngeal intima are dilator muscles. These are especially well-developed in sucking insects and forms the pharyngeal pump.

The esophagus is usually narrow but posteriorly may be dilated to form the crop where food is stored. During storage the food may undergo some digestion in insects whose saliva contains enzymes or which regurgitate digestive fluid from the midgut. In some species the intima of the crop forms spines or ridges which probably aid in breaking up solid food into smaller particles and mixing in the digestive fluid. The hindmost region of the foregut is the proventriculus, which may serve as a valve regulating the rate at which food enters the midgut, as a filter separating liquid and solid components, or as a grinder to further break up solid material. Its structure is, accordingly, quite varied. In species where it acts as a valve the intima of the proventriculus may form logitudinal folds and the circular muscle layer is thickened to form a sphincter. When a filter, the proventriculus contains spines which hold back the solid material, permitting only liquids to move posteriorly. Where the proventriculus acts as a gizzard, grinding up food, the intima is formed into strong, readily arranged teeth, and a thick-layer of circular muscle covers the entire structure. Posteriorly the foregut is invaginated slightly into the midgut to form the esophageal (=stomodeal) invagination. Its function is to ensure that food enters the midgut within the peritrophic membrane. It also appears to assist in molding the peritropic membrane into the correct shape in some insects.

Midgut

The midgut (=ventriculus = mesenteron) is of endodermal origin and, therefore, has no cuticular lining. In most insects, however, it is lined by a thin peritrophic membrane made of chitin (and, in some species, protein) whose function is to prevent mechanical damage to the midgut cells and, perhaps, entry of microorganisms

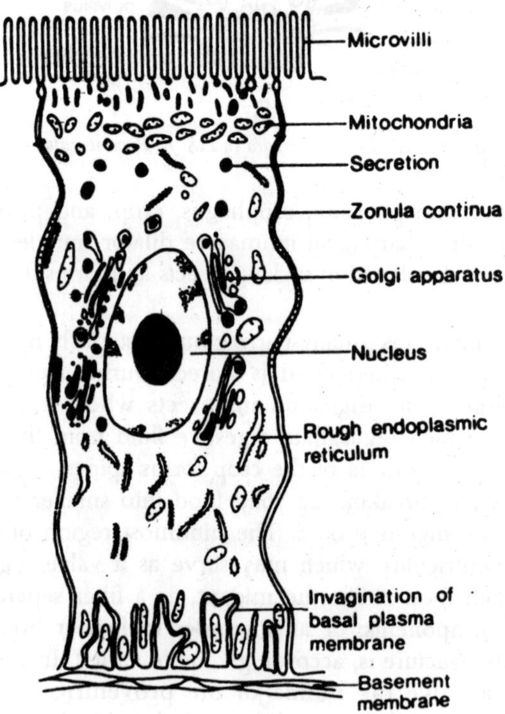

Fig. 2.17. Midgut epithelial cells–columnar.

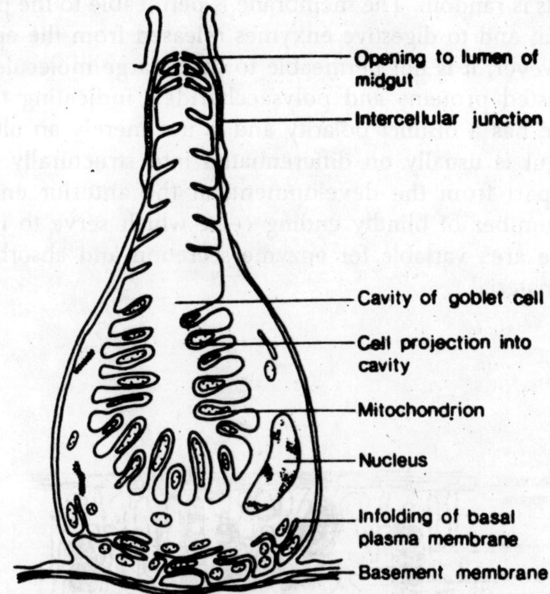

Fig. 2.18. Midgut epithelial cells–globlet cells.

into the body cavity. The peritrophic membrane is generally absent in fluid-feeding insects, for example, Hemiptera, adult Lepidoptera, and bloodsucking Diptera. The peritrophic membrane may be formed either by delamination of successive concentric lamellae throughout the length of the midgut (in Odonata, Ephemeroptera, Phasmida, some Orthoptera, some Coleoptera, and larval Lepidoptera), or by secretion from special cells at the anterior end of the midgut (in Diptera, and perhaps Dermaptera and isoptera), or by a combination of both methods as seems to occur in Dictyoptera, other Orthoptera, Hymenoptera and Neuroptera.

In Diptera the esophageal invagination presses firmly against the anterior wall of the midgut sot hat the originally viscous secretion of the peritrophic membrane-producing cells, as it hardens, is squeezed to form the tubular membrane. The membrane is made up of a meshwork of microfibrils between which is a thin proteinaceous film. The microfibrils have a constant 60° orientation to each other in membranes produced by delamination, thought to result from their secretion by the hexagonally close-packed microvilli of the epithelial cells. In peritrophic membranes produced form specialized anterior midgut cells, the orientation of the microfibrils is random. The membrane is permeable to the products of digestion and to digestive enxymes released from the epithelial cells. However, it is not permeable to other large molecules, such as undigested proteins and polysaccharides, indicating that the membrane has a distinct polarity and is not merely an ultrafilter. The midgut is usually on differentiated into structurally distinct regions apart from the development, at the anterior end, of a variable number of blindly ending ceca, which serve to increase the surface area variable for enzyme secretion and absorbtion of digested material.

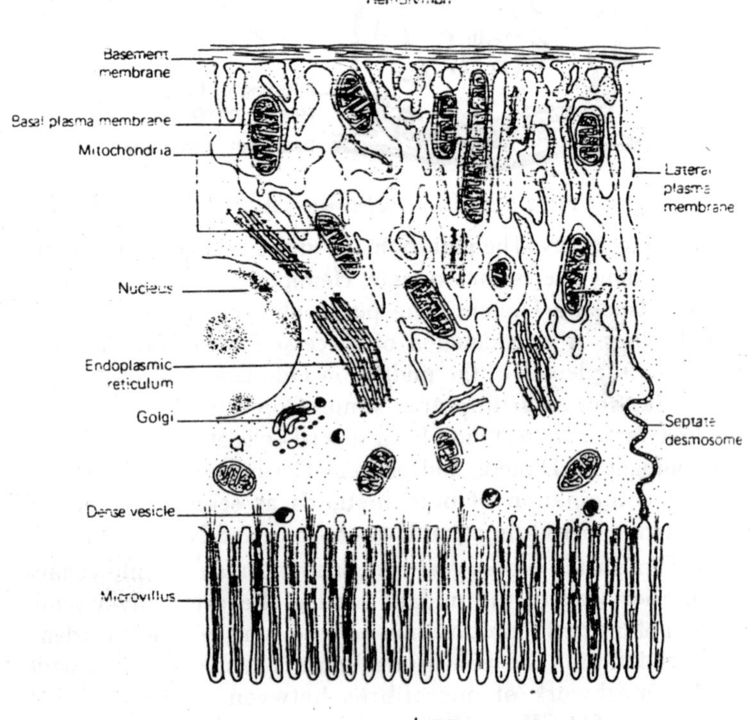

Fig. 2.19. Generalized midgut epithelial cell.

In many Heteropter, However, the midgut is divided into three or four easily visible regions. In the chinch bug (*Blissus leucopterus*) four such regions occur. The anterior region is large and saclike, and serves as a storage region (no crop is present). The second region serves as a value to regulate the flow of material into the third region where digestion probably occurs. Ten fingerlike ceca filled with bacteria are attached to the fourth region, which may be absorptive in function. The role of the bacteria is not known. In many Homoptera, which feed on plant sap, the midgut is modified both morphologically and anatomically so that excess water present in the food can be removed, thus preventing dilution of the hemolymph. Though details vary among different groups of Homoptera, the anterior end of the midgut (or, in some species, the posterior part of the esophagus) is brought into close contact with the posterior region of the midgut (or anterior hindgut), and the region of contact becomes enclosed within a sac called the "filter chamber." Such an arrangement facilitates rapid movement of water by osmosis from the lumen of the anterior midgut across the wall of the posterior midgut and possibly also the Malpighian tubules. Thus, relatively little of the original water in the food actually passes along the full length of the midgut.

Despite the lack of structural differentiation within the midgut of most insects, functional differentiation occurs, and this is reflected histologically. Specialization of certain anterior cells for peritrophic membrane production in Diptera was noted earlier. In addition, differentiation into digestive and absorptive regions occurs in many species. In tsetse flies the cells of the anterior midgut are small and are concerned with absorption of water from the ingested blood. They produce no enzymes and digestion does not begin until food reaches the middle region whose cells are large, rich in ribonucleic acid, and produce enzymes. In the posterior midgut the cells are smaller, closely packed, and probably concerned with absorption of digested food. In some species different regions of the midgut are apparently adapted to the absorption of particular food materials. In *Aedes* larvae the anterior midgut is concerned with fat absorption and storage, whereas the posterior portion absorbs carbohydrates and stores them as glycogen.

Hindgut

The hindgut is an ectodermal derivative and, as such, is lined with cuticle, though this is thinner than that of the foregut, a feature

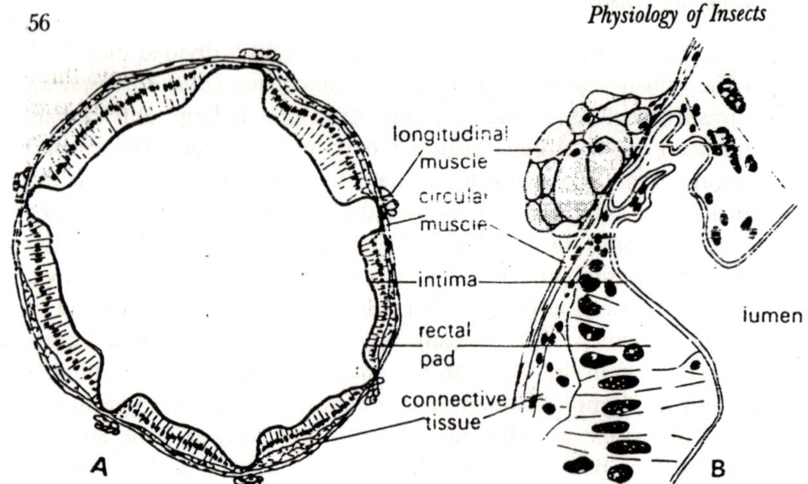

Fig. 2.20. A–Transverse section of the rectum of Chorthippus. B–Section of rectum more highly magnified.

related to the absorptive function of this region. The epithelial cells which surround the cuticle are *flattened* except in the *rectal pads* where they become highly columnar and filled with mitochondria. Muscles are only weakly developed and, usually, the longitudinal strands lie outside the sheet of circular muscle. In the hindgut the following regions usually can be distinguished: pylorus, ileum and rectum. The pylorus may have a well-developed circular muscle layer (pyloric sphincter) and regulate the movement of material from midgut ot hindgut. Also, the Malpighian tubules characteristically enter the gut in this region. The iluem is generally a narrow tube which serves to conduct undigested food to the rectum for final processing. In some insects, however, some absorption of ions and/or water may occur in this region. In a few species production and excretion of nitrogenous wastes occur in the iluem.

In many wood-eating insects, for example, species of termites and beetles, the ileum is dilated to form a "fermentation pouch" housing bacteria or protozoa which digest wood particles. The products of digestion, when liberated by the microorganisms, are absorbed across the wall of the ileum. The most posterior part of the gut, the rectum, is frequently dilated. Though for the most part thin-walled, the rectum includes six to eight thick-walled rectal pads whose function is to absorbions, water, and small organic molecules. As a result, the feces of terrestrial insects are expelled as a more or less dry pellet. Frequently, the pellets are ensheathed within the peritophic membrane which continues into the hindgut.

The Process of Digestion

The digestion of food by insects, or any other animal, involves the action of enzymes, which are secreted by various cells that comprise the tissues of the digestive tract. The sites of enzyme secretion vary so with different species that there is no basis for generalizations. The important point is that enzymes are produced, and the conditions in the different parts of the digestive tract are so adjusted and controlled that the enzymes are able to act.

Enzyme Actions

Enzymes are often defined as biological catalysts that are capable of initiating reactions as well as accelerating them. Enzymes usually enter into the chemical process that they catalyze, and they may be used up in the course of their function. Enzymes are proteins or compounds in which a protein molecules is the principal component. Frequently nonprotein groups are essential to their activity. If these non-protein groups are attached to the molecule, they are called prosthetic groups; but if the non-protein groups exist as entities not part of the protein molecule, they are coenzymes.

Types of enzymes

Enzymes are classified according to the type of reaction they catalyze. In the broadest possible terms enzymes are divided into two groups. These are the hydrolyzing enzymes, which include the principal enzymes of digestion, and the desmolyzing enzymes, which include all the others. Enzymes are named according to the substrate on which they act by adding the suffix *ase*. Thus a proteinase hydrolyzes a protein, a peptidase splits a peptide linkage, and a lipase splits fat into fatty acids and glycerol. The use of common nomenculatre may be confusing, but it is fortunate that the study of the enzyme systems of insects is sufficiently recent not to be hampered significantly by the use of common names.

Chemical breakdown

Digestive enzymes are hydrolases, and as such they split complex molecules into simpler ones by hydrolysis. Hydrolysis can be defined as a double decomposition reaction involving the splitting of water into its ions with the formation of a weak acid, a weak base, or both. Because all enzymes are proteins, they respond to physical and chemical factors that affect proteins. These factors can become critical in the function of an enzyme and are of special importance in the process of assaying enzyme activity. The results of an enzyme assay are almost always expressed as the rate of the reaction that is catalyzed, so that an assay of the activity of a digestive enzyme is made on the basis of a quantitative estimate of the formation of one of the hydrolysis products at successive time intervals. With animals as smal sainsects, this proedure imposes severe technological problems.

The usual procedure involves grinding fresh tissue by some physical means to produce a uniform homogenate called a *brei.* Both skill and advanced knowledge of anatomical structure are required to obtain homogenous tissue samples from insects. The enzyme potency of many of the tissues is extremely high, so that accidental inclusion of an unknown tissue (a section of Malpighian tube in a fat body sample, for example) can lead to spurious results. A common method for circumventing this problem has been to reduce whole insects, or arbitary divisions of insects (heads, thoraces, etc.), to a brei and to use this brei for the assay. This technique is bound to give results, but the value of these results is comparable to finding gold in alluvial drift. Doubtlessly the material sought is present somewhere, but there is very little direction toward the source. Quantitative assay is also hampered by the necessity for determining sample size. Wet weights are notoriously inaccurate and difficult to determine, particularly at the submilligram level.

A common substitute for wet weight is a chemical weighing procedure in which total protein in the sample (or a comparable sample from the same brei) is used to compare the amount of enzyme present in successive determinations. If all other conditions are favourable, the rate curve of an enzyme-controlled reaction shows a gradual rise with an increase in temperature until an optimum is reached. Above this optimum temperature, there is a sharp cut-off caused by the denaturation of protein of the enzyme. At suboptimal temperatures, an enzyme-controlled reaction is slow, but at these low temperatures the enzyme is more stable and lasts longer. When the temperatures are optimal or slightly above, the reaction proceeds more rapidly, but the enzyme is spent more quickly. The temperature optimum is not a fixed point.

As the temperature rises, several things happen simultaneously. The rate of the reaction increases, and at the same time the enzyme breaks down, so that if the measurements are made over short periods, the optimum determined may be somewhat higher than if incubation is carried out over a longer period. Proteins are composed of amino acids linked together. A single amino acid, or a chain of them, has a double-ended characteristic in that a carboxyl group is on one end and an amino group is on the other end. This means that two opposite reactive groups are exposed, and the degree of influence of one or the other is emphasized by the direction in which the molecule dissociates. Molecules of this type are variously known as *hybrids, ampholytes,* or *zwitterions.* The degree of dissociation in one direction, or the other, depends upon the concentration of hydrogen ions in the system.

The degree of dissociation is expressed as the negative logarithm of the hydrogen ion concentration, or pH. The rate of an enzyme reaction is controlled by the degree of dissociation of the enzyme protein and the direction in which it goes. The pH at which the undissociated part of the molecule is at its greatest concentration (the dissociation is at a minimum) is the *isoelectric point.* Even amino acid and protein has an isoelectric point. The curve that relates pH to enzyme activity is bell shaped, and the optimum shifts up and down the pH scale, depending upon the nature of the enzyme protein.

Pepsin and trypsin, for example, are both proteinases, but pepsin functions at an acid pH and trypsin is active at neutrality or in a slightly alkaline medium. Because enzymes enter into the reactions that they catalyze, the rate of the reaction can be controlled by the relative concentrations of the enzyme and substrate. If the quantity of enzyme is more or less fixed, the reaction velocity increases with an increase in substrate concentration up to the point at which no more enzyme molecules are available to combine with the substrate. When this level is reached, the rate of the reaction levels off. The rate of an enzyme activity can be altered in some cases by products of its own reaction. These factors have their greatest effect in vitro. Enzyme studies in vivo are difficult to perform quantitively because the function of the various vital systems of the insect tends to stabilize enzyme reactions.

Enzymatic Actions

Insects produce digestive enzymes, and it is a safe generalization that the enzymes that must be present in the digestive tract to take care of the food normally consumed are produced either by digestive secretion or by symbionts. With this as the starting point, it is possible to arrive at a procedure for the determination of the digestive enzymes of almost any insect, but first it must be recognized that there are a number of types of carbohydrates, proteins, and fats and that the enzymes may be very specific.

Carbohydrates

Carbohydrates exist as monosaccharides, oligosaccharides, or polysaccharides, depending upon the number of simple sugar molecules in the structure. The carbohydrates are split first into their monosaccharide components, usually hexoses, before they are assimilated. Starch is hydrolyzed through a series of dextrins to yield glucose and maltose. Three specific enzymes act in sequence on starch to produce α-glucose as the end product. α-glucose hexose that exists as a six-membered oxygen-bearing ring structure (pyranose) that has two (α and β) spatial configurations. The α configuration usually has more biological activity than the β. One of the enzymes that act on carbohydrate is *α-amylase* (the others are *α-oligosaccharase* and *maltase*), and this group of three enzymes is sometimes referred to in loose terminology simply as *amylase.* The same enzymes that hydrolyze starch also split glycogen. Cellulose, which makes up the bulk in plant tissue, consists of a series of β-glucoside units in long chains. This long-chained structure makes the material undigestible to most animals, but some insects that bore in wood and ingest relatively large quantities of cellulose either secrete a cellulose in the digestive tract or depend upon symbionts for the digestion of cellulose.

Proteases

Proteases are enzymes that act principally by hydrolytic splitting of the protein molecules at the peptide linkages. Proteases differ according to the size of the molecule they split, the groups adjacent to the peptide linkages that they attack, the pH at which they function, the presence of metallic prosthetic groups, and the effects of activating and inhibiting factors. In animals most proteases are of the L-series, but his may not be entirely the case with all insects. The proteases are classified as *endopeptidases* and *exopeptidases,* depending upon the location of the linkages in the protein molecule they attack. *Endopeptidases* include such specific enzymes as *pepsin, trypsin, chymotrypsin* and *cathepsin.* Pepsin acts in an acid medium (a pH from 1.0 to 5.0) and is rarely, if ever, found in insects. Pepsin attacks linkages adjacent to an aromatic amino acid and at neutrality is inactivated.

Trypsin, a proteinase found in insects as well as in other animals, is most effective in a slightly alkaline medium (a pH from 7.0 to 9.0). It is secreted as a trypsinogen that is activated autocatalytically or by specific enzyme such as enterokinase. Trypsin characteristically attacks peptide linkages adjacent to arginine or lysine.

Chymotrypsins–there are at least four–originate in the pancreas of mammals. chymotrypsins act in alkaline media and attack peptide linkages with adjacent aromatic amino acids, particularly tyrosine, phenylalanine, tryptophan, and to a lesser extent methionine. Chymotrypsins are inhibited by organic phosphates isopropyl fluorophosphate and by benzol derivatives of several amino acids. As the result of studies, several cathepsins are known to exist as intracellular proteases of the liver, kidney, and spleen of mammals, and cathepsins are also known to exist and function extracellularly in a number of invertebrates. These enzymes act most effectively in an acid medium (a pH from 4.0 to 6.0). The various cathepsins resemble pepsin, trypsin, and chymotrypsin in their substrate requirements. Proteinases that act in the pH range between 4.0 and 6.5 are arbitarily classified as cathepsins. Some cathepsins require reducing substances for activation, but some do not. The identification of this group of enzymes in insects is incomplete.

Exopeptidaes are enzymes that attack the terminal peptide linkages (substrates with free polar groups), most of which arise as the result of proteinase activity. Usually exopeptidases have a metal in the molecule or are activated by the presence of a metallic ion. The exopeptidase group is further divided into carboxypeptidases aminopeptidases and dipeptidases.

Carboxypeptidases remove a terminal amino acid with a free carboxyl group, the *aminopeptidases* split a peptide linkage next to a terminal amino acid that has a free amino group, and the *dipeptidases* are active in splitting the linkage that holds two amino acids together to form a depeptide. The pancreatic carboxypeptidase of mammals has been shown to contain one zinc atom per molecule, the aminopeptidases require manganese, and the dipeptidases are activated by cobalt, manganese, or zinc, depending upon the nature of the enzyme. Most of these exopeptidases act in a slightly alkaline medium.

Lipases and Esterases

Neutral fats are by definition esters of one or more higher fatty acids and glycerol, and the enzymes that accomplish a hydrolytic split of the acids and the alcohol are called lipases. Fatty acids also form compounds (esters) with other alcohols or with sterols. The enzymes that split esters are called esterases. Thus *lipase* is a more or less specific esterase. Both the lipases and the esterases act effectively throughout a broad pH range and are relatively non-specific. They usually act best in an alkaline medium. Insects can utilize (although they may not require) fats, and in some cases intact fats may be assimilated by cells, in which they are further metabolized. Both lipases and esterases are found in insects, either in the digestive tract or in various tissues.

The *esterases* have received a great deal of attention during the last 10 years because of their interaction with organophosphate insecticides. A great deal is still to be learned, both from the qualitative and quantitative standpoint, abot esterases and their functions in insects. When evaluating the literature on digestive enzymes, it should be remembered that enzymes found form an assay fo the brei of the whole digestive tract and its contents may be somewhat different from those enzymes elaborated intr the digestive tract. It is important to differentiate the contents from the tissues. This has not always been the practice, and the resulting data may be misleading.

The Digestion of Unusual Food

Of all the animals, only insects can feed successfully on and digest such refractory materials as keratin–the protein of wool, hair and feathers; and only a few insects can digest cellulose without the intervention of symbionts. These subjects have been given considerable attention because they are exotic. Cleveland is usually credited with the experimental proof that woodfeeding insects (termites) are able to survive on nearly pure cellulose land that they are dependent upon the activity of symbiotic protozoa for the production of cellulose. He used high oxygen tensions to defaunate his test insects, and some of the luster was removed from his experiments because the defaunating treatment also altered the growth rate of the termites.

Even so, the fact remains that termites are dependent upon their symbionts for their nutrition. The digestion of keratin offers another problem. This material is characterized by a high, but varying, content of cystine, and the sulfur of this amino acid forms disulfide bonds between adjacent polypeptide chains. The disulfide bridges contribute greatly to the stability of the keratin molecule, and the destruction of these bridges renders keratin more readily digestible. A number of factors contribute to the digestion of keratin by various of the Mallophaga, Dermestidae, and Lepidoptera (Tineidae). Disulfide linkages are less stable in an alkaline pH such as exists in the insect gut, but this is a minor factor in keratin digestion.

Linderstrom-Lang and Duspiva demonstrated that one essential difference between the larvae of the genus *Tineola* (clothes moth) and insects unable to digest wool is the unusually strong reducing potential in the midgut of the larvae. This reducing potential varies from – 200 mv. to –250 mv., and under these conditions reduction of the disulfide bonds to sulfydryl grops occurs so that a cystine moiety becomes two molecules of cysteine. The reduced keratin formed in this way can be digested by the proteases of the digestive tract. The origin of the reducing potential is not known. Waterhouse has presented a more recent discussion of this process. The digestion of wax by the larvae of two species of Lepidoptera that infest stored honeycomb has been a subject of attention because, like keratin, beeswax is a refractory material. Waterhouse has summarized the extensive literature on this subject, and Niemierko and Wlodawer have discussed the subject in a long series of journal papers. From the available data, it is most logical to conclude that lipolytic enzymes that digest beeswax arise from bacterial symbionts and not from the larval tissue.

Final proof of this idea must a wait the strictly axenic culture of these wax-digesting insects. Waterhouse has discussed the problems and progress in this line of endeavour. Haydak proved that larvae of *Calleria* do not require beeswax for normal growth and development over many generations; however, Young observed imporved growth of *Galleria* larvae over short periods when beeswax, certain fatty acids, or even paraffin wax was added to the diet. Thus we see that digestion in insects takes place in a specialized organ consisting of: a foregut, which may be modified into a storage unit or crop; a midgut where most of the digestion and assimilation takes place; and a hindgut, which serves both as a duct and a reabsorbing organ. Enzymes are produced by secretions of the salivary glands and by specialized cells in the midgut. A large number of insects possess highly evolved adaptations to specialized diets, including some highly refractory materials. Often, the explanation of these adaptations leads back to the presence of intestinal symbionts that produce the enzymes required for the process. Digestion in insects is an elementary process that is usually taken for granted; however, it others fascinating study for further research.

Factors Affecting Enzyme Actions

According to House (1974) three factors markedly affect digestion in insects; pH buffering capacity, and redox potential of the gut. The pH determines not only the activity of digestive enzymes, but also the nature and extent of microorganisms in the gut and the solubility of certain materials in the gut lumen. The latter affects the osmotic pressure of the gut contents and, in turn, the rate of absorption of molecules across the gut wall. Analyses of the pH in various regions of the gut have been made for a wide range of species, and various authors have attempted to correlate these with the feeding habits or phylogenetic position of an insect. At best, these correlations are only broadly correct, and many exceptions are known. Swingle observed that, in most insects studied by him, the gut is slightly acidic or slightly throughout its length. Further, the pH generally increases from foregut to midgut, then decreases from midgut to hindgut. Though the latter is true for most phytophagous species, in many ominivorous and carnivorous species the pH of the hindgut is greater than that of the midgut. Many variables affect the pH of different regions of the gut. Generaliy, the pH of the crop is the same as that of the food, though in some species it is consistently less than 7 due to the digestive activity of microorgansims or regurgitation of digestive juice from the midgut.

The pH of the midgut differs among species but tends to be constant for a given species because of the presence in this region of buffering agents. In a few species there are local variations in pH within the midgut, which are perhaps related to changes in digestive function from one part ot another. The hindgut typically has a pH slightly less than 7, presumably resulting from the presence of the nitrogenous easte product, uric acid. The hindgut contents of some phytophagous species may be quite acidic due to the formation of organic acids from cellulose by symbiotic microorganisms. The relatively constant pH found in the midgut results from the presence in the lumen of both inorganic and organic buffering agents. In some species, inorganic ions, especially phosphates, but including aluminium, ammonium calcium, iron, magnesium, potassium, sodium, carbonate, chloride, and nitrate, seem to offer sufficient buffering capacity. In other species organic acids, including amino acids and proteins, tend to supplement or replace the buffering effect of the inorganic ions.

The *redox potential,* which measures ability to gain or lose electrons, that is, to be reduced or oxidized, respectively, is an important factor in digestion in some insects. Normally, the redox potential of the gut is positive, which is indicative of oxidizing (aerobic) conditions. However, in species able to digest keratin the redox potential of the midgut fluid is strongly negative. It has been suggested that such an anaerobic (reducing) environment is necessary to effect splitting (reduction) of the disulfide bonds. Subsequenlty, a "normal" protease facilitates hydrolysis of the polypeptides. A reducing environment was assumed to be provided by the presence in the gut of reducing agents such as ascorbic acid, glutathione, or riboflavin. However, Gilmour (1965) has ponted out that the protease present in the gut is unusual in that it functions under anaerobic conditions. Further, the cystine released by hydrolysis could be converted intracellularly to cysteine, which itself could yield hydrogen sulfide. Both these substances, which are known to be present in the gut during digestion of keratin, are reducing agents and would facilitate breaking of disulfide bonds. Thus, the negative value for redox potential is, on this proposal, consequent upon, rather than necessary for, the digestion of keratin.

Regulation of Enzymatic Secretion

Apparently, digestive enzymes are not stored in the midgut cells but are liberated immediately into the gut lumen. In many insects which feed more or less continuously, digestive enzyme synthesis and secretion seem not to be regulated but continue even during nonfeeding periods, including starvation. In other species production of enzymes is clearly related quantitatively to food intake and secretagogue, neural, and hormonal control mechanisms have been suggested by different authors. Unfortunately, for most species, the evidence presented in support of one mechanism or another is equivocal. In a secretagogue system enzymes are produced in response to food present in the midgut.

Presumably the amount produced is directly influenced by the concentration of food in the lumen. Where neural or hormonal control of enzyme production has been proposed, the amount of food passing along the foregut, measured as the degree of stretching of the gut wall, is believed to result in the production of a proportionate amount of enzyme. In systems under neural control, information passes directly to the midgut via the stomatogastric nervous system. Where hormonal control exists, feeding stimulates the release of neurosecretion which travels via the hemolymph to the midgut cells. Insects do not appear able to regulate enzyme production quantitatively, that is, to alter the relative proportions of enzymes which they produce in accord with a change in diet. This suggests that a midgut epithelial cell produces a complete package of digestive enzymes.

Symbiotic Digestion

Microorganisms (bacteria, fungi, and protozoa) may be present in the gut, but for only a few species has there been a convincing demonstration of their importance in digestion. In many insects microorganisms appear to have no role, as the insects can be reared equally well in their absence. In other species microorganisms may be more important with respect ot an insect's nutrition than digestion *per se.* Where a role for microoganisms in digestion has been demonstrated, the relationship between the microorganisms and insect host is not always obligate, but may be facultative or even accidental. Bacteria are important cellulose-digesting agents in many phytophagous insects, especially wood-eating species whose hindgut may include a fermentation pouch in which the microorganism are housed. In other species, for example, the wood-eating cockroach *Panesthia,* bacteria in the crop are essential for cellulose digestion.

In larvae of the wax moth *Galleria,* bacteria normally present in the gut undoubtedly aid in the digestion of beeswax, yet bacteriologically sterile larvae produce an intrinsic lipase capable of degrading certain wax components. Finally, many insects feed on decaying vegetation and must, therefore, ingest a large number of saprophytic bacteria which, temporarily at least, would continue their degradative activity in the gut. In this sense therefore, though the relationship is accidental, the microorganisms are assisting in digestion. In lower termites and some primitive wood-eating cockroaches (*Cryptocercus,* flagellate and ciliate protozoa occur in enormous numbers in the hindgut.

The relationship between the insects and protozoa is mutualistic; that is, in return for a suitable, anaerobic environment in which to live, the protozoa phagocytose particles of wood eaten by the insects, fermenting the cellulose and releasing large amounts of glucose (in *Cryptocerus*) or organic acids (in termites) for use by the insects. Fungi rarely play a direct role in the digestive process of insects, though it is reported that yeasts capable of hydrolyzing carbohydrates occur in the gut of some leafhoppers (Homoptera). However, a mutualistic relationship has evolved between many fungi and insects in which the funig convert wood into a more usable form, while the insects serve to transport the fungi to new locations. Some ants and higher termites, for example, culture ascomycete or basidiomycete fungi in special regions of the nest called fungus gardens. Chewed wood or other vegetation is brought ot the fungus garden and becomes the substrate on which the fungi grow, forming hyphae to be eaten by the insects. Certain wood-boring insects, for example, bark beetles (Scolytinae) inoculate their tunnels with fungal cells when they invade a new tree. The fungal mycelium which develops, along with partially decomposed wood, can then be used as food by the insects.

Transportation of Digestive Food

Absorption is the transportation of digested food from the site of digestion to other part or parts of the body. Contrary to the views of some early entomologists, the foregut, specifically the crop region, is not the site of absorption of digestive products. The majority of absorption occurs in the midgut, especially the anterior portion, including the mesenteric ceca, and to a lesser extent the hindgut. The later region is, however, primarily of importance as the site of water or ion resorption in connection with osmoregulation, though in insects which have symbiotic microorganisms in the hindgut it may also be an important site for absorption of small organic molecules, especially carboxylic and amino acids. Most absorption of organic molecules across the midgut wall is passive, that is, from a higher to a lower concentration, though the rapid rate at which some molecules are absorbed suggests that their movement is facilitated by special carriers. The absorption rate is enhanced by a steep concentration gradient maintained between the midgut lumen and hemolymph. This may be achieved by absorption of water from the gut lumen so that the hemolymph becomes more dilute or by rapid conversion of the absorbed molecules to a more complex form.

3

VENTILATORY SYSTEM

Respiration is the energy liberating process and in most of animals the respiratory organs take in and give off the respiratory medium, and it is only the oxygen which is transported to various organs of the body. But in group insecta it is the atmospheric air which is directly supplied to body organs to make the exchange of gases possible between tissue fluid and respiratory organs. For this a system of internal tubes–the tracheal tubes is formed to carry oxygen is carried directly to its sites of utilization and the blood is not concerned with gas transport. The tracheae open to the outside though segmental pores, the *spiracles,* which generally have some closing mechanism which permits water loss from the respiratory surfaces to kept to a minimum. The spiracles open in response to a low concentration of oxygen or a high concentration of carbon dioxide in the tissues. Diffusion alone can account for the gaseous requirements of the tissues of most insects at rest, but in larger insects or during activity demands on oxygen are greater. To meet these demands the insect pumps air in and out of the tracheal system by expanding and collapsing air-sacs which are enlarged by movements of the body. These movements are controlled by endogenous rhythms within the central nervous system.

In some insects which have low oxygen requirements the spiracles may open in such a way as to permit the entry of oxygen, but they prevent the exist of water and carbon dioxide except in occasional bursts. This is regarded as a water conserving mechanism. In some insects living in moist environments where water loss is not a problem gaseous exchange may take place through the permeable cuticle.

A System of Tracheal Tube

In insects there is a system of tracheal tubes which is developed from integument by a process of invagination for respiratory purposes occurs in Onychophora and in some terrestrial forms at least of all the major groups of Arthropoda. The organs probably, as claimed by Ripper (1931), are formed independently in most of hte several groups in which they are developed. Some of the Arachnida, a few isopod crustaceans, most of the chilopoda, and the majority of insects have branched tracheae. Most of the diplopoda are provided with segmental clusters of unbranched respiratory tubules, and one family of the protura has unbranched tracheae arising only on the thorax. The Arachnida have tracheae and respiratory pouches known as "*book lungs,*" so named because their walls are produced into parallel lamellate folds. In the Onychophora the respiratory invaginations take the form of groups of small tubules irregularly scattered over the inner surface of the body wall.

Embryonic Origin of the Tracheal Tubes

During embryonic development the tracheal arise as more or less solid ingrowths of cell from ectoderm. The enclosing between them only a potential lumen. the external orifices of the depressions become the *spiracles;* the internal tubular parts are the rudiments of the tracheae. The tracheal pits extend inward and branch into ramifying tubes that eventually extend to all parts of the body in

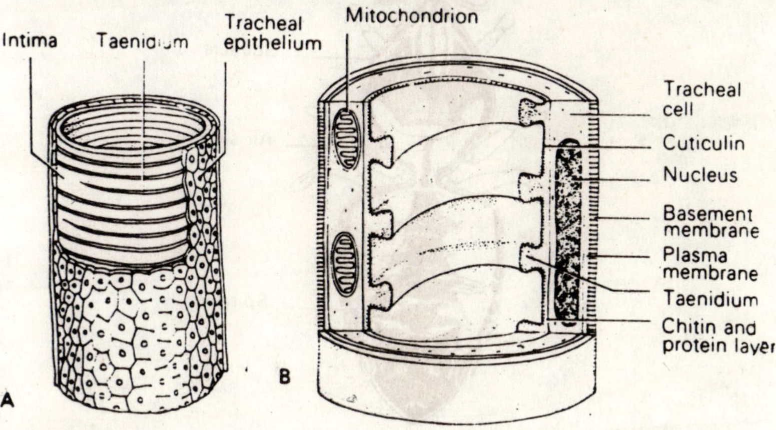

Fig. 3.1. Tracheal structure. A–Portion of trachea. B–Fine structure of trachea.

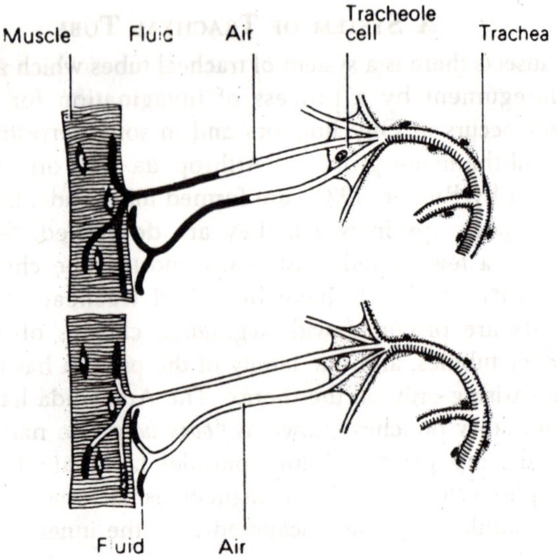

Fig. 3.2. Tracheal structure. Tracheoles in close contact with muscle.

insects haivng a fully developed tracheal system. Judging from the position of the abdominal spiracles in adult insects, it would seem that the primitive position of the tracheal invaginations is in the lateral parts of the dorsum of the body segments just above the

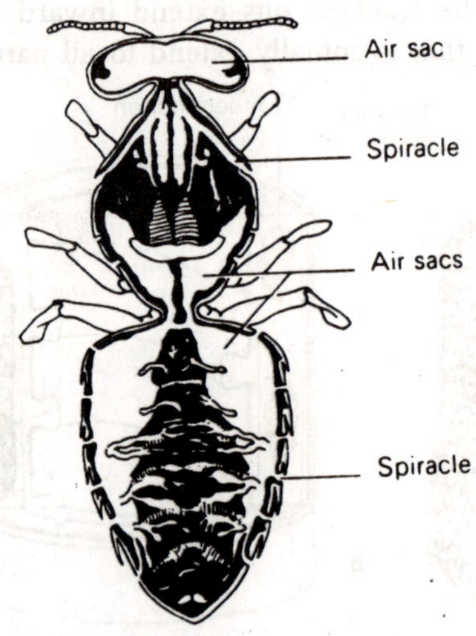

Fig. 3.3. Air sacs in the honeybee.

limb bases. In embryos the spiracular rudiments commonly are thus located.

Spiracles

Spiracles are the external openings of tracheal tubes. It is possible that the primitive insects had a pair of tracheal invaginations in each of the 17 somites of the gnathal, thoracic, and abdominal regions of the body; but there is no suggestion of tracheae ever having been formed in the protocephalon. Direct proof of the existence of tracheal invaginations, however, has been found only on 14 segments, which are the second maxillary segment, the 3 thoracic segments, and the first 10 abdominal segments. Tracheal invagination of the second maxillary segment are said by Nelson (1915) to be formed in the embryo of the honey bee, where they appear on the anterior part of the segment above the bases of the labial rudiments.

These tracheal pits of the second maxillary segment, Nelson says, give rise to the tracheal system of the head but are soon closed and leave no trace of their existence in the head of the adult bee. In some of the Sminthuridae (Collembola) a pair of spiracles is situated on the sides of the neck close behind the head. Davies (1927) believes that these apparent *cervical spiracles* belong to the prothorax because of their position relative to the muscle attachments on the posterior margin of the head capsule. When we consider, however, that the submerginal ridge of the head on which these muscles are inserted marks the intersegmental line between the two maxillary segments and that the membranous neck is

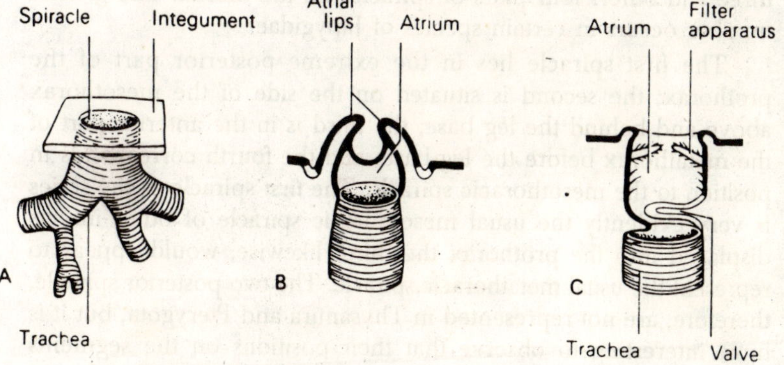

Fig. 3.4. Types of spiracles. A–Simple, nonatriate type. B–Atriate spiracle with lip closure mechanism. C–Atriate spiracle with filte apparatus and valve closure mechanism.

derived in part from the segment of the labium, it becomes evident that the neck spiracles of *Sminthurus* lie in the region of the second maxillary segment. These spiracles, therefore, may be persisting examples of the second maxillary spiracles, known otherwise only as temporary tracheal openings in the embryo the honey bee.

Prothoracic spiracles are present in the embryonic stage of some insects, but these spiracles are completely obliterated before hatching. The embryonic prothoracic spiracles have been described by Cholodkowsky (1891) in *Blattela,* and by Wheeler (1889) in *Leptinotarsa.* The thorax of postembryonic stages of all insects, except Diplura, never has more than two pairs of spiracles, and these two pairs are formed in the embryo on the anterior part of the meso thorax and the metathorax, respectively. In many insects, however, the thoracic spiracles migrate forward during development and come thus to have a definitive position in the secondary intersegmental membranes or in the posterior parts of the segments preceding.

The mesothoracic spiracles particularly are subjected to this anterior migration and hence often occur in larval or adult insects on the sides of the prothorax, for which reason they are frequently called the "*prothoracic*" *spiracles.* The *anterior* (*mesothoracic*) *spiracles* are usually the larger of the thoracic spiracles, those of the *metathorax* being generally small, and sometimes rudimentary, as in certain larvae. In adult Pterygota the thoracic spiracle have a "pleural" position, but it is probable that their definitive location between the pleural sclerites is a result of the secondary upward extension of the subeoxal plates of the leg bases on each side of the spiracles. The Diplura differ from other insects in that some species have three and others four pairs of spiracles on the thorax. The greater number occurs in certain species of Japygidae.

The first spiracle lies in the extreme posterior part of the prothorax; the second is situated on the side of the mesothorax above and behind the leg base; the third is in the anterior part of the metathorax before the leg base; and the fourth corresponds in position to the mesothoracic spiracle. The first spiracle of the series is very evidently the usual mesothoracic spiracle of other insects displaced into the prothorax the third, likewise, would appear to represent the usual metathoracic spiracle. The two posterior spiracle, therefore, are not represented in Thysanura and Pterygota, but it is most interesting to observe that their positions on the segments correspond exactly to the usual position of the spiracles in Chilopoda.

Moreover, these posterior thoracic spiracle of Diplura fall in line with the series of abdominal spiracles. Hence, we might conclude that the anterior thoracic spiracles of Diplura represent the two thoracic spiracles present in other insects, and that the posterior spiracles, belonging morphologically to the same series as the abdominal spiracles, have been eliminated from the thorax of other insects. If, then, the anterior thoracic spiracles have had an independent origin, we have here an explanation of the curious fact that the structure of these spiracles in pterygote insects is almost always different from that of the abdominal spiracles. In *Campodea,* according to Grassi (1886), the anterior metathoracle spiracles are absent, but the same spiracles are frequently rudimentary or absent in Pterygota. On the abdomen there are usually eight pairs of spiracles, and this is the maximum number of abdominal spiracles in postembryonic stages of all insects, but the number may be variously reduced.

Cholodkowsky (1891) reports the existence of a pair of tracheal invaginations on each of the first nine abdominal segments in the embryo of *Blattella* (*Phyllodromia*); and Heymons (1897) finds in the *Lepisma* embryo, in addition to nine distinct pairs of abdominal spiracles, masses of ecotdermal cells on the tenth segment at points corresponding to the spiracular invaginations on the preceding segments, which he takes to be rudiments of a tenth pair of abdominal spiracles. The first abdominal spiraces are often situated close to the thorax, but their abdominal relation is shown by the fact that they always lie posterior to a line through the base of the third phragma, which is an intersegmental fold between the metathorax and the first abdominal segment.

Arrangement of the Tracheal Tubes

As the primary tracheal invaginations grow into the body of insects, they divide a short distance from their origins into major and minor branches, and the latter eventually ramify to all the tissues. In insects having a well-developed tracheal system some of the branches from consecutive and opposite spiracles unite to form longitudinal trunks and transverse commissures. In general, the mature tracheal system attains an organization having a pretty definite fundamental pattern. It is probable that in a primitive stage each somite of the body was independently tracheated from its own pair of spiracles, and that the connection of the segmental systems by longitudinal turnks is a secondary condition evolved to give more efficient aeration.

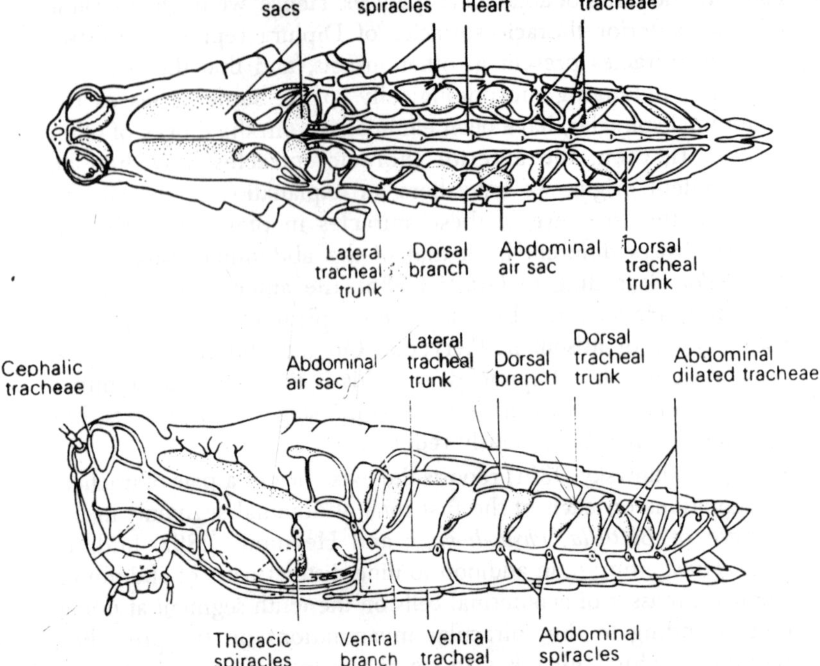

Fig. 3.5. Representative types of tracheal systems–Dorsal and lateral views of the open type, grasshopper.

In general it is found that in each half of each segment there are three principal tracheae given off from the longitudinal trunk in the neighbourhood of the connection of the latter with the spiracle. Hence, we may suppose that primarily a short *spiracular trachea* extended inward from the spiracle and gave off three main branches. Of the latter, one is *dorsal trachea* (b) going to the dorsal musculature of the body wall and to the dorsal blood vessel; another is a *ventral trachea* (c) supplying the ventral musculature and the ventral nerve cord, and sending a branch into the leg in the leg bearing segments; the third is a median *visceral trachea* (d) having its principal ramifications on the walls of the alimentary canal, with branches to the fat body, in the appropriate segments, to the gonads and the genital ducts. The plurisegmental longitudinal trunks are formed by the union of anterior and posterior branches from the spiracular tracheae of consecutive segments.

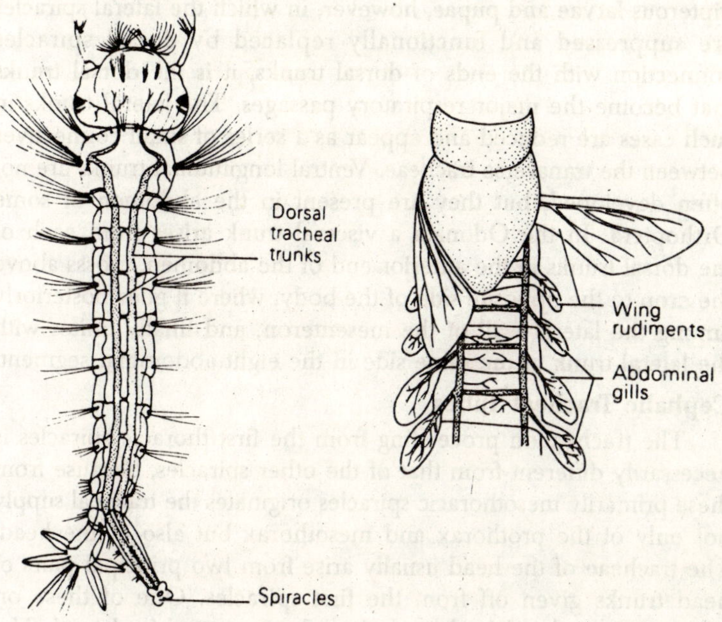

Fig. 3.6. Representative types of tracheal systems. A–Open type with two posterior spiracles, mosquito larva. B–Closed type with no functional spiracles and cuticular ventilation, mayfly nymph (only posterior part of thorax and anterior part of abdomen shown).

The lengthwise trunks most generally present are a pair of *lateral longitudinal turnks,* on on each side of the body, connecting all the spiracular tracheae from the first thoracic spiracle to the last abdominal spiracle. But there is often present also a pair of *dorsal longitudinal trunks,* connecting the dorsal tracheae of successive segments, and sometimes a pair of *ventral longitudinal* uniting the ventral tracheae. In some insects, finally there are *visceral longitudinal trunks* on the sides of the alimentary canal. By anastomosis of the dorsal or the ventral tracheae in each segments there are frequently formed *commissural trunks* continuous from one side of the body to the other.

Thus, there may be present a *dorsal tracheal commissure* crossing above the dorsal blood vessel or a *ventral tracheal commissure* below the ventral nerve cord. The lateral longitudinal trunks are usually the largest tracheae in the insect. Because of their size and their immediate connection with the spiracles, these trunks become generally the chief avenues of air circulation in the body. In dipterous larvae and pupae, however, in which the lateral spiracles are suppressed and functionally replaced by dorsal spiracles connection with the ends of dorsal trunks, it is the dorsal trunks that become the major respiratory passages. The lateral trunks in such cases are reduced and appear as a series of small connectives between the transverse tracheae. Ventral longitudinal trunks are not often developed, but they are present in the abdomen of some Orthoptera. In the Odonata a visceral trunk arises from each of the dorsal trunks in the anterior end of the abdomen, crosss above the crop to the opposite side of the body, where it goes posteriorly among the lateral wall of the mesenteron, and finally unites with the lateral trunk of the same side in the eight abdominal segment.

Cephalic Tracheal Tubes

The tracheation proceeding from the first thoracic spiracles is necessarily different from that of the other spiracles, because from these primarily mesothoracic spiracles originates the tracheal supply not only of the prothorax and mesothorax but also of the head. The tracheae of the head usually arise from two principal pairs of head trunks given off from the first spiracles. One of these on each side is a *dorsal head trunk,* the other a *ventral head trunk.* The actual number of tracheae entering the back of the head form the thorax, however, may be increased by an immediate branching of the two primary trunks, and it is not clear that the principal trunks themselves are in all cases strictly homologous branches. The dorsal head trunk sends branches to the antennae, the compound eyes, the mandibles, the brain, and the adductor muscles of the mandibles. The ventral turnk branches to the first and second maxillae and to the mandibular adductors and forms an anastomosis with the dorsal trunk.

A similar distribution of the head tracheae, Lehmann says, occurs in *Machilis* and in Ephemerida. In *Lepisma,* according to Sule (1927), a short cephalic trunk springs from the dorsal branch of the first (mesothoracic) spiracle and soon divides into a *trachea cephalica dorsalis* and *a trachea cephalica ventralis.* The first branches to the prothoracic tergum, the dorsum of the head, the brian, the optic lobes, the uppr parts of the eyes, and the antennae. The second branches to the prosternum and neighbouring organs, the posterior part of the head, the salivary glands, the inner region of the eyes, and the gnathal appendages. From this it would appear that in general the procephalic part of the head is trancheated

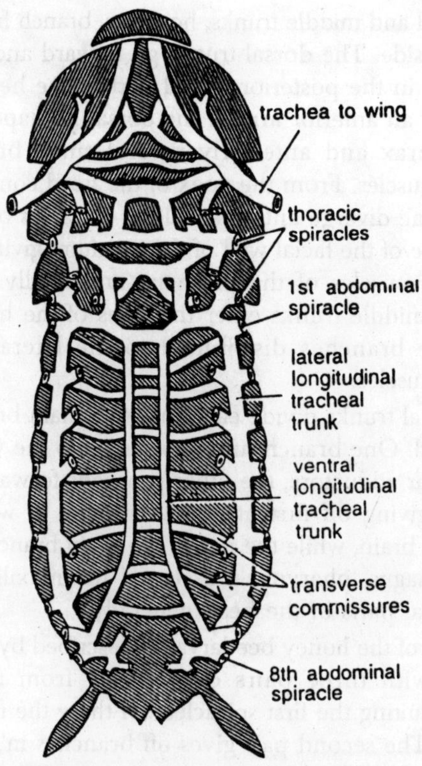

Fig. 3.7. P. americana. Tracheal system in dorsal view.

from the dorsal head trunk, and the gnathal region from the ventral trunk, though Lehmann finds that the mandibles receive their tracheae from the dorsal trunk.

The ventral trunk, Lehmann, says, is the first head trachea formed in the embryo of *Dixippus*, the dorsal trunk being an outgrowth from it. Studies on the head tracheation of other insects show a considerable diversity in the distribution of the branches from the principal trunks. Thus, according to Alt, the ventral head trunks of *Dytiscus* supply almost all the trancheation of the head muscles and give branches to all the appendages, including the antennae and mandibles, as well as to first and second maxillae. In the caterpillar three large tracheal trunks enter the head from each anterior spiracle, one being dorsal, another ventral, and the third having a middle position between the other two.

The dorsal and middle trunks, however, branch from a common base on each side. The dorsal trunks go upward and unite to from a commissure in the posterior dorsal part of the head. Each gives off posteriorly an anterior arm of the dorsal X-shaped commissure of the prothorax and anteriorly several small branches to the mandibular muscles. From the apex of the head commissure a pair of long tracheae diverge anteriorly along the arms of the V-shaped epistomal ridge of the facial wall of the cranium, giving off branches to the dorsal muscles of the pharynx, and finally ending in the labrum. The middle trunks enter the sides of the head and break up into large branches distributed to the lateral parts of the mandibular muscles.

The ventral trunks divide each into two main branches as they enter the head. One branch turns upward into the ventral parts of the mandibular adductors; the other proceeds forward beneath the oesophagus, giving off first at dorsal trachea, of which a branch penetrates the brain, while the rest of the main branch is distributed to the oesophagus, pharynx, muscles of the maxilla and labium, and the ventral parts of the head generally.

The head of the honey bee larva, as described by Nelson (1924), is supplied with three pairs of tracheae from the transverse commissure uniting the first spiracles. Of these the mesal pair goes to the brain. The second pair gives off branches in the upper part of the head to aorta and the brain, but the main trunks go ventrally and ramify to the maxillae, the mandibles, the antennal rudiments, and the labrum. The lateral third pair goes to the salivary glands. In the embryo of the bee, Nelson (1915) says, the invaginations from the temporary second maxillary spiracles give off each four primary tracheal branches, one going posteriorly, one dorsally, and the other two anteriorly. The posterior branches connect with the tracheal system of the thorax, the dorsal branches from opposite sides unite to form the anterior commissure, and the anterior branches on each side become the two principal pairs of head tracheae.

Thoracic Tracheal Tubes

Thorax is the middle part of the body. The tracheal system of the thorax is often complex and differs much in different insects. In its simpler forms, however, it departs little from the more generalized plan of segmental tracheation in the abdomen, except for the supply of tracheae to both the prothorax and the mesothorax from the mesothoracic spiracles, and in the frequent reduction or obliteration of the metathoracic spiracles. A good example of generalized tracheal system of the thorax is seen in a caterpillar. The large lateral trunks are continuous from the abdomen to the mesothoracic spiracles, which are situated on the sides of the dorsum of the protoorax. The metathoracic spiracles are rudimentary, but the site of each is connected with the lateral trunk by a small tracheal strand.

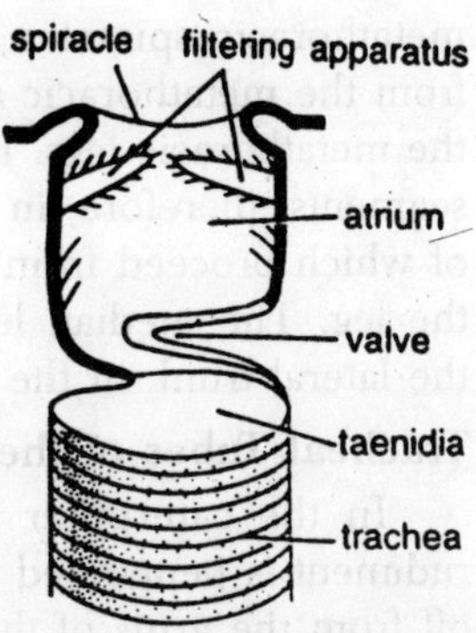

Fig. 3.8. A spiracle with atrium, filtering apparatus and valve.

The principal somatic and visceral branches are given off from the lateral trunk in the neighbourhood of the spiracles, and in each segment there is a well-developed ventral commisure crossing the anterior part of the sternal region. The tracheal system of the thoracic legs of the caterpillar is of particular interest because in some respects it illustrates the leg tracheation typical of most insects. Each leg has two tracheae, one lateral, the other median. In the prothorax the lateral trachea of the leg is derived from the ventral commissure; the median one comes directly from the branches of the first, or mesothoracic, spiracle. In the mesothorax and the metathorax, the lateral leg trachea is a branch from a tracheal loop formed apparently by the union of tracheae from the spiracles preceding and following that is, from the mesothoracic and

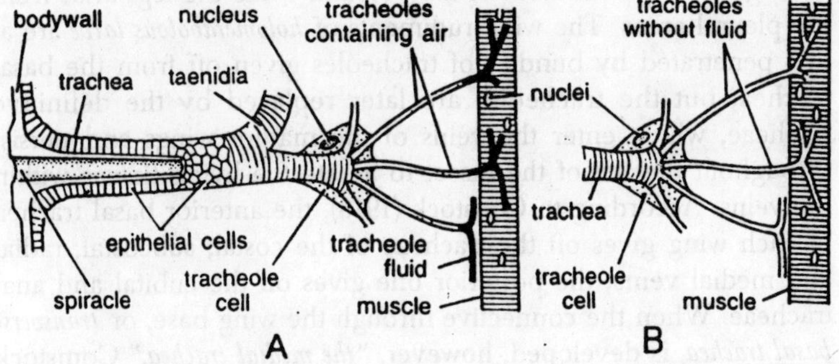

Fig. 3.9. Periplaneta. Role of tracheoles in gaseous exchange. A–Tracheoles with fluid at rest. B–Tracheoles without fluid after work.

metathoracic spiracles in the case of the mesothoracic legs, and from the metathoracic and first abdominal spriacles in the case of the metathoracic legs. Each lateral leg trachea of the wing-bearing segments, therefore, in its entirely has the form of a Y, the arms of which proceed from successive spiracles, while the stem enters the leg. The median leg trachea in these segments springs from the lateral trunk in the neighbourhood of the spiracle following.

Tracheal Tubes of the Wings

In the caterpillar it is to be seen that each internal wing rudiment is penetrated from opposite ends by two tracheae given off from the arms of the lateral Y-shaped leg trachea of the same segment. The two basal wing trachea appear to become continuous through the wing rudiment in older larvae. It has been shown by Chapmen (1918), from a comparative study of the basal connections of the wing tracheae, that the wing tracheation here exemplified in the caterpillar respresents the primary tracheation of the wing in all insects. There are, of course, many deviations from the typical condition and many developments along different lines of specialization; but all such modifications, Chapman shows, may be derived from the fundamental simple plan, in which two tracheae proceed from the convergent arms of the lateral leg trachea and enter the wing base.

According to Kennedy (1922a), the anterior branch is the original wing trachea. There is probably no morphological significance in the origin of the wing tracheae from the Y-shaped leg tracheae, since this particular tracheation of the legs occurs only in the wing-bearing segments. The spiracles and the wings belong to the lateral areas of the dorsum, while the legs arise from the pleural areas. The wing rudiments of *holometabolous larve* are at first penetrated by bundles of tracheoles given off from the basal trachea, but the tracheoles are later replaced by the definitive tracheae, which enter the veins of the mature wings and persist throughout the life of the insect to aerate the living tissues within the veins. According to Constock (1918), the anterior basal trachea of each wing gives off the tracheae of the costal, subcostal, radial and medial veins; the posterior one gives off the cubital and anal tracheae. When the connective through the wing base, or *transverse basal trachea,* is developed, however, "*the medial trachea,*" Comstock says, "tends to migrate along the transverse basal trachea toward the cubito-anal group of tracheae" and thus becomes more closely associated with the posterior wing tracheae in the wings of more specialized insects.

Abdominal Trancheal Tubes

With insects having a fully developed tracheal system, the lateral tracheal trunk extend posteriorly to the last pair of spiracles, which are usually those of eighth abdominal segment. In the caterpillar the principal abdominal branches are given off from the longitudinal trunks in the neighbourhood of the spiracles. Transverse ventral commissures uniting the larteral trunks occur in each of the first seven segments, and in the eighth segment there is a dorsal commissure. The ganglia of the ventral nerve cord, except the last, are tracheated by branches springing from the ventral commissures of their proper segments, regardless of the positions of the ganglia. The composite last ganglion, lying in the sixth segment, for which there receives its tracheae. The leg tracheae of the abdomen the lateral tracheae of the prothoracic legs counterparts in the mesothorax and metathor of adomen, going principal to the walls of the alimentary canal, arise from the lateral trunks near the spiracles. The heart is tracheated by terminal branches of the dorsal tracheae.

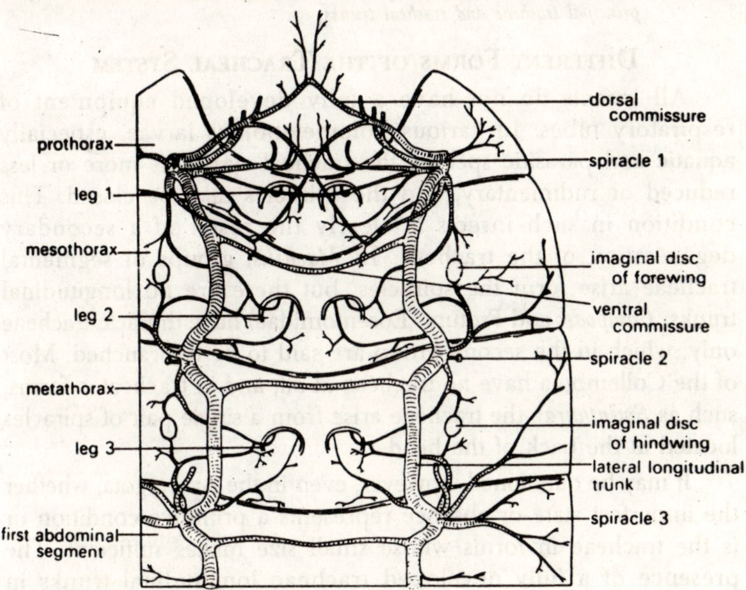

Fig. 3.10. Tracheation of the thorax and first abdominal segment of a caterpillar, dorsal view.

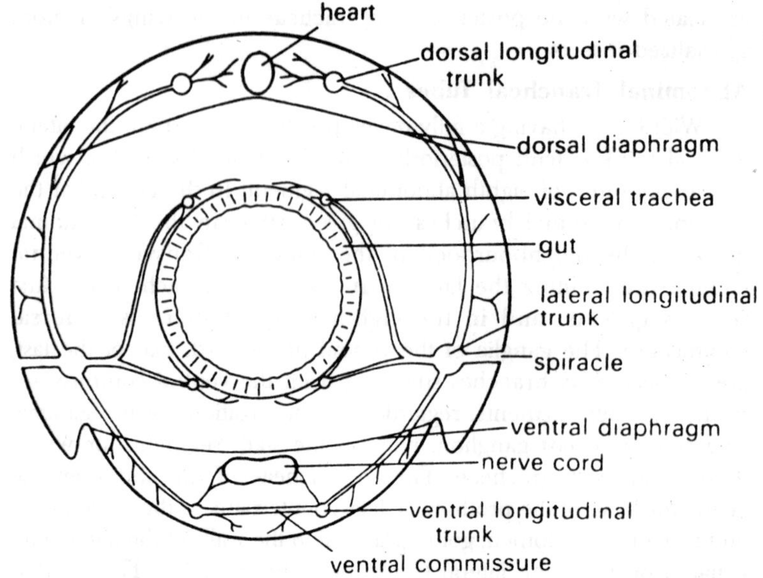

Fig. 3.11. Diagrammatic cross-section of the abdomen of an orthopteran showing the principal tracheae and tracheal trunks.

Different Forms of the Tracheal System

All insects do not have a fully developed equipment of respiratory tubes. In various holometabolous larvae, especially aquatic and parasitic species, the tracheal system is more or less reduced or rudimentary, and the spiracles may be closed. This condition in such insects is clearly the result of a secondary degeneration of the tracheae. In *Machilis,* groups of segmental tracheae arise form the spiracles, but there are no longitudinal trunks, *Campodec* and Protura (Eosentomidae) have thoracic tracheae only, which in the second group are said to be unbranched. Most of the Collembola have no tracheae at all, and in tracheatec forms, such as *Sminthuras,* the tracheae arise from a single pair of spiracles located at the back of the head.

It may be questioned, however, even in the Apterygota, whether the imperfect state or absence represents a primitive condition or is the tracheae in forms whose small size makes sufficient. The presence of a fully developed tracheae longitudinal trunks in Japygidae suggests that in its usual form is an inheritance from com modern Apterygota. An interesting example of specialization

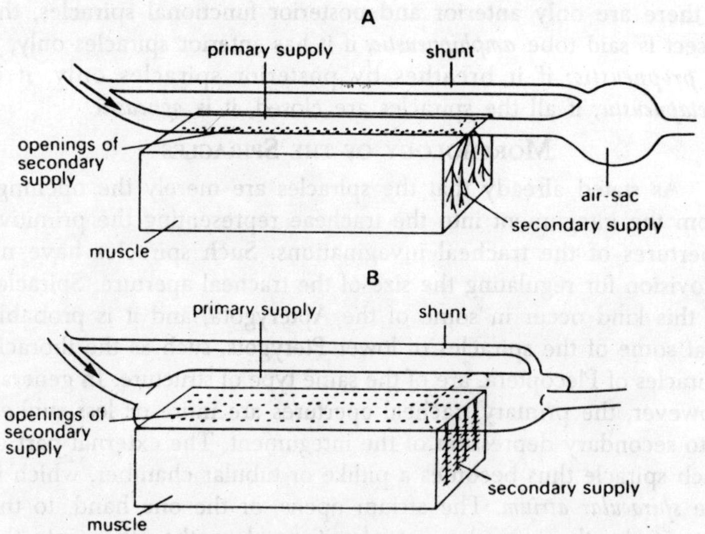

Fig. 3.12. Diagram of the tracheal supply to flight muscles in which the primary supply is (A) a trachea, or (B) an air-sac. Arrows indicate the inward flow of air.

seen in the larvae and pupae of most Diptera, longitudinal trunks become the principal respiration. The dorsal trunks of all dipterous through spiracles situated at one end or at both ends. These dorsal spiracles appear to be secondary respiratory orifices, since in some cases there are present also the usual lateral spiracles, though the latter are closed and remain rudimentary during the larval and pupal stages and are not funcionally restored until the imaginal stage. All dipterous larvae have a pair of posterior drosal spiracles, and some have in addition an anterior pair of posterior dorsal spiracles, and some have in addition an anterior pair on the porothorax. The pupae have only th anterior dorsal spiracles.

In a tipulid larva the lateral and dorsal tracheal trunks are equally developed; but in most other dipterous larvae, as is well shown in a mosquito larva or a muscoid maggot, the dorsal turnks are reduced to inconspicuous connectives between the roots of the transverse tracheae along the line of the closed and rudimentary lateral spiracles. According to distribution of the functional spiracles, several types of respiratory conditions may be distinguished. The *holopneustic* type is the generalized one in which the insect is provided with the usual bilateral series of 10 pairs of open spiracles; if there are only anterior and posterior functional spiracles, the insect is said tobe *amphipneustic;* if it has anterior spiracles only, it is *propneustic;* if it breathes by posterior spiracles only, it is *metapneustic;* if all the spiracles are closed, it is *opneustic.*

Morphology of the Spiracles

As stated already that the spiracles are merely the openings from the integument into the tracheae representing the primitive apertures of the tracheal invaginations. Such spiracles have no provision for regulating the size of the tracheal aperture. Spiracles of this kind occur in some of the Apterygota, and it is probable that some of the spiracles of lower Pterygots, such as the thoracic spiracles of Plecoptera, are of the same type of structure. In general, however, the primary tracheal apertures are more or less sunken into secondary depression of the integument. The external part of each spiracle thus becomes a pitlike or tubular chamber, which is the *spiracular atrium.* The atrium opens, or the one hand, to the exterior by the secondary *atrial orifice* and on the other, into the trachea by the primitive *tracheal orifice.* The walls of the atrium are often rugose and may be strengthened by transverse circular ridges, but such structures are not true taenidia, which pertain to the walls of the tracheae only. The atrial walls are also commonly clothed with hairs or other cuticular processes, such as occur on the external body wall.

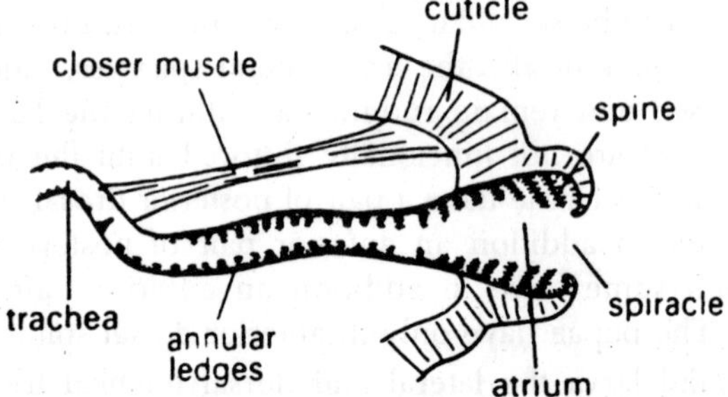

Fig. 3.13. Longitudinal section of the spiracle of a louse, Haemotopinus, showing the dust-catching spines and ledges.

In some cases the atrium is subdivided into an outer and an inner chamber, which differ in diameter or in the structure of their walls, but in general the atrium is quite distinct from the spiracular trachea. The lips of the atrial orifice may be flush with the surface

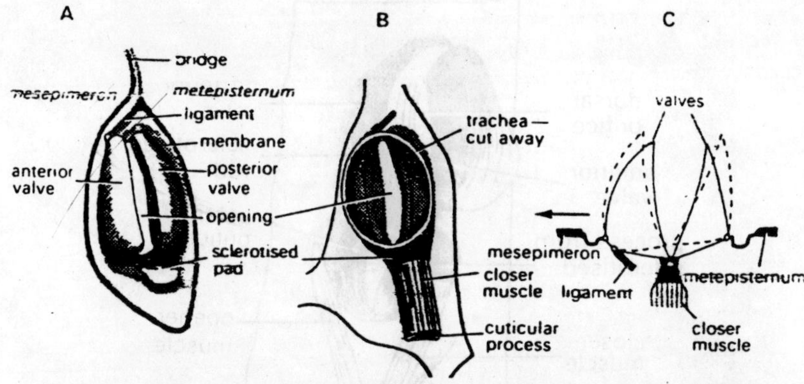

Fig. 3.14. The second thoracic spiracle of a locust. A–External view. B–Internal view. C–Diagrammatic transverse section showing how movement (indicated by arrow) of the mesepimeron causes the valves to open wide.

of the integument, raised in a marginal flange or short tube, or produced into a pair of valve-like plates, which are sometimes movable by special muscles. The opening is often contained in a small sclerotic plate of the body wall forming a distinct spiracular sclerities, or *peritreme.* Atriate spiracles are usually provided with a mechanism for regulating the passage of air to and from the spiracular trachea. This mechanism is generally called the *closing apparatus,* though it serves both to open and to close the spiracle.

The structure of the closing appartus differs much in different insects, and it is often quite different between the thoracic and the abdominal spriacles of the same species. Two principal types of occulusor mechanism, however, may be distinguished, with numerous modifications under each. The first type is a device of one kind or another for closing the outer lips of the atrium. The second is a mechanism for regulating the size of the tracheal aperture at the inner end of the atrium. Accessory structures are often present in the outer part of the atrium in spiracles of the second type, which simply guard the atrial orifice. Such structures commonly have the form of opposing rows of tapering processes of the atrial wall thickly clothed with interlacing hairs, the whole mass of which forms a *filter apparatus* that freely permits the passage of air, but which prevents the entrance of foreign particles or water into the atrium. Closing apparatus of the spiracles if absent in some of the higher insects, but the lack of the mechanism in such cases is probably a secondary condition.

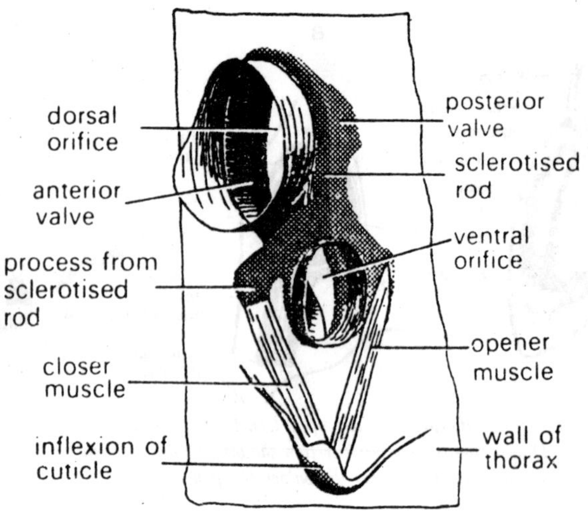

Fig. 3.15. The first thoracic spiracle of a locust, internal view.

The thoracic spriacles are more variable in structure than are the abdominal spiracles, and in general the lip type of closing apparatus is characteristic of them; the abdominal spiracles more consistently have the inner type of closing mechanism. Unusual modifications of the spiracular structure occur in certain holometabolous larvae, as in the so-called "*biforous*" spiracles of celeopterous larvae and the dorsal spiracles of *dipterous larvae.* It will be possible to give have only a brief description of the principal varieties of spiracular structure, illustrated by a few typical examples.

External Closing Apparatus of Spiracles

The type of spiracular structure in which the closing apparatus is formed by the lips of the atrial aperture is well-illustrated in the thoracic spiracles of Acrididae. In *Dissosteira carolina* the first, or *mesothoracic, spiracle,* lying in the membrane between the prothorax and the mesothorax, is an obliquely vertical slit in the peritremal sclerite with strongly protruding anterior and posterior lips. The anterior lip is a rigid elevation of the anterior edge of the anterior edge of the atrial aperture; its inner face, however, is soft and deeply grooved parallel with the outer margin. The posterior lie is a weaker and freely movable flap, but it has a sharp, strongly sclerotized marginal band, which, when the spiracle is closed, fits into the groove of the anterior lip. The atrium of this spiracle is

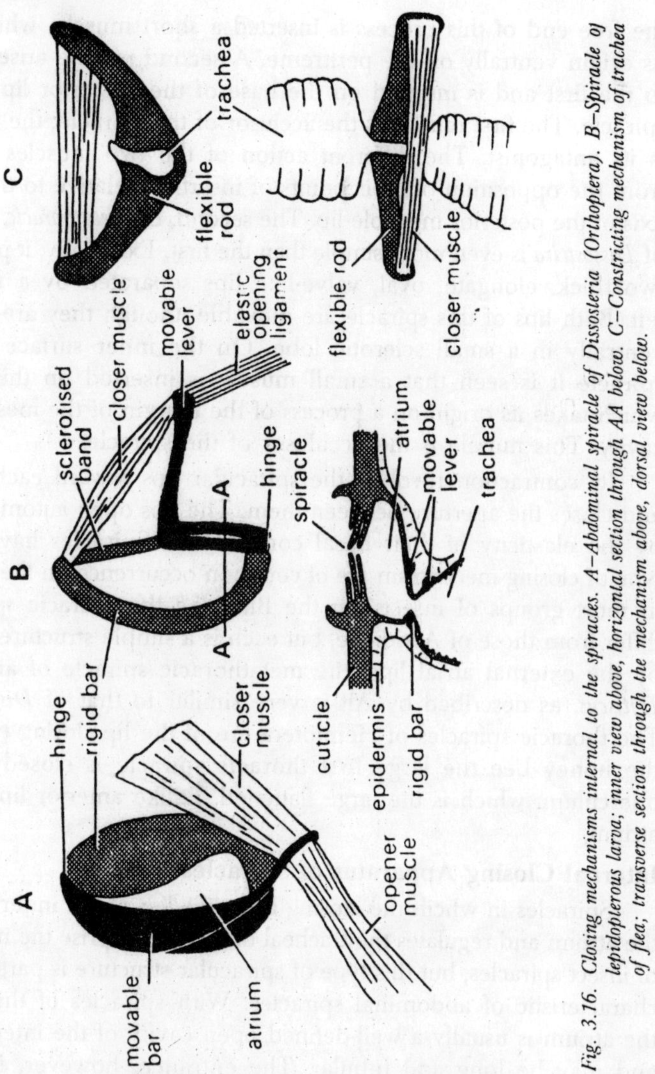

Fig. 3.16. Closing mechanisms internal to the spiracles. A—Abdominal spiracle of Dissosteria (Orthoptera). B—Spiracle of lepidopterous larva; inner view above, horizontal section through AA below. C—Constricting mechanism of trachea of flea; transverse section through the mechanism above, dorsal view below.

the shallow cavity between the lips. From it there are given of two tracheae, a large dorsal one and a smaller ventral one.

In the septum between the two tracheal opening is a strong bar projecting anteriorly and ventrally from the posterior lip. Upon the feed end of this process is inserted a short muscle, whch has its origin ventrally on the peritreme. A second muscle arises close ot the first and is inserted on the base of the posterior lip. Upon the free end of this process is inserted a short muscle, which has its origin ventrally on the peritreme. A second muscle arises close to the first and is inserted on the base of the posterior lip of the spiracle. The first muscle is the acclusor of the spiracle; the second is its antagonist. The different action of the two muscles results from the opposition of their points of insertion relative to the long axis of the posterior movable lip. The second, or *metathoracic, spiracle* of *Dissosteira* is even more simple than the first. Externally, it presents two thick, elongate, oval, valve-like lips separated by a vertical left. Both lips of this spiracle are movable, though they are united ventrally in a small sclerotic lobe. On the inner surface of the spiracle it is seen that a small muscle is inserted on this lobe, which takes its origin on a process of the margin of the mesocoxal cavity. This muscle is the occulusor of the spiracle.

Its contraction revolves the spiracular lips toward each other and closes the aperture between them. The lips open automatically by the elasticity of their basal connections. Spiracles having lip type of closing mechanism are of common occurrence on the thorax in most groups of insects. In the Blattidae the thoracic spiracles differ from those of Aerididae, but each is a simple structure closed by the external atrial lips. the metathoracic spiracle of an adult *Dytiscus,* as described by Alt is very similar to that of *Dissosteira.* The thoracie spiracles of Hemiptera are of the lip-closing type. In the honey bee the large first thoracic spiracle is closed by an operculum, which is the large flattened, lidlike anterior lip of the atrium.

Internal Closing Apparatus of Spiracles

Spiracles in whcih the closing appratus lies at the inner end of the atrium and regulates the tracheal opening comprise the majority of insect spiracles, but this type of spiracular structure is particularly characteristic of abdominal spiracles. With spiracles of this kind, the atrium is usually a well-defined open cavity of the integument and may be long and tubular. The entrance, however, is often guarded by a filter apparatus, usually in the form of two rows of matted brushes projecting from opposite walls of the atrium (b). The lips of the atrial aperture have various forms, but they are never movable and they take no active part the closing of the spiracle. The size of the atrial orific varies much regardless of the size of the atrium; sometimes it is contracted to a small pore opening into a relatively large artial chamber, and in special cases it is closed. Two common sub-type of structure are found among spiracles of the inner closing type.

In one sub-type of occlusor mechanism is a simple *pincheock apparatus,* consisting of two sclerotic bars in opposite walls of the atrium just before the mouth of the trachea, with a muscle stretched between their projecting ends. The contraction of the muscle brings the bars together and thus closes the tracheal entrance. Usually a second muscle arising on the body wall inserted on the end of the anterior bar opposite the attachment of the occlusor muscle and acts as a dilator of the spiracle. The abdominal spiracles of Blattidae have a closing apparatus of this kind, but the free end of the anterior bar is prolonged as a manubrium to give stronger effect to the muscles.

The dilator muscle in Blattidae arises anteriorly on the deflected lateral lobe of the tergum that contains the spiracle. In the Acrididae the closing mechanism of the abdominal spiracles is a modification of the blattid type, in which the posterior bar is absent, and the anterior bar is represented by the entire anterior wall of the atrium, which is movable and produced ventrally in the manubrium. The occulusor muscle arises on the tergal wall immediately behind the spiracle, and the long dilator muscle arises on the lateral edge of the sternum. In the second sub-type of *occulusor apparatus* in spiracles of the inner closing type the effective organ is a valve. The valve consists of a fold of the inner end of one wall of the atrium, of a mechanism for inflecting the fold over the tracheal mouth. An occlusor apparatus of this kind is the common form of closed apparatus in the abdominal spiracles of holmetabolous insects, essential elements of the closing structure include a crescentic or semicircular elastic bar, the so-called *closing,* the ends of which are produced outside the atrial walls as two thick conical pore; second, a soft, convex fold, the *closing,* projecting into the atrial lumen form the wall opp the bow; and, third, a closing muscle stretched like a bows between the ends of the bow.

The closing band, or valve, is us on the posterior wall of the atrium and is located just before the of the trachea. The contraction of the occlusor in pulls on the two ends of the bow and forces the valinward until it entirely closes the tracheal orifice. In some spiracle the closed valve overlaps externally the bow on the opposite of the aperture. The opening of the spiracle may be entirely by the elasticity of the bow, but usually a dilator muscle, a ventrally on the body wall, is inserted on the lower process of the opposite the attachment of the occlusor muscle. A simple closing apparatus of the form just described occurs abdominal spiracles of many insects, but numerous departures from typical structure are found in the holometabolous orders. A modifiction results from the suppression of the ventral process bow, and the development of a point of flexure by the dorsal process and the upper end of the bow.

The pull closing muscle on the dorsal process then brings the base of the and the closing band against the inner edge of the bow to tracheal aperture. This type of structure, found with variations Coleoptera, might also be supposed to be a derivative of the pin type of mechanism, but its effective element is a valve, typical valvular spiracle. The valve mechanism is highly de in Lepidoptera by the extension of the dorsal muscle process into lever. A dorsal dilator muscle or strand of elastic tissue arising on the body wall and inserted on the lever, is present pillars, in addition to a ventral muscle, which is here inserted atrial wall.

The structural details and action of the closing apparatus of pillar spiracle are shown in Fig. representing the first thoracic in which the structure is the same as that of the abdominal except that the position of the parts is reversed, the lever and being anterior in the thoracic spiracles and the bow posterior is supported on a looped bar in the membranous valve, and the latter, when closed, is received into a deep concavity of the posterior atrial wall. Distal to the valve the atrial aperture is protected by a filter apparatus composed of two opposing mats of thick, brushlike processes projecting from the anterior and posterior walls of the atrium.

Spiracles with Double Openings

In the larvae of various families of Coleoptera there occur spiracles of a type known as "biforous." the term originally implified that each spiracle had two external openings, but it is now commonly extended to other spiracles similar in appearance, but it is now commonly extended to other spiracles similar in appearance, but having only one opening or probably, in some cases, one at all. Good examples of true biforos spiracles are found in the larvae of Elasteridae. The functional openings of these spiracles are secondary formations, since the primary atrial orifice is closed except during ecdyses. A largal spiracle of *Alaus oculatus* presents externally an ovate peritremal area having anteriorly a dark sclerotic thickening and posteriorly two elongate covergent plates. The thickening marks the site of the closed atrial orifice.

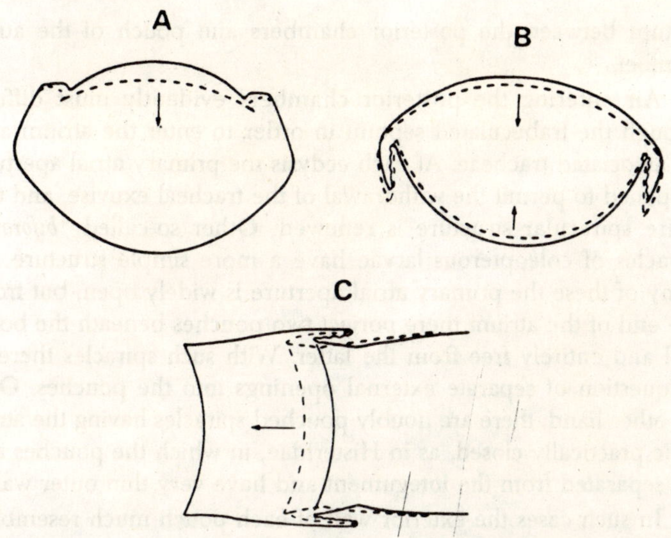

Fig. 3.17. Diagrammatic representations of types of abdominal ventilatory movements. Dashed lines indicate the contracted positions, arrows the directions of movement. (A) and (B) in transverse section, (C) in longitudinal section.

The convergent plates have each a clear median area traversed by an axial line. Internally, the spiracle consists of a closed atrial chamber, which gives off posteriorly a wide membranous pouch, the external wall of which consists of a thin doubly convex membrane strengthened by branching and interjoining trabeculae. External to the pouch, and projecting beyond it posteriorly, are two shallow cuticle-lined chambers beneath the convergent external plates of the spiracle. A manipulation of the spiracle of *Alaus oculatus* demonstrates beyond question that these chambers can be widely opened along the axial lines of their outer walls and gives every reason to believe that the opening are natural clefts though in the usual condition their lips are closely appressed. Roberts (1921) has shown that in section the spiracles of *Agriotes* are cleft along the median lines of the posterior chambers. It has often been claimed, however, that the chambers are closed cavities, and that observed openings are artifacts. The spiracles of *Alaus oculatus* thus appear to consists of a closed atrial chamber provided with a broad posterior diverticulum, and of two open, secondary posterior chambers, the inner walls of which are adnate with the outer wall of the atrial pouch, forming a thin, doubly arched, trabeculated septum between the posterior chambers and pouch of the atrial chamber.

Air entering the posterior chambers evidently must diffuse through the trabeculated septum in order to enter the atrium and the associated tracheae. At each ecdysis the primary atrial aperture is opened to permit the withdrawal of the tracheal exuvise, and the entire spiracular structure is renewed. Other so-called "*biforous*" spiracles of coleopterous larvae have a more simple structure. In many of these the primary atrial aperture is widely open, but from one end of the atrium there porject two pouches beneath the body wall and entirely free from the latter. With such spiracles there is no question of separate external openings into the pouches. One the other hand, there are doubly pouched spiracles having the atrial orific practically closed, as in Histeridae, in which the pouches are not separated from the integument and have very thin outer walls.

In such cases the exterior wall of each pouch much resembles that of the posterior chambers of *Alaus* and may even be marked by a faint median line. Much discussion has centered around the question as to whether these pouches are open to the exterior or are closed. Steinke (1919) seems to concede that they may be open in some cases, but he rightly says it is a very difficult thing to prove. If they are closed, as they appear to be, the exchange of respiratory gases must take place by diffusion through their very delicate outer walls. In the larvae of *Donacia* the dorsally situated

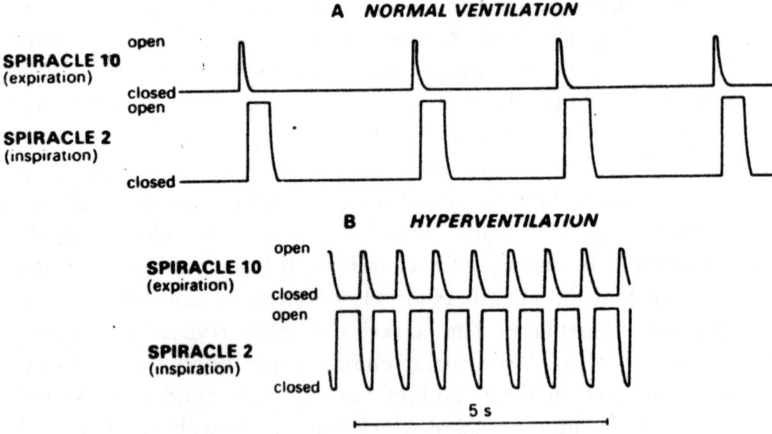

Fig. 3.18. Diagram illustrating the activity of the spiracles of Schistocerca in ventilation. A–Normal ventilation. B–Hyperventilation.

spiracles of the eighth abdominal segment have each a pair of slender tubular pouches extended into a long free spine-like process of the peritreme. The respiratory spines of these posterior spiracles are used to penetrate the vascular tissues of water plants for obtaining air. According to Boving (1910), the spines are inspiratory in function and are imperforate; expiration takes place through the open lateral spiracles and thorugh the atrial apertures of the posterior spine-bearing spiracles.

It is evident that there is needed a thorough comparative study of biforous spiracles in Coleoptera. From the foregoing discussion it appears that there may be two types of closed spiracles here included, one in which the atrium gives off a pair of imperforate diverticular, into which air diffuses through the external integument, the other (Elateride) in which the diffusion surfaces are concealed in secondary open invaginations of the body wall.

Spiracles of Dipterous Larvae

Spiracles of a unique type of structure occur in the larvae and pupae of Diptera directly connected with the ends of the dorsal tracheal trunks. The position, structure, development, tracheal connections, and the temporary nature of these dorsal spiracles suggest that they are secondary respiratory structures havng no relation to the lateral spiracles, which are closed or suppressed during immature stages and functionally restored in the adult. The anterior larval or pupal spiracles take the form of perforated lobes or tubes or of trumpetlike horns arising from the posterior part of the prothorax. The posterior larval spiracles are usually contained in a pair of prominent plates situated on the eighth segment or the composite terminal segment of the adbomen, where they are generally exposed, though they may be concealed in a shallow cavity or elevated on a respiratory tube. The posterior spiracles typically have one, two, or three openings.

In tipulid larvae the posterior spiracluar plates were formerly supposed to consist of a mesh of fine rods branching from a central disc, admitting air through the interstices, but Gerbig (1913) has shown that peripheral area of the spiracle is imperforate, and that the functional opening is a median slit in the central disc obscured by its overlapping lips. In first-instar larvae the spiracular apertures are plainly open. The external part of each prothroacic dorsal spiracle of cyclorrhaphous larvae has the form of a small lobe, usually branched or digitate, with numerous pores communicating with the strium. The posterior spiracles present each two openings in the first instar of the larva and usually three in the second and third instars. The spiracular apertures open into a large atrial chamber connected with the end of the corresponding dorsal tracheal trunk.

At the first and the second moult the entire spiracular structure is formed anew and takes on a different form characteristics of the ensuring instar. Investigators do not agree as to whether the new atrial chamber is an outgrowth of the one preceding or an ingrowth from the integument, but in either case the old chamber serves for the discharge of the tracheal intima and is than closed, while the new formation becomes the functional breathing orifice for the succeding instar. The site of the earlier spiracle is marked by a scar on the surface of the integument, which remains connected with the base of the new atrial cavity by a strand of cuticular tissue. At the third molt of the larva the dorsal spiracles are not renewed. The lateral imaginal spiracles of these flies appear first on the fourth instar of the larva (formed within the puparium) just before the transformation to the pupa.

Morphology of the Tracheae

The histological structure of tracheae is essentially the same, as that of the body surface from which they are derived. The matrix layer of a trachea is an epithelium of flat polygonal cells continuous wth the epidermis around the spiracle. On the outside is a basement membrane, and on the inside a strong cuticular intima. The characteristic feature of an insect trachea is its closely ringed appearance resulting from the presence of folds or thickenings of the intima in the form of minute circular or spiral ridges, the *taenidia,* which project on the inner surface.

In some insects the inner walls of the tracheae are covered by short spicules or clothed with simple or branched hairs arising from the taenidial ridges. The taenidia are generally not continuous through any considerable length of the trachea but form a succession of ridges, each of which makes a few turns around the tracheal wall and then terminates. When a trachea is broke, the torn edge usually pulls out in a long spiral band, which, it will be observed in most cases, is not a single taenidium but a strip of the tracheal wall containing several taenidia. The taenidia of the large dorsal, trunks of some dipterous larvae, however, appear to be simple,

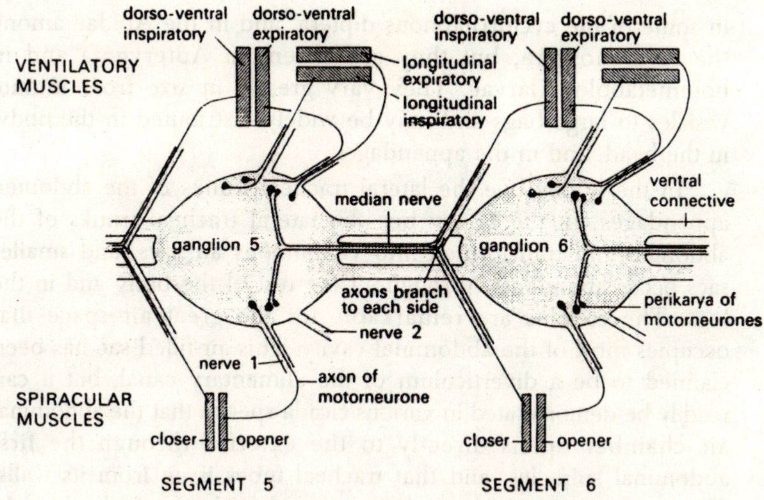

Fig. 3.19. Diagrammatic representation of the motor control of abdominal ventilation (top half) and spiracular closure (lower half) in Schistocerca. The positions of the perikrya within a ganglion are only approximations.

uninterrupted rings, since a single taenidial "thread" may be easily removed from the broken end of trunk.

While in general the taenidia are continuous around the walls of a trachea and serve to keep the tube open, those of the dorsal trunks of a tabanid larva, and presumably in other diterous larvae, are all jointed, or broken by points of flexibility, in a definite line along each side of the trachea. A trachea having this structure, when devoid of air, collapses of a flat band. Dunavan (1929) has observed that a collapsing and also a shortening take place in the dorsal tracheal trunks of a living *Eristalis* larva during respiration. In insects having a mechanical respiration, Krogh (1920a) distinguishes *respiration tracheae* (that is, *ventilation tracheae*), which are oval in cross section and easily compressible, from *diffusion* tracheae, which are rigid and cylindrical.

The Tracheal Air Sacs

The tracheal tubes are seldom of a uniform or an evenly tapering diameter; generally they are widened in some places and narrowed at others. If a widened part of a trachea forms a conspicuous enlargement in the course of the tube, the dilation is called a *tracheal air sac.* Air sacs are present in certain members of most of the pterygote orders and reach their greatest development in some of the cyclorrhaphous diptera, and in the Apidae among the Hymenoptera, but they are absent in Apterygota and in holometabolous larvae. They vary greatly in size from minute vesicles to large bags and may be widely distributed in the body, in the head, and in the appendages.

In the honey bee the lateral tracheal trunks of the abdomen appendages. In the honey bee the lateral tracheal trunks of the abdomen and transformed into voluminous air sacs, and smaller sacs occur abundantly throughout the rest of the body and in the legs. The cicadas are remarkable for the great air space that occupies most of the abdominal cavity. This air-filled sac has been claimed to be a diverticulum of the alimentary canal, but it can readily be demonstrated in various cicada species that the abdominal air chamber opens directly to the exterior through the first abdominal spiracles, and that tracheal tubes issue from its walls. The air sacs respond in a greater degree than do the tracheal trunks to increased and decreased pressure in the body resulting from the movements of respiration and thus give a more efficient ventilation to the tracheal system during breathing.

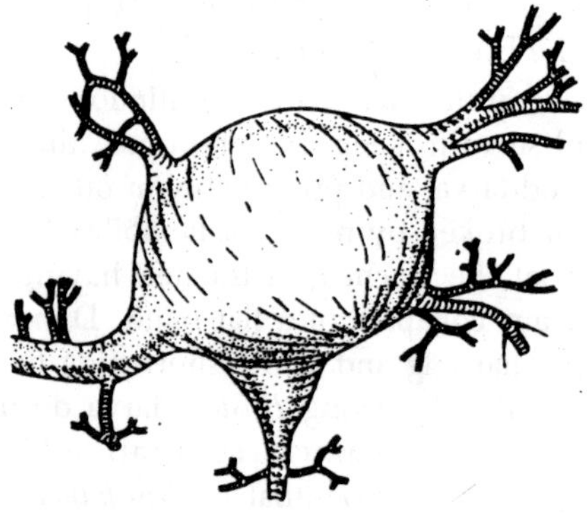

Fig. 3.20. An air sac.

The air sacs in some insects are particularly responsive to the respiratory movements because their walls lack the taenidial ridges characteristic of the tracheae. It is generally assumed that intima of the air sacs as well as that of the tracheae is a chitinous membrane. Tests made by van Wisselingh (1898) on the air sacs and tracheae of the house fly, and by Chapbell (1929) on the air sac and attached tracheae of both the house fly and the honey bee failed to show the presence of chitin; but Koch (1932) claims that with more delicate methods of technique the presence of chitin can be demonstrated in the tracheal intima of both these insects.

THE TRACHEOLES

The final link between the end branches of the tracheae and the cells of the body tissues is formed by minute tubules called *tracheoles.* The tracheoles are said to differ from the tracheae in that they are contained within single cells. They are cuticular canals, generally less than a micron in diameter, lacking taenidal ridges, formed in elongate and usually branched cells of the tracheal epithelium. When the tracheoles are first developed they have no opening into the lumen of the trachea, but with the removal of the tracheal intima at the succeding moult the lumina of the tracheoles becomes continuous with the cavity of trachea. In other words, a tracheole, apparently, is a tubular outgrowth of the newly forming tracheal intima formed within a single cells of the tracheal epithelium. A tracheole, therefore, is probably not a truly intracellular structure but resembles the duct of a unicellular gland, which penetrates the cell body as an invagination of the cell wall.

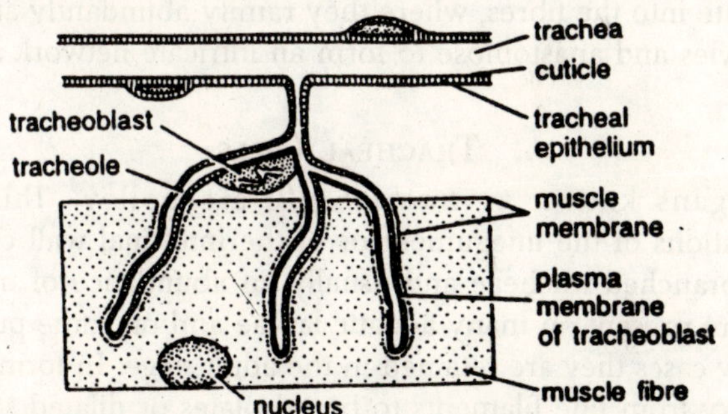

Fig. 3.21. An intracellular tracheole (diagrammatic).

During the formative stage the tracheole becomes coiled within its cells, but as the cell elongates the tracheole straightens out, until finally it extends a long distance form its point of origin, and the attenuated cell matrix around the tube becomes scarcely perceptible. The tracheoles are usually given off in clusters from the tracheae. In some cases they appear to be simple tubules, but generally they are dichotomoulsy branched. The terminations of the tracheoles have not been studied in many insects, but their final branches have been found to anastomose in a fine capillary network over the tissue cells, in which there may be united groups of tracheoles from several different tracheal sources. Von Wistinghausen (1890) has described the tracheole capillary net of the silk glands of caterpillars, and E. Holmgren (1896) finds a similar network of anastomosing tracheoles not only on the silk glands but also on the flat cells, the Malpighian tubules, and the walls of hte mesenteron in caterpillars.

According to Holmgren (1896a), the canaliculi of the tracheole net are formed in a different set of cells from those in which the primary tracheoles are generated. The tracheole ending usually lie on the surfaces of the cells, but they are said in some cases to dip beneath the cell surface and thus appear to live within the body of the cell. It is probable, however, that the tracheoles do not ordinarily penetrate the cell cytoplasm. Where they are seen to lie within the circumference of an epithelial cells, Holmgren says, they are contained in a pouch or sheath of the basement membrane. The tracheole net of most muscles is also superficial. In the wing muscles, however, according to Athanasiu and Dragoiu (1913, 1915), the tracheae branch profusely between the fibers, and the tracheoles penetrate into the fibres, where they ramify abundantly among the sarcostyles and anastomose to form an intricate network about the latter.

TRACHEAL GILLS

Organs known as *tracheal gills* are hollow, thin-walled evaginations of the integument or of the intestinal wall containing finely branched tracheae and usually an abundance of tracheoles. They are present on many aquatic larvae and on some pupae, and in a few cases they are retained in the adult stage. In form, tracheal gills vary from fine filaments to broad plates or dilated sacs. They may be situated on any external part of body, including the head, the thor and the abdomen, or in the rectal part of the proctodaeum, but they usually confined to the exterior of the abdomen. Typical *filamentous tracheal gills* occur on the larvae of Plecopte some Ephenmeridae, most Trichoptera, the neuropteran *Corydalus coritus,* on several species of aquatic lipidopterous larvae of the get *Nymphula,* and on the pupa of the dipterous genus *Simulium.*

The slender tapering appendicular process borne on the sides of abdominal segments of sialid larvae and of gyrinid and certain coleopterous larvae are usually regarded also as having a respiratory function because each is penetrated by a tracheal branch from lateral tracheal trunk. Familiar examples of plate-like tracheal lateral tracheal trunk. Familiar examples of plate-like tracheal are those that occur along the sides of the abdomen of many epheme larvae and at the end of the abdomen of larvae of zygopterous Odona. An interesting description of the filamentous gills of lepidoptera larvae is given by Welch (1922), who finds that each gill filament contain a tracheal branch from the main lateral trunk of the tracheal system, a that the inner surface of the gill is covered by innumerable tracheal lying parallel with one another. Nearly five hundred gill filaments be present on a single individual of *Nymphula obscuralis.*

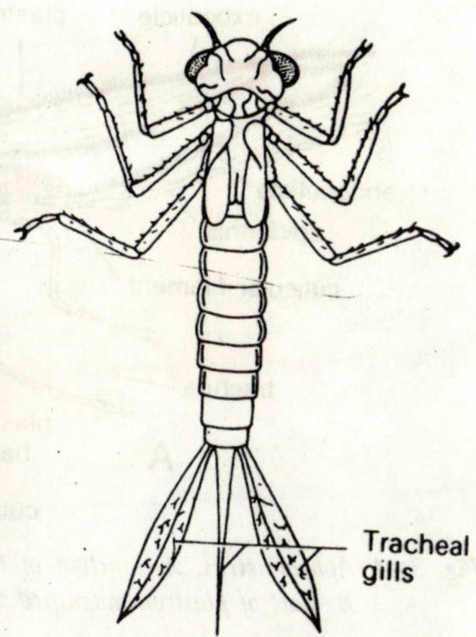

Fig. 3.22. Terminal abdominal tracheal gills in a damselfly nymph.

The *terminal gills* of zygopterous larvae are borne by the epiproct and the paraprocts. Usually they have the form of elongate plates, but certain species they are vesicular. Most highly developed

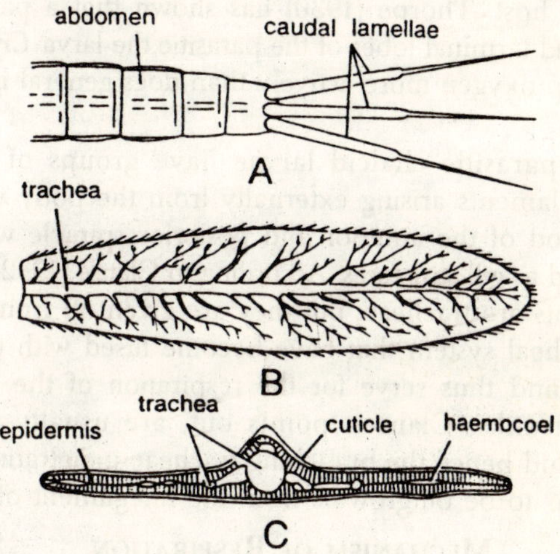

Fig. 3.23. Tracheal gills of larval Zygoptera. A–Posterior end of larva of Coenagrion in dorsal view. B–One lamella in lateral view. C–Caudal lamella of Synlestes in T.S.

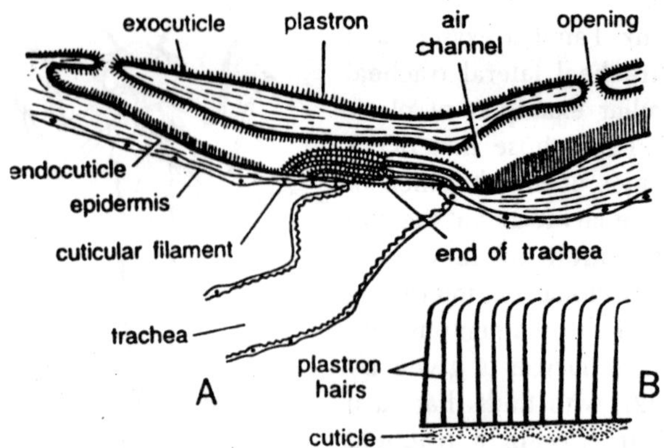

Fig. 3.24. Aphelocheirus. A–Junction of trachea with system of channels in the cuticle. B–Part of plastron magnified to show hairs.

of all tracheal gill structures are the rectal gills of the larvae of the anisopteous Odonata. These gills consists of six sets of invaginations of various shapes projecting in longitudinal rows, from the inner wall of the anterior part of the rectum, together forming the so-called "*branchial basket.*" The respiratory lobes a richly tracheated from the dorsal and visceral longitudinal trunks of the tracheal system. Though parasitic larvae generally have no special respiratory equipment for breathing the oxygen dissolved in the blood of the host. Thorpe (1930) has shown that a pair of long, well-tracheated terminal lobes of the parasitic the larva *Cryptochaetum iceryae* take up oxygen more actively than does general integument of the insect.

Certain parasitic chalcid larvae have groups of branched trachealike filaments arising externally from the body wall in the neighbourhood of the anterior and posterior spiracle which have been regarded as gill structures. According to Clause (1932), however, these filaments are tracheae, but they are given off from trunks of the host tracheal system that have become fused with the wall of the parasite and thus serve for the respiration of the latter. The connecting trunks, Clausen points out, are usually broken in dissections, and hence the branching tracheae penetrating the host tissues appear to be outgrowths from the integument of the larva.

Mechanism of Respiration

Respiration through tracheae branching to all parts of the body may be accomplished entirely by the diffusion of gases within the tracheae; but probably the majority of adult insects produce a partial ventilation of the tracheal system by means of movements of the body wall. In the second case, tracheal breathing has many features in common with lung breathing and involves the presence of a mechanism for producing and controlling the respiratory movements.

Respiration by Gas Diffusion in the Tracheae

Since many insects, especially larval forms, do not make any perceptible breathing movements, it is evident that respiration in such cases must be accomplished largely or entirely by the diffusion of gases through the tracheae. It is possible, however, that with some larvae the movement of the body or particularly the successive contraction of the lateral body muscles overlying the longitudinal tracheal trunks may cause an irregular passage of air through these trunks. Likewise it may be supposed that in the larvae of Diptera the pulsations of the heart might effect a compression of the large dorsal trunks lying to each side of the

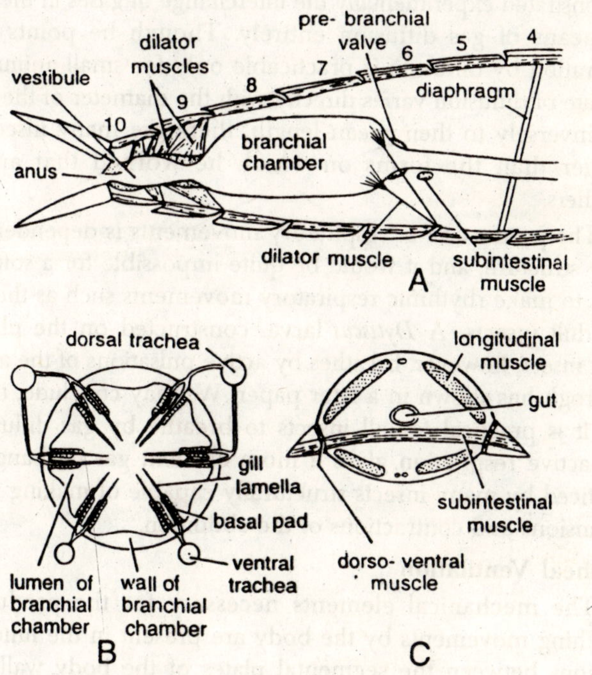

Fig. 3.25. Abdomen of dragonfly larva in L.S. B–Branchial chamber in cross section. C–T.S. of sixth abdominal segment.

heart might effect a compression of the large dorsal trunks lying to each side of the heart.

As already mentioned, it has been observed by Dunavan (1929) that the dorsal trunks of an *Eristalis* larva both shorten and collapse during respiration, though Dunavan was not able to discover the means by which the activity of the tracheae is produced. In general, however, there can be no doubt that diffusion accounts for the major part of gas transfer through the tracheae of insects that make no specific respiratory movements, and, even in inseccts that actively breathe, it is only the larger tracheae that are ventilated; the peripheral respiration is always by means of gas diffusion. The part played by diffusion in the respiration of insects has been conclusively shown by Krogh (1920, 1920a). Using tenebrionid larvae and the larva of *Cossus* as subjects having an open tracheal system, and aeschnid larvae as examples of aquatic insects with a closed tracheal system and breathing by means of gills, Krogh demonstrated experimentally the interchange of gases in the tracheae by means of gas diffusion entirely. Though he points out that respiration by diffusion is practicable only for small animals, since the rate of diffusion varies directly with the diameter of the tracheae and inversely to their mean length, there are many insects much smaller than the forms on which he worked that are active breathers.

The production of respiratory movements is dependent on the body structure, and it would be quite impossible for a soft-skinned larva to make rhythmic respiratory movements such as those made by adult insects. A *Dytiscus* larva, constructed on the plan of an adult insect, however, breathes by active pulsations of the abdomen, as Krogh has shown in a later paper. We may conclude, therefore, that it is practical for all insects to breathe by gas diffusion, but that active respiration gives a more efficient gas exchange and is practiced by many insects structurally capable of making rhythmic expansions and contractions of the abdomen.

Tracheal Ventilation

The mechanical elements necessary for the production of breathing movements by the body are present in the fundamental relations between the segmental plates of the body wall and the somatic muscles. The possibility of breathing, therefore, is possessed by all insects with sclerotic plates in the body wall, and, after the acquisition of tracheae, mechanical respiration needed only the development of control centers in the nervous system.

The respiratory movements

The movements of respiration affect principally the abdomen. They are produced by the somatic muscles and by the elasticity of the body wall. Expiration results from a dorsoventral compression of the abdomen or, in some cases, also from a longitudinal contraction of the abdomen. The effectors in the first case are the lateral tergosternal muscles; in the second they are the intersegmental longitudinal muscles. Both movement may occur together in the same insect. Expiration may be accomplished entirely by the elasticity of the body wall; but, in insects that breathe strongly, some of the vertical and longitudinal muscles are generally converted into dilators and protractors of the abdomen by a change in their mechanical relations to the plates on which they are attached. Thus, with insects in which the abdominal terga overlap the edges of the sterna, some of the external lateral muscles become dilators if their tergal attachments, and in such cases, the effectiveness of the dilators is usually increased by the dorsal extension of their sternal attachments on lateral apodemal arms of the sternum.

Similarly, a protractor apparatus is formed by a transposition of the anterior ends of the external drosal and ventral muscles to the posterior margins of the terga and sterna, respectively, so that these muscles become antagonistic to the internal longitudinals. The effectiveness of the protractors likewise may be increased by the forward extension of their points of insertion on anterior tergal and sternal apodemes. A comprehensive study of the breathing movements of insects has not been made in recent years, and our best source of information on the subject is still the work of Plateau (1884). By means of lantern projections of the shadows of living insects, Plateau made observations on the respiratory movements of insects representing the principal orders. From his results he distinguished three principal types of respiratory mechanism, based on the structure of the abdomen and the manner of breathing.

In the *first type*, the sterna are usually firm and strongly convex and move but little in respiration; the terga, on the other hand, are mobile and noticeably rise and fall with each inspiration and expiration. Insects that breathe in this manner include Heteroptera and Coleoptera. In the *second type,* the terga are large and overlap the sterna laterally, usually concealing the membranous lateral zones of the segments. Both the terga and the sterna approach and separate in this type of structure, but the movements of the sternal are the more pronounced. Here Plateau included the Odonata, Acirdidae, aculeate Hymenoptera and Diptera. He observes, however, that is Phryganiidae and Hymenoptera the dorsoventral movements of the abdomen are accompanied by more pronounced movements in a longitudinal direction. The *third type* of respiratory mechanism is found in insects having the terga and sterna separated on the sides of the abdomen by ample membranous areas.

During breathing the terga and sterna approach and separate, while the lateral membranes correspondingly bulge outward or are drawn inward. Insects having this type of structure include Tettigoniidae, Neuroptera, Trichoptera and Lepidoptera. The rate and amplitude of the breathing movements are characteristically different in different insects and vary also in each individual according to the strength of external stimuli and according to the activity of the insect. Lee (1925), for example, records the average rate of breathing for females of *Melanoplus femur-robrum* as being 5.8 a minute at 49°F, and increasing to 26.6 at 80°. Herber and Slifer (1928), however, find much variation in the breathing of quiescent grasshoppers when observations are continued for a considerable length of time, the variations affecting not only the rate of breathing but also the depth of the abdominal pulsations. Thus, they report for a male of *Melanoplus femur-robrum,* observed for an hour, a fluctuations from 21.5 to 67.5 seconds in the time occupied by 10 respiratory movements.

The Tettidgoniidae appear to be more active breathers than the Acrididae, and during stridulation the breathing of the males is especially pronounced. The Phasmidae, on the other hand, are very slow breathers. According to Stahn (1928), the European walkingstick *Dixippus morosus* when at rest makes only 1.4 to 2.3 expirations a minute, though all stimulating influences cause an increase in the respiratory rater. The breathing movements of *Dixippus* are said to affect both the abdomen and the thorax.

Course of the air in the tracheae

During recent years there has been much discussion onthe question of a differential function of the spiracles as inspiratory and expiratory orifices, and on that of the direction of the air currents in the longitudinal tracheal trunks. Experiments made by Lee (1925) on the respiration of grasshopper (Acrididae) and observation that the thoracic and first two abdominal spiracles open during the expansion of the abdomen and close during contraction, while the last six abdomen spiracles open and close with the reverse movements, seemed to show that in normal breathing by grasshopper inhalation takes place through the anterior spiracles, and exhalation through the posterior spiracles.

Lee's results were disputed by MacKay (1927); and McArthur (1929), after making similar experiments on several species of Acrididae, arrived at the following conclusions: The first four spiracles of the grasshopper are usually inspiratory, and the last six expiratory, but the action of the spiracles is variable under both normal and abnormal conditions; the mechanism of the spiracular valves is capable of reversing the times of opening and closing of the spiracles relative to the respiratory movements of the abdomen; the direction of air currents through the tracheae can thus be reversed, or the air can be forced into any one of several possible paths by the internal control of the spiracular valves. Subsequent investigations have confirmed in general Lee's original claim that the air stream goes posteriorly through the body of Acrididae with a more efficient appratus than that used by preceding writers, McGovran (1931), experimenting on *Chortophaga viridifasciatus,* reports that the respiratory movements produce a pulsatory movement of air through the tracheal trunks, and that inspiration is principally into the thorax, while expiration is principally by way of the abdomen. An adult female, at 28°C, passed an average of 0.222 cubic centimeter of air through the body per minute-per gram body weight. finally, the work of Fraenkel (1932) gives essentially the same resutls on Orthoptera. The thoracic and first two pairs of abdominal spriacles, Fraenkel says, open during inspiration and close during expiration, while the other six pairs of abdominal spiracles show a reverse action relative to the respiratory movements of the body.

Furthermore, Fraenkel demonstrated experimentally in *Schistocerca gregaria* a movement of the respiratory air posteriorly in the tracheae. Quantitative measurements showed from 5 to 20 cubic millimeters transported per second, or from 7.5 to 24.4 cubic millimeters with each expiratory movement. The work of von

Buddenbrock and von Rohr (1923) on the respiration of *Dixippus morosus* led these investigators to the conclusion that the tracheal air stream goes *forward* in the walkingstick, the thoracic spiracles being expiratory and the abdominal spiracles inspiratory, except that a small quantity of air may sometimes issue from the next to the last pair of abdominal spiracles. Stahn (1928) obtained the same results in experiments on *Dixippus*, but he observes that the expiratory stream appears often to be interrupted by expiration through the abdomen during passive breathing.

According to Du Buisson (1926), the action of the spiracular valves in *Dixippus* is variable. During ordinary breathing, he claims, inspiration takes place through all the spiracles, but the reverse may occur, or, again, the movements are disordered and have no rhythm. In earlier studies on *Stenobothrus* and *Locusta,* Du Buisson (1924, 1924a) claimed that in these insects also inspiration usually takes place through all the spiracles, but that expiration is ordinarily by way of the thoracic spiracles only. Under unusual conditions, however, he says, *Stenobothrus* may keep the thoracic spiracles continuously closed, and expiration then takes place through the abdominal apertures.

That insects have no definitely fixed direction of breathing is also the conclusion of Demoll (1927), deduced from experiments on *Melolontha*. By subjecting either the thorax or the abdomen of an intact beetle to nascent chlorine, he found that the insect was quickly killed. If the wings were cut off, however, and the thorax protected from the gas while the abdomen was exposed to it, the insect was unaffected, since the open tracheae of the wing stumps, together with the spiracles of the thorax, afforded a sufficient means of respiration and allowed the insect to keep the abdominal spiracles closed. From the diversity of the results obtained by different investigators we may conclude that there is no law governing inhalation and exhalation through special sets of spiracles applicable alike to all insects, and that the direction of the respiratory currents may alternate even in the same individual; but it appears that respiration has a usual though not a fixed course in each species, which presumbaly is characteristic also of the family, and probably of the order in most cases.

The function of the air sacs

The greater diameter of the air sacs as compared with that of the tracheae makes the walls of the sacs relatively weaker, and the air sacs are, therefore, more responsive than the tracheae to variations of pressure in the surrounding blood or other tissues, created by the alternating respiratory movements of the body wall. Particularly is this true of air sacs, such as those of the honey bee, which have no taenidia in their walls. In their response to pressure changes the air sacs resemble lungs; but inasmuch as peripheral tracheae are given off from them, their action is more accurately stated by Betts (1923), who says, "function of the air sacs is that of the bag of a bellows," or, as Demoll (1927) puts it, they guarantee an intensive ventilation of the tracheae during breathing. By a device for making direct observations on the action of the tracheal sacs under varying pressures, Demoll demonstrated that the sacs are compressed with increasing pressure around them, a part of their air content being thus driven into the pressue-resisting tracheal tubes, and that, with decreasing pressure, they are inflated. There is no direct evidence that the air sacs function as storage chambers for air. In special cases they serve to give atmospheric pressure against the inner surfaces of tympanal organs or, in certain aquatic species, to maintain buoyancy in the water.

The respiratory stimuli

Most studies on the respiratory stimuli of insects appear to be based on the assumption that carbon dioxide is not carried by the blood, and that, therefore, the respiratory movements must be regulated by the relative amounts of carbon dioxide and oxygenin the tracheal air, and experiments have shown, in fact, that such is the case. Temperature also influences the rate of breathing, but its primary effect is presumably on the processes of metabolism. The mechanism of breathing response to increased or decreased activity on the part of the insect has received little attention experimentally; but inasmuch as it has been shown that a part of the carbon dioxide produced by metabolism may be eliminated by other means than the tracheae, it is probable that most of it is thrown off from the tissues into the blood. If so, it then becomes possible that the ordinary respiratory regulation in insects, as in vertebrates, is brought about by fluctuations of the hydrogen-ion concentration of the circulating medium.

The stimulus for the fundamental rhythmic movements of respiration, Fraenkel (1932a) concludes, arises within the controlling nerve centers and has no peripheral source. The first attempts at determining the regulatory value of gases on the respiratory movements, of insects are those of Babak and Foustka (1907). From experiments on the breathing reactions of libellulid larvae to alterations in the carbon dioxide and oxygen pressure of the water medium, these investigators concluded that the rate and amplitude of breathing are dependent on the oxygen supply that reaches the nervous system through the tracheae, but that carbon dioxide or carbonic acid can scarcely be a regulatory stimulus for respiration, since it is effective only in excessive amounts. Stahn (1928), however, claims that the experimental methods of Babak and Foustka were not reliable for determining of he effects of small quantities of carbon dioxide. Experimenting with *Dixippus morosus,* Stahn found that small increases in the carbon dioxide content of the inspired air are reflected in the rate of the breathing movements, and that the effects of an excess of carbon dioxide are remarkably parallel with the effects of deficiency of oxygen.

In brief, Stahn concludes that the primary stimulating agent for increased breathing activity is carbon dioxide in small excess over the amount in ordinary air, the lower threshold being 0.2 percent of carbon dioxide in the inspired air, and the effective maximum about 0.3 to 3 percent. A slight decrease in the oxygen content, however, has the same effect as an increase of carbon dioxide, the maximum effectiveness of oxygen as a control stimulus being from 20 to 15 percent. A strong and apparently toxic accleration of breathing occurs when the oxygen content falls below 8 percent, or when the carbon dioxide content exceeds 12 to 15 percent. The respiratory effects of temperature have been studied by Walling (1906), who found that normal grasshoppers (Acrididae) making on the average 40 contractions of the abdomen a minute at 14°C increase the rate of breathing to 110 contractions a minute as the temperature, during a period of 4 hours, is increased to 54°C; at still higher temperatures the rate declines, and respiration ceases at 59°C. Lowered temperature has an opposite effect. At 5°C grasshoppers breathe faintly, if at all, from five to six times a minute, though breathing by normal individuals may not cease until the temperature falls to 0°C.

It will be noticed that the breathing rate of grasshoppers at ordinary temperatures given by walling is considerably higher than the figures of Lee (1925) quoted above. Nothing is known definitely as to how the varying carbon dioxide and oxygen pressure in the tracheal air makes itself effective as a respiratory stimulus. It has been supposed that there may be sensory nerves connecting the tracheae with the respiratory nerve centers; but a sensory innervation of the tracheae has not been observed, and Stahn suggests that the respiratory centers may be stimulated directly by the condition of the air that diffuses from the tracheoles into the nerve ganglia.

Nervous Control

There is no specific respiratory center in the nervous system of insects for the production and regulation of the breathing movements. Each ganglion of the ventral nerve cord of the abdomen contains an independent respiratory center controlling the movements of its segment, but it appears that the thoracic ganglia also play a part in the production or control of the respiratory movements. Experimental results in some cases are possibly somewhat confused by the fact that the ganglion proper to a segment may lie in some other segment, and the thorax often contains one or more of the abdominal ganglia. Though Matula (1911) claimed that in *Aeschna* larvae the activity of the ventral ganglia is under the control of a cerebral breathing center, his conclusion was disproved by Wallengren (1913), who showed that headless larvae are still sensitive to the oxygen tension of the water. On the other hand, Wallengren found that dragonfly larvae having the prothorax removed give no response to external respiratory stimuli, from which observation he concluded that the prothoracic ganglion plays an important role in the respiratory regulation.

At an earlier dat H.Z. Ewing (1904) had shown that in grasshoppers each ganglion of the ventral nerve cord contains a respiratory center and will activate the breathing movements of its segment when the latter is removed from the rest of the body. The more recent work of Stahn (1928) on the respiration of *Dixippus morosus* and *Aschna* larvae and of Fraenkel (1932a) on *Schistocerco* confirm the view that the head contains no respiratory nerve center but Stahn concludes that there must be distinguished in the body ganglion primary and secondary respiratory centers. The first lie in the abdominal, the metathoracic, and the mesothoracic ganglia, and possibly also in the prothoracic ganglion; the second is contained in the prothoracic ganglion.

The primary centers of *Dixippus*, as shown in insects with both head and prothorax removed, are responsive only to large dosage of carbon dioxide (12 to 15 percent) in the inspired air or to large decreases in the oxygen content (10 to 8 percent or less). The secondary center of the prothoracic ganglion, on the other hand, is a center finer adjustments, since insects from which the prothorax has not been removed are responsive to much smaller increases of carbon dioxide (12 to 15 percent), and to much smaller decreases in the oxygen content of the inspired air (down to 10 to 8 percent). In *Aeschna* larvae, Stahn says, the respiratory control is almost entirely taken over by the secondary centers of the prothorax.

Respiratory Mechanism in the Tracheoles

In living insects the tracheoles are filled to a varying extent from their distal ends with liquid. The composition of this liquid is unknown, but it is of such a nature that it can be absorbed through the walls of the tracheoles. It has been shown by Wigglesworth in mosquito larvae and in certain other insects that the amount of liquid in the tracheole branches distributed to a muscle is inversely affected by the activity of the muscle. From various experiments Wigglesworth concludes that the absorption of the liquid from the tracheoles is a direct result of increased pressure resulting from the formation of metabolities surrounding the ends of the tracheoles, to which the tracheole walls are impermeable.

Following metabolic activity in the muscles, or supposedly in any other tissue, therefore, the liquid is absorbed from the tracheoles, and air extends toward their extremities, where it comes into closer proximity to the cells requiring oxidation. By several experiments Wigglesworth sought to demonstrate his theory. In the first place, it was found that asphyxiation of mosquito larvae causes at first a violent muscular reaction, which is followed by penetration of air from the tracheae into the tracheoles going to the muscles. Lactic acid, Wigglesworth showed, is produced by the mosquito larva during asphyxiation. In a second set of experiments the liquid was absorbed from the tracheoles following infiltration into the body of 10 to 5 percent solutions of sodium chloride, and of lactic acid and potassium lactate at different strengths, the last being effective down 2 percent.

Again, it was found that the same effect was produced allowing the body liquid from one larva to diffuse into a second. Find testing the effect of gases, Wigglesworth showed that carbon dioxide and hydrogen cause muscular contraction followed by extension of dioxide in the tracheoles, until the insect is narcotized. Treatment penetration of the followed again by a rise of the liquid in the tubules. Poisonous have the same effect as non-poisonous base, but "the extent to air moves down the tracheoles depends upon the degree of muscles activity which precedes the death of the insect." Soon after death liquid rises in the tracheoles. From these experiments it seems Wigglesworth contends, that the absorption of the oxygenated and conversely the penetration of air into the tracheoles going to muscles is caused by the metabolic activity in the muscles. The respiratory effect of the movement of the liquid in the tracheae is that the air of the tracheae is quickly brought into closer relation the cells of a tissue as metabolism in the latter creates a'need for it.

Presumably, under normal conditions, oxygen is continued dissolved in the tracheole liquid and with the latter is taken into cells. It is possible that the entire oxygen supply of the tissues the latter by way of the tracheoles; but it does not seen possible all the carbon dioxide produced can be absorbed into the tracheae matter how intimately the latter may cover the cell surfaces or penetrate the cell bodies.

Other Functions of the Tracheal Tubes

Apart from respiration the tracheal system has a number of other functions. The whole system, and in particular the air-sacs, lowers gravity of the insect. In aquatic insects, but not in terrestrial ones, it also gives some degree of buoyancy and in the larvae of *Chaoborus* (Diptera) the trachae from hydrostatic organs enabling the buoyancy to be adjusted. Air-sacs, being collapsible, allow for the growth of organs within the body without any marked changes in body form. Thus at the beginning of an instar the tracheal system of Locusta (Orthoptera) occupies 42% of the body volume. By the end of the instar it only occupies 3.8% due to the growth of the other organs causing compression of the air-sacs.

When adult *Drosophila* emerge the air-sacs are collapsed, but subsequently they expand and at the same time there is a marked reduction in blood volume. Availability of oxidisable substrate, rather than shortage of oxygen, is likely to be a limiting factor in the activity of flight muscles and the air-sacs may indirectly imporve the fuel supply by the permitting a reduction in blood volume with a consequent increase in the concentration of fuels. Possibly intramuscular ventilation, causing marked changes in the volumes of the flight muscles, may also improve the blood supply to the muscles. In some noctuids (Lepidoptera) tracheae form a reflecting tapetum beneath the eye, and tympanal organs are usually backed by an air-sac which, being open to the outside air allows the tympanum to vibrate freely with a minimum of damping. Expansion of the tracheal system may assist in inflation of the insect after a moult. Thus in dragonflies, spiracle closure, preventing the escape of gas from the tracheae, accompanies each muscular effort of the abdomen during expansion of the wings. Some insects, such as *Aeschna* (Odonata), have an extensive development of air-sacs, apparently having no respiratory function, round the pterothoracic musculature. They probably serve an insulating function, helping to maintain the temperature of the flight muscles. An important general function of tracheae and tracheoblasts is in acting as connective tissue, binding other organs together.

4

Respiration in Aquatic Environment

In many larvae with a closed tracheal system, that is, one in which none of the spiracles is functional, respiration occurs exclusively over the general body surface (*cutaneous respiration*) and the subepidermal tracheal plexus is well-developed. This plexus may be generally distributed or more or less confined to areas where the cuticle is thin to make the diffusion of gases possible. Many dipteran larvae possess small blood-filled sacs which have been referred to as "*blood-gills,*" but the general consensus of opinion is that most of these structures are not true gills and it has been suggested that their primary function is in the control of chloride ion uptake. In a large number of aquatic larvae with a closed tracheal system, gills, containing abundant ramifications of the tracheal system (*tracheal gills*), have evolved. On the other hand, a wide variety of aquatic insects possesses an open tracheal system. In many cases, they are entirely reliant on coming to the surface to breathe, or on tapping the air spaces of submerged aquatic plants. Some carry bubbles or films of air (*physical* or *gas gills*) when they are sumberged and their spiracles then communicate directly with this air.

A gas gill may be temporary, when the animal has to return periodically to the surface of the water to replenish it. Alternatively, it may be permanent (plastron), involving the development of either the body hairs or cuticular modifications of outgrowths from the vicinity of the spiracles (*spiracular gills*). Nevertheless, even in larvae which possess tracheal gills or breathe atmospheric air, cutaneous respiration is in many cases still important. For instance, Fraenkel and Herford (1938) have shown that, in well-aerated water, *Culex* larvae (which bear functional spiracles) take up about half of their requirements through the body wall.

Closed Tracheal System

General Body Surface

Insects that do not have tracheae, or that have an imperfect or a secondarily closed tracheal system and are not provided with other devices for respiration, must effect the outer exchange of gases directly through the integument. The best known examples of insects lacking tracheae are the Collembola, the majority of which have no breathing mechanism of any kind, and certain species of aquatic Chironomidae, in which tracheae, if present, are very imperfectly developed. Many parasitic insect larvae, living entirely submerged in the liquids or tissues of the host, also must respire through the soft body wall, though they may be provided with a well-developed tracheal system. Seurat (1899) observes that some internal parasitic hymenopterous larvae have a system of finely branching tracheal tubes covering the inner surface of the body wall, into which air is absorbed from that dissolved in the blood of the host. certain parasitic larvae are provided with filamentous processes of the body wall that appear to be gills. Though with the majority of free-living insects having a normally open tracheal system inspiration takes place largely through the tracheae, there is evidence to suggest that the expiration of carbon dioxide takes place at least in part by way of the integument. Thus Krogh (1913), finding that the carbon dioxide deficit in the tracheae of the hind leg of a grasshopper after expiration is always considerably lower

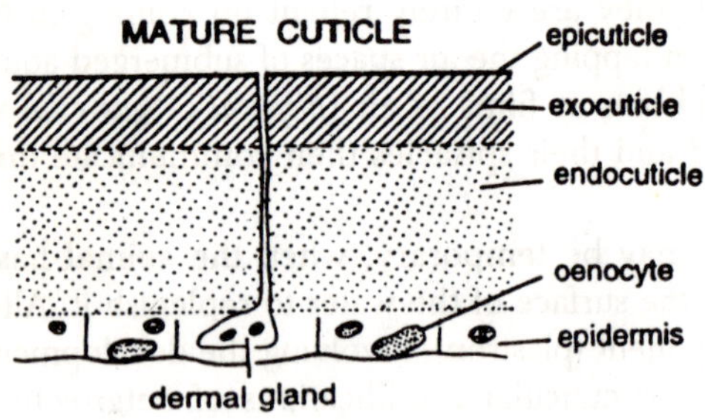

Fig. 4.1. A mature insect cuticle showing oenocytes.

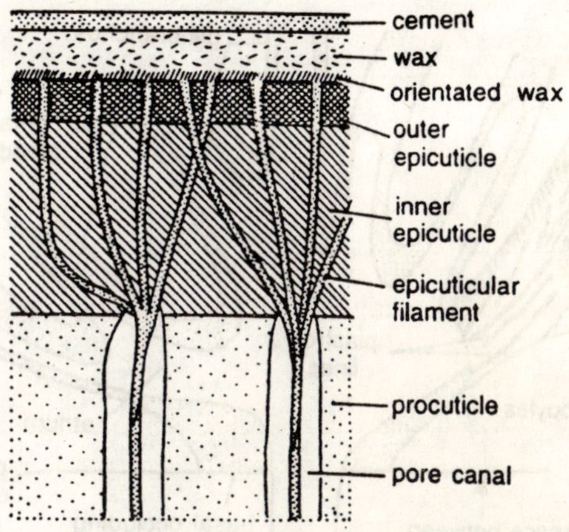

Fig. 4.2. Section through epicuticle (Diagrammatic).

than the oxygen deficit, concluded that a large part of the carbon dioxide formed in the tissues of the leg must be carried away by other means than the tracheae. The same idea has been expressed by other investigators and appears to be demonstrated experimentally for certain insects. Von Buddenbrock and von Rohr (1923), for example, claim that in *Dixippus morosus* one-fourth of the carbon dioxide produced is given off through the body wall, and Demoll (1927) reports finding no carbon dioxide in the tracheae of *Melolontha*.

On the other hand, Wrede (1926) found little evidence of carbon dioxide expiration through the skin of ordinary caterpillars. In aquatic caterpillars, however, such as certain species of *Bellura, Nymphula* and *Pyrausta* that lack gills, respiration under water, Welch (1922) has pointed out, must take place through the general body integument. It has not been shown that any particular part of the integument, when gills ae absent, serves for the elimination of carbon dioxide, but we may suppose tht diffusion would be most likely to take place through the less dense areas, such as the conjunctive membranes and other parts where sclerotization is weak or absent. In the Acrididae there are fenestralike, unpigmented areas of the body wall along the midline of the back above the heart chambers inthe abdomen, and above the aortic pouches in the thorax, which suggest tht they may be permeable to gases and perhaps serve for the elimination of carbon dioxide. In soft-skinned aquatic dipterous larvae carbon dioxide is usually found to be given off over the entire body surface.

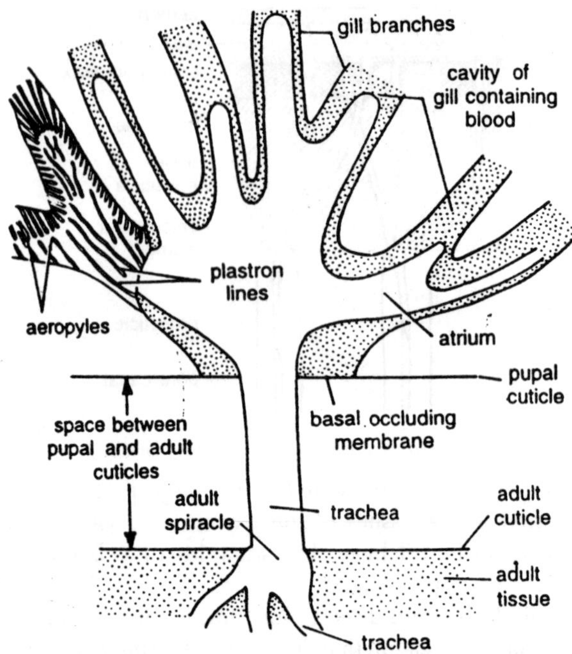

Fig. 4.3. Taphrophila. Spiracular gills of pharate adult.

Blood Gills

Various aquatic larval insects are provided with thin-walled, hollow diverticula of the integument or of the proctodaeum, which, in the absence of definite knowledge concerning their function, are usually termed "*blood gills.*" External process of this kind variously distributed on the body, occur on the aquatic larva of the bettle *Hygrobia* (*Pelobius*) *tarda,* on the aquatic larvae of certain species of *chironomus,* and on the aquatic caterpillar of *Catacylsta fulicalis.* The tapering fleshy processes arising near the anus of many tipulid larvae are sometimes cited also as examples of blood gills, and it is said that streams of blood may be seen to circulate through them, but each of these processes is penetrated by a trachea, and Gerbig suggests, therefore, that the organs serve in a double capacity of blood gills and tracheal gills. The "gills" of the *Hygrobia* larva, however, which consists of clusters of deliate filaments arising behind the bases of each pair of thoracic legs, contain no tracheae; and the same is true of the group of slender processes arising at the posterior end of the abdomen of *Chironomus* larvae, and of the filamentous appendages distributed over the body of the larva of *Cataclysta fulicalis*. These organs, therefore, may be supposed to be blood gills. The characterisitc tail of certain parasitic hymenopterous larvae has been supposed to be a gill.

A respiratory function has not been demonstrated in connection with this organ, but Wigglesworth (1931) points out that, when it has the form of a fluid-filled vesicle, its surface offers a respiratory possibility. Proctodaeal evaginations occur in some trichopterous larve, and in the larvae of certain Simuliidae, which are protractile from the anus and are generally regarded as organs for aquatic respiration. The structure of the slender proctodaeal evaginations of a trichopterous larva is described by Branch (1922), who finds that the organs are hollow tubular diverticula produced from the posterior ends of the six folds in the wall of the prerectal part of the intestine. They are capable of being extended from the anus, apparently by pressure exerted by the abdominal walls, and each is retractile by a branched muscle inserted within it, which arises on the body wall.

The similar processes of *Simulium* larvae, as described by Headlee (1906), consist of three simple or branched, soft, translucent filaments protractile through the anus from the ventral wall of the rectum, into which they may be retracted by a pair of muscles arising on the dorsum of the abdomen and inserted by branches on their bases. Since each of these processes contains masses of fine tracheal tubes filled with air, Headlee suggests that the organs perhaps function both as blood gills and as tracheal gills. Most of the supposed "*blood gills*" of insects have not been subjected to physiological tests for a respiratory function. Experiments on *Chironomus* larvae made by Fox (1921), using the infusorian *Bodo sulcatus* as an indicator, which is positively chemotactic to certain concentrations of dissolved oxygen, appear to show that oxygen consumption takes place through the general body wall and not through the so-called gill filaments.

Similar results were obtained from microspectroscopic tests. Carbon dioxide, according to Fox, is given off likewise through the general integument of the *Chironomus* larva. The structure and function of the gill-like anal lobes of mosquito larvae have been studied by Wigglesworth (1933, 1933a), who concludes that the organs serve for the absorption of water, since respiratory tests show that oxygen is absorbed on all parts of the body, though most actively at the bases of the anal lobes, and that carbon dioxide is given off equally from the entire body surface.

Tracheal Gills

Tracheal gills are thin outgrowth of the body wall which contain a rich tracheal plexus. They are almost exclusively confined to larval stages. However, they are found in some trichopteran pupae. Also, in some plecopterans and in the trichopteran *Hydropsyche,* they persist into the adult stage in a shriveled, non-functional condition. They occur most often on the abdomen, but are also found on the thorax and head. In some instance there has been considerable controversy over the function of tracheal gills, especially since in many cases the larvae live quite successfully after they have been removed. Also, their absence may cause virtually no different to the total oxygen consumption of the animal.

However, Koch (1936) has pointed out that although an abundant tracheal supply may not be a very good criterian on its own for determining function, if the structure has very good criterion on its own for determining function, if the structure has very low metabolic requirements, and thus takes up more oxygen from the environment than it needs for itself, then it will be an important site of gaseous exchange. In general, the thin cuticular covering of the gills, their rich supply of tracheal capillaries, and the absence of structures within them which possess a high metabolic rate all tend to argue in favour of this function.

(i) Lateral Abdominal Gills

Morphological and Occurrence. Pairs of segmental, abdominal gills are the most common arrangement. In their simplest form they are flat, platelike outgrowth of the body wall (lamellae), as found in many ephmeropteran larvae. In *Baetis* and *Cloeon* there are seven pairs. In the former, each gill is a single lamella, but in *Cloeon* some are double. In other ephemeropterans the gills are biramous with simple (*Leptophlebia*) or multifurcate (*Habrophlebia*) branching and further elaboration occurs in *Ecdyonurus* branching and further elaboration occurs in *Ecdyonurus* and other dorsoventrally flattened larvae, where many of the lamellae shield a tuft of filamentous gills.

In *Ephemera* and *Hexagenia,* the gills are biramous and lanceolate, except for the first pair which is vestigial. They are fringed with long filaments and are generally reflexed over the back. It has been demonstrated by Dodds and Hisaw (1924) that for different species of mayflies the area of the gills is inversely proportional to the oxygen content of their environment. I.. all other orders the gills are filamentous. Among the Odonata, lateral abdominal gills are found in just two zygopteran families, the Epallagidae and Polythoridae (Calopterygidae), the members of which also possess caudal gills. The gills are sijple in the Epallagidae, but are segmented in the Polythoridae. Abdominal gills are also found in just one family of plecopterans, the Eustheniidae.

Among the Neuroptera they are found in larvae belonging to the family Sialoidea (suborder Megaloptera) and may be simple, as in *Corydalis* and *Chauliodes* (eight pairs), or segmented, as in *Sialis* (seven pairs). In addition to the lateral gills, *Sialis* has a single, segmented, terminal on the ninth segment. In the other suborder of the Neuroptera, the Planipennia, only the larvae of *Sisyra* are known to have gills and here they consists of seven pairs of jointed filaments. Finally, among Coleopteran, larvae of the Gyrinidae have ten pairs, and larvae of *Hydrocharis* and *Berosus* have seven pairs of hair-fringed filamentous gills. Considerable elaboration occurs in some genera of mayflies. Thus, in *Cacnis,* which has gills on the first six segments, the first pair are reduced, while the second pair are very large and cover the remainder, forming a terancital chamber. In *Prosopistoma* the wing sheaths and the highly developed pro- and mesothoracic terga help to form a branchial chamber which completely encloses the five pairs of gills.

(ii) Caudal Gills

Morphology and occurrence. The presence of three (one median and two lateral) caudal gills is a diagnostic feature of zygopteran larvae. In a few cases, such as the semiterrrestrial and terrestiral species of *Megalagrion* from Hawaii, the gills are somewhat reduced, but otherwise are generally well-developed. In young larvae the gills are ferular, with the median and one quardrilateral and the lateral ones triangular (triquetral) in sector. In the Calopteryginae the lateral gills retain this shape. However, together with the medialones, they usually become lamellate or, in a few families, saccoid. MacNeill (1958, 1960) has shown that in many species the gills are divided into two more or less distinct zones (proximal and distal). In such cases the distal zone is not discernible in early instars, but it increases in length by successive outgrowths of an organized structure present within the proximal part. Often, the additons produce a banded appearance (chevron bands). This is termed "protrusive growth" and zgopteran larvae can be divided into two types on the basis of whether protrusive growth is absent (simplest type) or present (duplex type).

In duplex gills the distal zone has a thinner cuticle than the proximal zone and a relatively small number of spines. In early instars there is no easily identifiable division between the two zones, but later on in lamellate gills it may consist of an abrupt change inthe type of armature, a small node, or a large constriction. Duplex saccoid gills only occur in three genera of the family Protoneuridae and in all cases the distal zone is extremely small. It has been suggested that the caudal gills (or lamellae) of zygopteran larvae are used for swimming, or at least that they aid swimming by acting as rudders while Roster (1886) suggested that they served for both respiration and locomotion. However, swimming is a compartively rare event, even in the better swimmers. It is achieved by lateral movements of the body and is virtually unimpaired by removal of the gills. Furthermore, saccoid gills would appear to hinder, rather than aid, swimming movements.

It has ben suggested that the primitive method of respiration in dragonfly larvae is by means of lateral abdominal gills, as are found in a few species, but the observation of Robert (1958) that *Epallage fatime* lives in streams which sometimes dry up, leaving stagnant pools, has led him to suggest that lateral abdominal gills may be a secondary adaptation to conditions of low oxygen tension. Nevertheless they are undoubtedly fairly archaic structures. MacNeill (1960) argues strongly that protrusive growth, and hence the duplex gill, is an evolutionary development which increases the respiratory efficiency of the caudal gills, both by increasing their surface area and by providing a region with an especially thin cuticle. An alternative means of increasing the surface area of the gills is seen in the development of the saccoid form, which presumably renders protrusive growth unnecessary. This may explain why duplex saccoid gills are only found in a few species and why, when they do occur, the distal region is always very small.

In most cases, however saccoid gills would be extremely vulnerable and they are only found in those species which live under rocks and in rock crevices in fairly fast-flowing water, where they are reasonably well-protected. Removal of the caudal gills in *Coenagrion* results in the development of a rich supply of tracheoles in the body wall, which is also indicative of a respiratory function. It is tempting to suggest that in zygopteran larvae general cutaneous respiration is still important because of the vulnerability of external caudal lamellae; whereas anisopteran larvae with invulnerable internal gills, have little need of cutaneous respiration.

(iii) Rectal Gills

Morphology and occurrence. Rectal tracheal gills are confined to the anisopteran dragonflies. They lie in the anterior region of the rectum, which has become enlarged to form a branchial chamber. Basically there are six main gill folds, each of which runs the full length of the branchial chamber. Each is supported by a series of short cross-fold on either side. This has been called the simplex system by Tillyard (1971), who has provided a classification of the various types. In some families (e.g. Cordulegastrinae) the gills folds consists of an undulated membrane which is not subdivided (*undulate type*), but in most Gomphinae the edges of the gills are divided into long papillate (*papillate type*). A number of families have the so-called duplex system, in which only the cross-folds function as gills. This also can be subdivided on the basis of characteristics of the individual gill.

In the implicate type (only found in the Brachytronini), consecutive lamellate gills, arranged on either side of a reduced main fold, overlap each other. In the folliate type (confined to the Aeshnini) the bases of the gills are constricted. This type can be further divided. Thus, *Aeshna* has crinkled lamellate gills (*normale foliate*) while in *Anax* the distal portion of each gills is swollen and bears a number of papillae (*papilo-folicate*). Finally, there is the lamellae type, restricted to the Libellulidae. Here, the main folds have completely disappered and each gill, as the name suggests, is a simple lamella with a broad base. In the Synthemini there are only 12 lamellae in each longitudinal row (archilamellate), but all others have many more (neolamellate). In all cases, the gills are covered by an extremely thin layer of cuticle. This encloses an epithelial syncytium through which run the tracheal capillaries.

The capillaries do not end blindly, but are looped. At their bases the walls of the gills are separated by extensions of the hemocoele which contain the tracheae surrounded by hypobranchial tissue and the epthelium of one or both walls has become well-developed to form a basal pad(s). Immediately posterior to the branchial chamber there is a highly muscular region of the rectum called the vestibule. In contrast to the branchial chamber, the circular muscles of the vestibule are extremely well-developed and it has a number of dilator muscles attached to it. The branchial chamber and vestibule are guarded by valves. Water is pumped in and out of the branchial chamber via the anus and vestibule.

(iv) Other Gills

Morphology and occurrence. A number of plecopterans have tufts of finger-like gills which occur in a variety of places on the head, thorax, and abdomen, and also on the coxae. Similar tufts are found on the ventral surface of the abdomen of *Corydalis* and *Neuromus*. In all of those trichopteran larvae which construct portable cases, as well as in a few of those that do not, filamentous gills occur in segmental groups in various positions on the abdomen. *Macronema zebratum*, for example, has about 60 gills.

In case-bearing caddis larvae there is an increase in the number of gill filaments with increase in body size. Larvae without cases do not show this relationship, but there is an indication that these animals have an inverse relationship between the number of filaments and their normal environmental pO_2. *Phalacrocerca* (Diptera) and *Nymphula stratiotalis* and *N. maculalis* (Lepidoptera) also possess filamentous gills, on most body segments in the former and laterally on the meso- and metathoracic and abdominal segments in *Nymphula*. *Ymphula maculalis* has over 400 gills. In *Dicranota* (Diptera), there are two ventral pairs on the last abdominal segment. In two genera of mayflies, *Jolia* and *Oligoneuria*, tracheal gills are fond on the head. Finally, *Peltodytes* (Coleoptera) has a number of jointed filaments on the dorsal surface of the thorax and abdomen, and another coleopteran (*Hydrobia*) has ventral gills near the base of each leg and on the first three abdominal segments.

Open Tracheal Systems

Insects with functional spiracles face the primary danger of flooding of their tracheal system. There are two main categories into which these insects can divided. On the one hand, there are those which need to maintain contact with the surface or with the air spaces of plants, and at best can only make short dives; on the other, there are those which carry air (gas gills) around with them and which have evolved varying degrees of independence of atmospheric oxygen.

Water Insects Utilizing Atmospheric Oxygen

(i) Surface Breathers

The principle problems faced by surface breathers are how to prevent water from entering the spiracles and how to break through the surface film. They are solved by having part of the cuticle, or a group of hairs, around the spiracle with a greater affinity for air than for water and with which the water thus has a high "*contact angle.*" Thus, when the structure comes into contact with the surface film "the cohension of the water is greater than the adhesion to the body" and the spiracular region remains dry. Such regions of the cuticle or hairs are said to be "hydrofuge." In this case of a ring of hairs around the spiracle, their hydrofuge nature will result in them curving over the spiracle when the animal is submerged

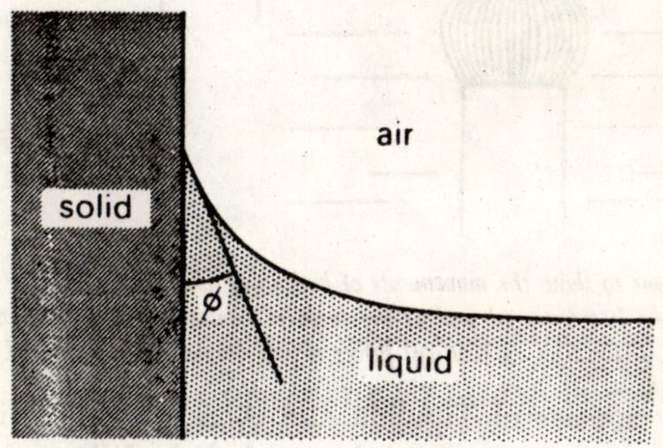

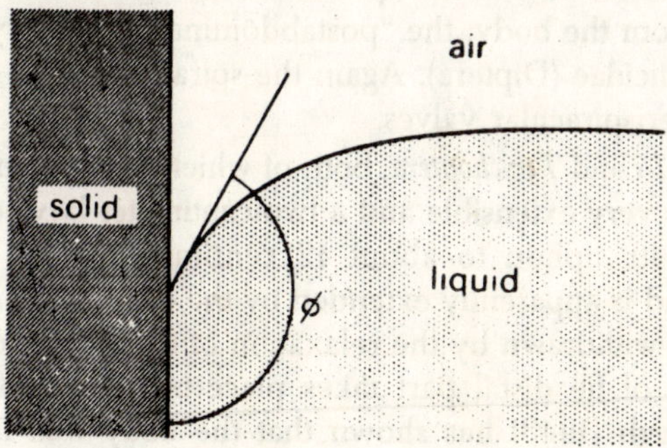

Fig. 4.4. Diagrams to illustrate low and high contact angles.

and spreading out because of the surface tension effect when the animal is at the surface. Aquatic diptecan larvae provide examples of hydrofuge areas of cuticle around the spiracles. This is achieved by the oily secretion of the peristigmatic glands in this region, first described towards the end of the last century. Their function in this respect was first suggested by Gazagnaire (1886) and they have since been described in many species.

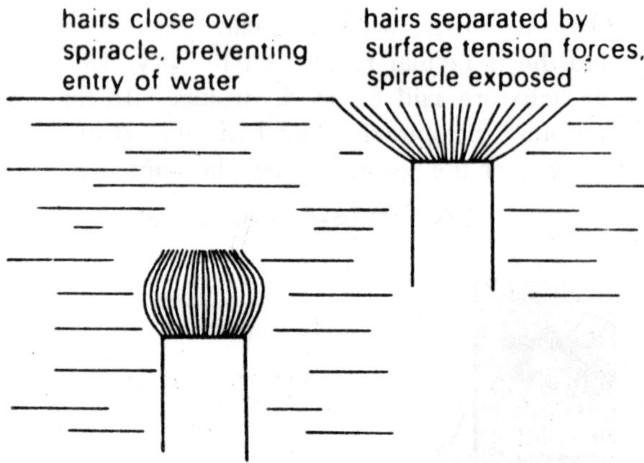

Fig. 4.5. Diagrams to show the movements of hydrofuge hairs surrounding a spiracle when the insect is submerged and at the surface. The movement of the hairs is entirely passive, depending on physical forces acting between the hairs and the water.

In many species of Tipulida and Anisopodidae (Diptera) the spiracles open onto a postabdominal disk. When the animal submerges, the disk is withdrawn and the surrounding fleshy lobes close over it to prevent entry of water. A number of dipterna larvae have their functional spiracles situation on the end of a projection from the body; the "postabdominal respiratory siphon," as in the Culicidae (Diptera). Again the spiracles can be closed by the fleshy perispiracular valves.

In *Eristalis* and *Ptrychoptera,* both of which are cottom dwellers, the siphon is very extensible and a two-centimeter larva of *Eristalis* can extend its siphon to about 12 centimeters. The siphon is telescopic and is apparently extended by the contraction of circular muscles and withdrawn by the relaxation of these muscles and the invagination of he distal part takes place by means of retractor muscles. Krogh (1943) has shown that the body wall is virtually impervious and that respiration only occurs via these posterior

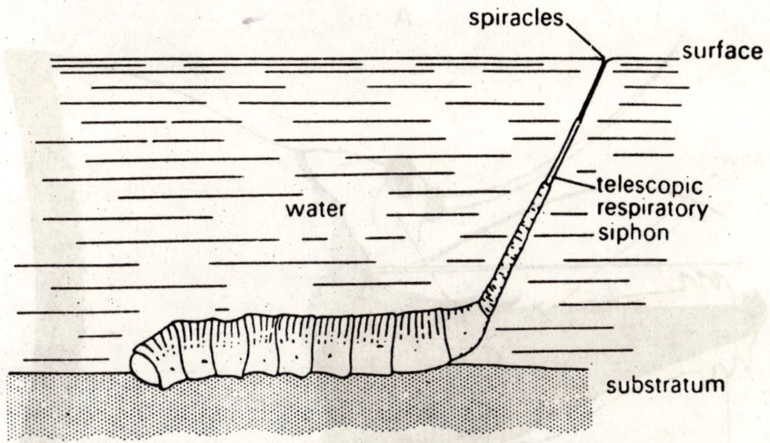

Fig. 4.6. Larva of Eristalis with the respiratory siphon partly extended.

spiracles. Hydrofuge hairs surrounding the spiracles have been described in the larva of *Hedriodiscus truquii* (Diptera) which lives in thermal springs but otherwise seem to be fairly well restricted to insects with gas gills, such as *Notonecta* (Hemiptera). Also, in some insects which have a gas gill, communication with the atmosphere is made via a respiratory tube. Thus, in *Nepa* (Hemiptera) the region around the posteriorly situated spiracles is developed to form a tube, and in *Hydrophilus* (Coleoptera) the antennae have become modified to perform a similar function. A double log plot of the rate of oxygen uptake against body weight for larvae of *Hedriodiscus truquii* showed a linear relationship up to a weight of 40 mg. Beyond this, the respiratory rate did no increase.

(ii) Drawing oxygen from plant air spaces

Larvae, pupae and young adults of the *Donaciinae* (Coleoptera) live in mud on the roots of aquatic plants, an environment which has a low oxygen content and Houlihan (1969) points out that the larve are able to tolerate a high environmental pCO_2. According to Beadle and Beadle (1949), this is characteristic of many aquatic insects. The functional spiracles of the larva at the end of a long, pointed posterior siphon, which it uses to penetrate the gas spaces in the roots. Boving (1910) observed that, prior to pupation, the larva constructs a cocoon on a root. While doing this, it bites a hole in the root and leaves a space in the floor of the cocoon above the hole, and it was suggested by Ege (1951b) that the pupa

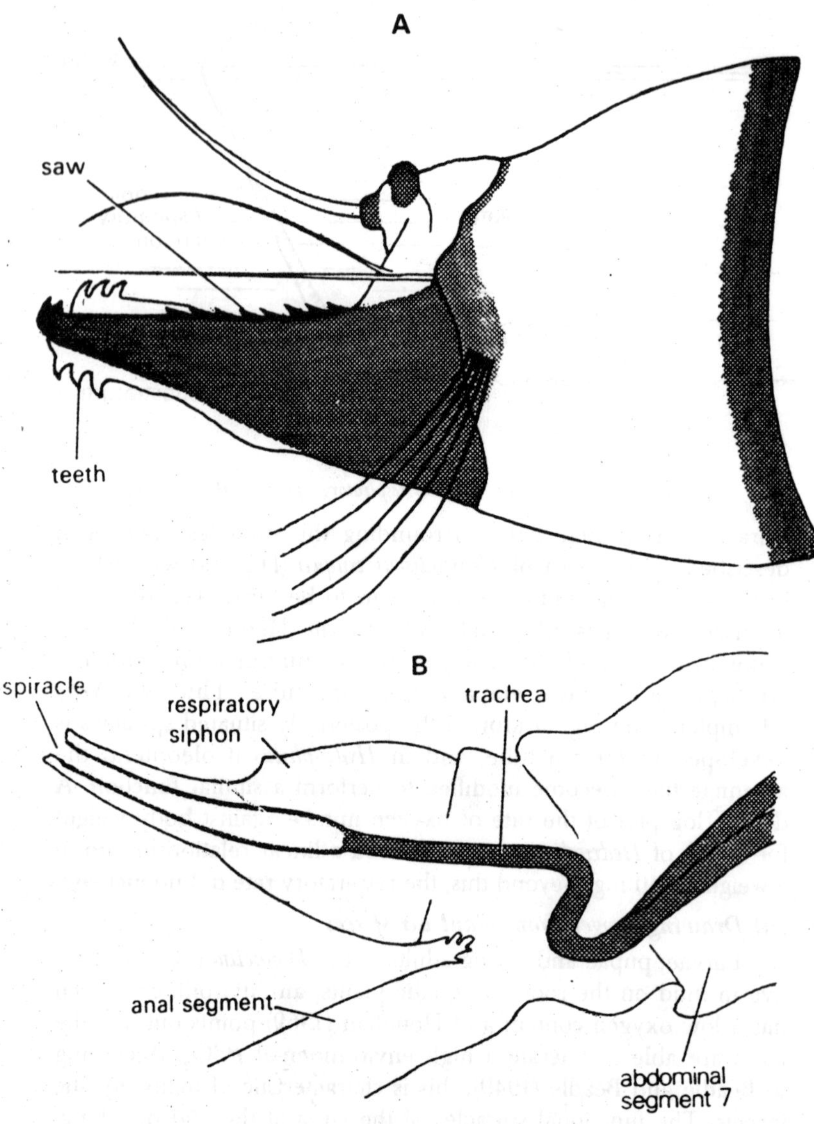

Fig. 4.7. A–Postabdominal respiratory siphon of Mansonia larva. B–Section of posterior end of larva to show tracheae and the terminal spiracle.

and adult continue to obtain their oxygen from the plant. The hole in the root eventually becomes covered over by a scar, buy oxygen still passes through this into the cocoon.

In *Donacia simplex* pupation occurs around October. The pupal stage is short, but the adult stays in the cocoon until June. Houlihan (1970) demonstrated that the respiratory rate (ml O_2/mgm/hour) shows little change with increase in temperature over the range 4-14°C in adults taken from the cocoons in Junuary, but those removed progressively later in the year started to show a marked increase in oxygen uptake above 8°C. Emerged, inactive adults showed a similar increase in the rate of oxygen uptake with increase in temperature, but the overall rate was much higher than in those taken from their cocoons. Houlihan (1979) suggests that the adults are thus adapted to a low oxygen tension, brought about by the formation of scar tissue which slows down the diffusion of oxygen into the cocoon.

Several other insects obtain their supply of oxygen in a similar manner. The larvae of the diperans *Chrysogaster* and *Notiphila* occupy the same type of habitat as *Donacia*: and they and the larvae of other dipterans, such as *Mansonia* and *Taeniorhynchus richiardii*, have similar respiratory siphons. In the pupae of *Notiphila* and *Mansonia* the siphons are thoracic. Less or haptrus (weevil) larvae also piece plants. Alternatively, some insects, such as *Elmis* (Coleoptera) and the pupating larva of *Hydrocampa* (Lepidoptera), bite into the air spaces.

Gas Gills

It has been known for some time that many insects carry air with them when they submerge and it was first suggested by Comstock (1887) that these bubbles extract oxygen from the water Brocher (1910) thought that in *Dytiscus* the air had a purely hydrostatic function and was nto involved in any way with respiration, but both Babak (1912) and Wesenberg-Lund (1912) considered it to be of both respiratory and hydrostatic importance. Ege (1915a) worked on this problem with a variety of species (dytiscids, corixids and *Notonecta*) and demonstrated that the air did indeed have an important role to play in respiration.

In all of the species at which he looked, the trapped air acted, not only as a store of oxygen, but also as a physical or gas gill during winter. This was also the case in the summer for *Hydrophorus* and the smaller species of *Corixa;* but in *Notonecta, Dytiscus,* and various other dytiscids only the stored, or primary, supply of oxygen was used if the animals were at all active. Ege (1915a) worked exclusively with animals in which the air "store" steadily diminished in size, ultimately depending on the temperature and the amount of activity, forcing the animal to return to the surface to replenish its supply. These are therefore compressible gills. On the other hand, many insects have involved a gas gill which is incompressible and this allows them to remain below the surface of the water indefinitely. The incompressible gas gill was called the "*plastron*" by Thorpe and Crisp (1947a), a term first employed by Brocher (1912b) to cover all forms of gas gill.

In both types of gas gill, when the oxygen tension (pO_2) in the gill falls below that of the water, oxygen will diffuse into it, the rate depending on the difference in tension. Nitrogen is extremely important in these gills and is primarily responsible for their functioning. The initial gas concentrations in the gas gilm will be the same as in air (i.e., about 20% oxygen, <1% carbon dioxide and 79% nitrogen). Carbon dioxide is extremely soluble in water and diffuses out of the store as it is produced by the animal. Thus, gas gills can be thought to solely in terms of oxygen and nitrogen.

(i) Compressible Gas Gill (Temporary Air Stores)

Structure and occurrence. Compressible gas gills are normally held in place by body hairs and/or consist of bubbles in the subelytral space. Thus, among the Coleoptera, some members of the Dryopoldea and the Hydrophilidae have a film of air held in place by a hair pile. The hairs are not so regular and densely packed as in the case of the plastron. In the Dryopoidea there are between 6 x 10^4 and 8 x 10^5 hairs/cm^2; in the Hydrophilidae 1 x 10^4 to 1 x 10^7/cm^2. In a number of hydrophilids there are two sets of hairs (long and short). The long hairs hold a bubble which decreases in volume fairly quickly when the animal dives, whereas the smaller hairs hold a gas film which is relatively permanent as long as the animal does not dive too deep. These two layers have been termed the "macroplastron" and "microplastron,"respectively. When the macroplastron has collapsed, the long hairs which supported it help to protect the inner layer from wetting. *Dryops* and some hydrophilids can crawl about on submerged plants, enveloped by an air bubble form which only the animal's extremites project. Many hemipterans carry a layer of air supported by an extensive hair pile and often supplymented by an air bubble in the subelytral space. *Gerris* and *Velia* both have a macroplastron and a microplastron supported by long and short hairs, respectively, and some adult trichopterans also have these two layers.

Among the Lepidoptera the larvae of *Palustra* and *Diacrisia* carry air in two lateral tracts of both large and small hairs. The larvae of some pyralids (e.g., *Cataclysta*) and *Hydrocampa* have longitudinally ribbed or flanged spines and, in the former at least, these are thought to retain an air film. In a wide range of other insects found on the surface or in the vicinity of water, hydrofuge hairs are present. They are not dense and probably serve to prevent wetting if the animal accidentally falls into the water. In others words they have more of a "rainproofing" effect.

(ii) Incompressible Gas Gills (Plastron Respiration)

Structure and occurrence. A plastron is a thin film of air which may be held in place by hairs or by cuticular modifications. The "hair plastron" is found on the adults of certain coleopterans. *Aphelocheirus aestivalis* (Hemiptera) and *Acentropus niveus* (Lepidoptera). The best-known example is that of *Aphelocheirus.* The adult has nine pairs of spiracles, two thoracic and seven abdominal. The first three have single openings into deep longitudinal grooves, while the others have numerous small openings along canals which radiate out from the center, forming a "rosette." The grooves, canals, and openings are all lined with fine hairs, as is also the whole of the ventral and much of the dorsal surface of the animal. There are between 2 and 2.5 x 10^8 hairs/cm^2. The animal is heavier than

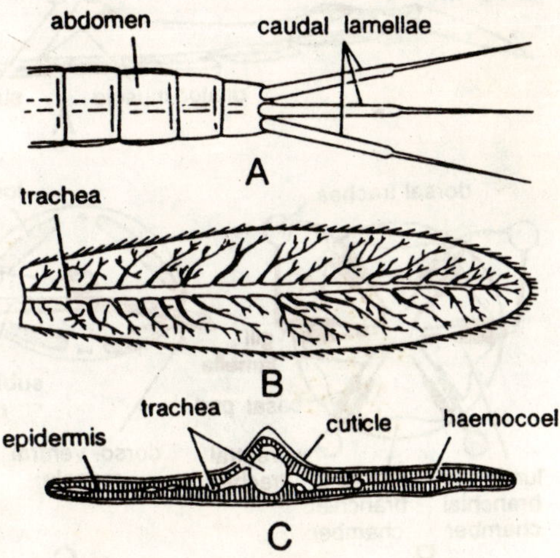

Fig. 4.8. Tracheal gills of larval Zygoptera. A–Posterior end of larva of Coenagrion in dorsal view. B–One lamella in lateral view. C–Caudal lamella of Synlestes in T.S.

water, in contrast to those animals which carry air bubbles, and avoids regions of low pO_2 by means of its tactile sensilla, visual sense, and specialized pressure receptors. The plastron of *Aphelocheirus* is externaly efficient and is only equalled in a few elmid beetles (*Stenelmis crenata* and *Cylloepus barberi*) and in the weevil *Phytobus velatus.*

Phytobius is covered by scales which touch or overlap each other over most of the body. The scales are covered with a dense hair pile (1.8 x 2.0 x 10^8 hairs/cm^2). All of these animals belong to group I of Thorpe can Crisp (1949) and Thorpe (1950) which is characterized by a hair density in excess of $10^8/cm^2$. In the rest of the animals which possess a hair plastron, the hair tend to be longer and not so dense as in those just described, and they have been assigned to a second group. However, like the animals in Group I, members of group II do not normally need to keep returning to the surface and may remain submerged for months. They are characterized by a hair density of 3 x 10^6 x 1.6 x 10^7/

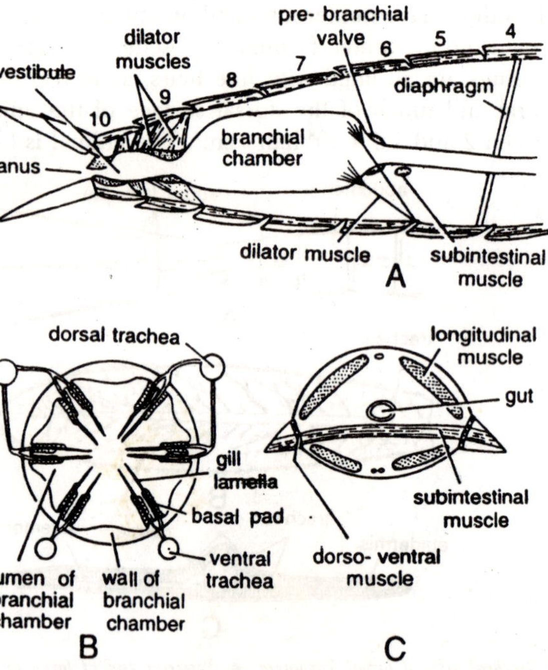

Fig. 4.9. Abdomen of dragonfly larva in L.S. B–Branchial chamber in corss section. C–T.S. of sixth abdominal segment.

cm^2. Included in Group II are most of the aquatic elmid beetles (e.g., *Elmis* and *Riolus*). *Elmis maugei* and *Riolus capresu* have eight pairs of spiracles which open into a lateral groove on each side. These grooves communicate with the subelytral space and the plastron.

Table 4.1. The Oxygen Tension (pO_2) in Gas Gills During a Dive

Air	*pO_2 about 20%*	
At start of dive	*pO_2 16.0-19.6%*	
Species	*Time after dividing (minutes)*	*pO_2 (%)*
Dytiscidae		
Colymbetes fuscus	1	15.4
Acilius sp.	2	11.9
Colymbetes sp.	5	9.2
Dytiscus marginalis	5	3.0
Ilybius sp.	20	0.7
Notonectidae		
Notonecta sp		
Dorsal space	1	11.0
	2	5.3
Canals (ventral surface)	1	15.4
	2	3.9
Corixidae		
Corixa geoffroyi		
Passive	10	9.4
	15	5.8
Active	5	6.5

The latter covers most of the lateral and lateroventral surfaces of the animal. The hairs hold a thin film of air, but a thicker layer is often present. This extra layer (macroplastron) is soon lost by the Ege effect when the animal submerges. It can only be maintained by frequent visits to the surface or by using some of the air in the subelytral space. However, under well-oxygenated conditions the microplastron is sufficient for the animal's needs. Also in this group is another beetle, *Haemonia mutica* (Donaciinae), the plastron of which covers the whole of the ventral surface and the antennae.

The hair density is greater on the latter. The hairs are bent at their tips, as in *Aphelocheirus,* but are much longer. The hamipteran *Hydrocyrius columbiae* (Belostomatidae) has film of hair held in place on the ventral surface by small hairs, bent at their tips, at a density of 1 to 2 x $10^6/cm^2$. Longer hairs, ending in flattened blades, lie over the surface of the short hairs. This ventral air film communicates via three pairs of *"bridges"* with a subelytral bubble. When the animal comes to the surface of the air supply is renewed via a posterior retractile siphon. All of the spiracles except the first communicate with the gas gill.

The lepidopteran *Acentropus niveus* has a wingless form which remains submerged for the whole of its adult life of 3 or more days. Its plastron consists of three- and four-pronged scales and it probably also belongs to Group II. Thorpe and Crisp (1949) have commented that in Group II the plastron is functionally but not structurally perfect.

The other way in which the plastron may be held in place is by means of cuticular modifications, and these are variable associated with spiracular gills. The spiracular gill is an outgrowth of the spiracular atrium and/or of the region immediately surrounding it. With the exception of those in some of the Chironomidae, all spriacular gills possess a plastron. They are

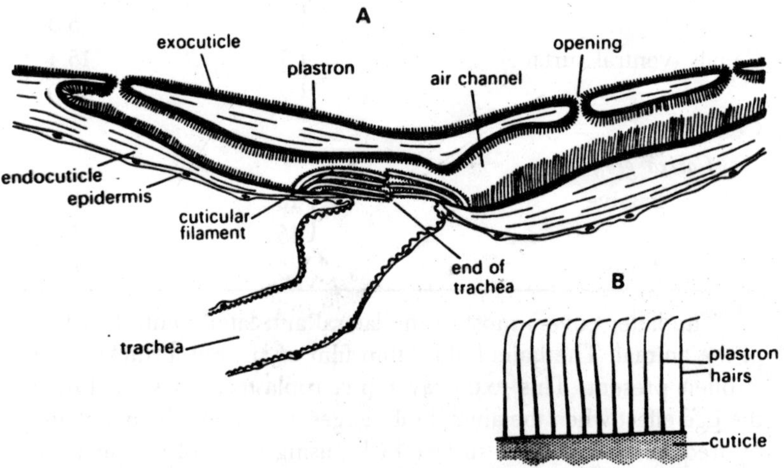

Fig. 4.10. A–Section through a spiracular rosette of Aphelocheirus showing the junction of the trachea with the system of channels in the cuticle. B–Part of the plastron highly enlarged, showing the form of the hairs.

confined to the pupal stages of certain Coleoptera and diptera, except in the Torridinicolidae, where the larvae also bear them and the Sphaeridae and Hydroscaphidae, where they are confined to the larvae.

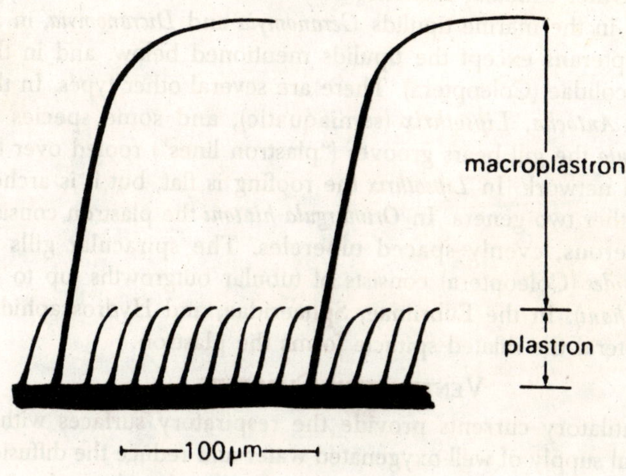

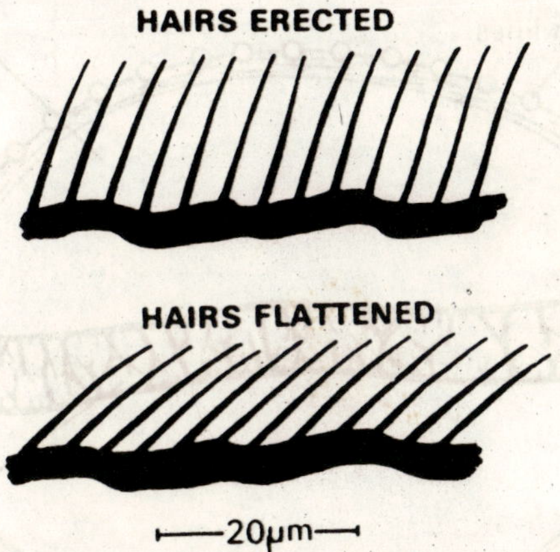

Fig. 4.11. Diagrams of the arrangement of hairs in (A) Hydrophilus and (B) Elmis, showing the mode of formation of a macroplastron.

In most cases the plastron consists of vertical struts which divide near their tips to provide horizontal branches which are fused with the branches of adjacent struts. The effect is to form an open hydrofuge network, supported by struts, enclosing an air-filled lumen. Other cuticular branches may travese the lumen. This type is found in the marine tipulids *Geranomyia* and *Dicranomyia,* in all other dipterans except the tipulids mentioned below, and in the Torridincolidae (Coleoptera). There are several other types. In the tipulids *Antocha, Lipsothrix* (semiaquatic), and some species of *Orimargula* the gill bears grooves ("plastron lines") roofed over by an open network. In *Lipsothrix* the roofing is flat, but it is arched in the other two genera. In *Orimargula hintoni* the plastron consists of numerous, evenly spaced tubercles. The spiracular gills of *Psephenoides* (Coleoptera) consists of tubular outgrowths (up to 40 in *P. gahani*). In the Eubriinae, Sphaeridae, and Hydroscaphidae (Coleoptera) the dilated spiracle forms the plastron.

Ventilatory Currents

Ventilatory currents provide the respiratory surfaces with a continual supply of well-oxygenated water and reduce the diffusion barrier imposed by the establishment of a boundary layer. In most cases, ventilatory currents are produced either by undulations of the body or by rhythmic movements of the gills themselves. Leg movements may also be used to produce currents, as in hemipterans.

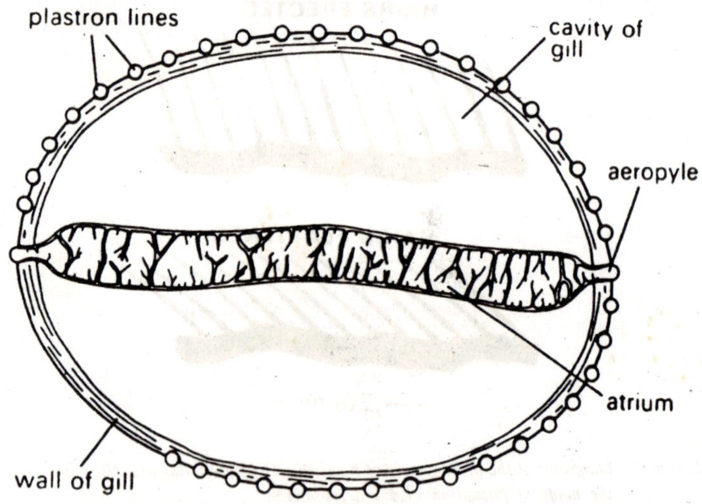

Fig. 4.12. Transverse section through a branch of a gill of pupal Taphrophila.

Thus in *Notonecta* de Ruiter *et.al.* (1951) demonstrated that leg movements and active or passive movement through the water increased the rate of oxygen uptake. Finally, anisopteran dragonfly larvae ventilate their branchial chamber by muscular pumping movements. However, in only a few cases do we have any detailed knowledge of the nature and mode of production of ventilatory currents.

Respiration in Ephemeropteran Larvae

In *Ecdyonurus dispar, Ephemera dancia,* and *Leptophlebia marginata* the ventilatory currents flow from the sides and/or below the animal across the gills toward the dorsal surface and then backward along the dorsal midline. However, this is not the case in *Cloeon dipterum,* where the current passes backward over the dorsal surface of the abdomen and outward between the gills, or in *Caenis horaria,* in which the ventilatory current passes between the gills and across the dorsal surface of the abdomen from one side to other. Common to all species is an anterior-to-posterior metachronal rhythm, each gill starting to move slightly in advance of the next posterior one, and in all except *Caenis* the gills on opposite sides of the same segment beat in unison. Each gill in *Ecdyonurus dispar* consists of an anterior lamella and a posterior bunch of filaments. The second to sixth pairs of lamellae produce the ventilatory current which irrigates their own surfaces and the bunches of filaments. Each lamella is slightly curved, with its posterior surface convex, and projects laterally when at rest.

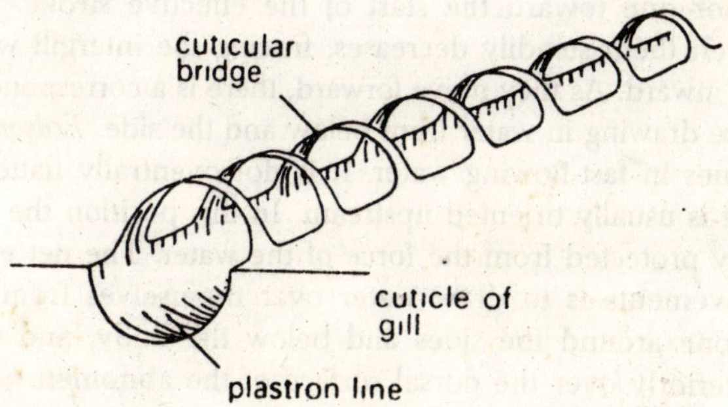

Fig. 4.13. Diagram of the cuticular bridges across a plastron line on the gill of Taphrophila.

During ventilation it moves backward and upward and then forward and downward, its tip travelling in an ellipse. The backward movement is the effective stroke and the convex posterior surface

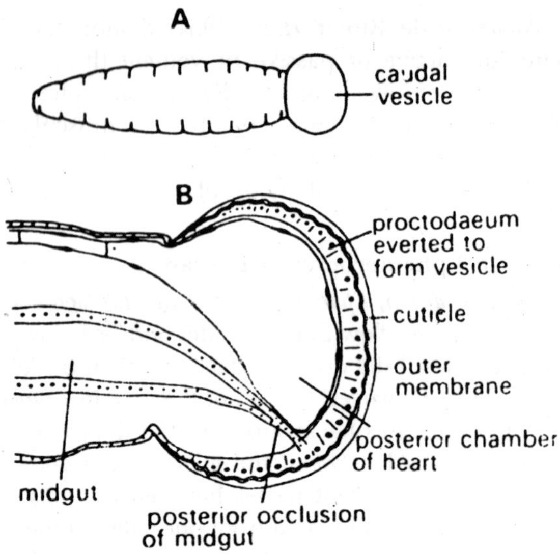

Fig. 4.14. A–Larva of Apanteles showing the caudal vesicle. B–Longitudinal section of the vesicle.

leads, gradually increasing its angle to the direction of movement. During the forward recovery stroke the lamella is "feathered" so that it makes virtually no angle with its path of motion. Thus, each lamella generates a current upward and backward. The leg between successive pairs of lamellae results in the lower border of each lamella coming into contact with the posterior surface of the next posterior one toward the start of the effective stroke. The angle between them steadily decreases, frocing the intergill water upward and inward. As they move forward, there is a corresponding suction phase drawing in water from below and the side. *Ecdyonurus* lives on stones in fast-flowing water. It is dorsoventrally flattened and its head is usually oriented upstream. In this position the gills are relatively protected from the force of the water. The net effect of their movements is to draw water over themselves from the fairly still zone around the sides and below the body, and then force it posteriorly over the dorsal surface of the abdomen.

Ephemera danica has a similar ventilatory current, except that water is only drawn toward the dorsal surface of the abdomen anteriorly. There are seven pairs of bifurcate gills, each narrow lamella being fringed with numerous filaments, but only the second to sixth produce the current. At rest, each gill branch is held reflexed over the dorsal abdominal surface. The effective stroke is a downward and inward movement, during which an undulatory wave passes from the base of the gill branch to its tip, thereby setting a current in this direction. Another undulation passes posteriorly across the gill surface, generating a posterior flow. The net effect is that each gill branch produces a current directed diagonally backward towards the dorsal midline. The gill branch then moves upward and outward.

It presents an appreciable angle to its own path of motion in both directions and so the basic oscillatory movemens probably contributes little or nothing to the current. The filaments of the two branches of each gill and of consecutive gills are arranged so that they overlap at all times and are so close together that there is no flow between them. However, the two second gills are twisted spirally such that, during the effective stroke, water is drawn in laterally at their bases and flows round to their inner surfaces.

The backward and downward stroke of consecutive pairs of gills results in a posteriorly directed current. The intersegmental phase lag is about one-eighth of an oscillation (i.e., 45°), which means that when the second pair is at one end of an oscillation, the sixth pair is at the other end. Larvae of *Ephemera* burrow is fairly fine sediment and Eastham (1939) suggests that the reason for the virtual absence of lateral currents is that they would damage the walls of the burrow. Like *Ephemera, Leptophlebia marginata* has seven pairs of bifurcate gills, the first and last of which are not used in the production of the ventilatory current. The gill branches are simple and lanceolate.

As in *Ecdyonurus* they draw water in towards the dorsal midline, but they differ from this animal in one noticeable respect. Successively more posterior gills are held at lower angles to the abdomen so that the anterior ones sample the water dorsolaterally, the posterior ones ventrolaterally. The lamellae move through an ellipse with the effective stroke directed backward and inward. During the effective stroke both lamellae of each pair present a large angle to their path of motion, the anterior one being inclined towards the midline and so producing a posteromedian current. During the return stroke, both lamellae are "feathered," as in *Ecdyonurus.*

However, the posterior lamella of each pair lags behind the anterior one and does nto reach so far forward during the return stroke. Thus, during the effective stroke the anterior lamella meets the posterior one at an angle which is open toward the midline. This angle is progressively reduced and the interlamellar water squeezed inwards. As in *Ephemera,* each gill is about one-eighth oscillation in advance of the one next behind it; in other words, there is a phase lag of 45°. *Cloeon dipterum,* like *Leptophlebia* lives in still water. However, the lamellate gills of *Cloeon* are all held laterally in the same plane. They draw water posteriorly along the dorsal midline and then force it outward and upward between themselves. The last pair do not beat, but serve to deflect the current laterally.

In *Ecdyonurus, Ephemeria,* and *Cloeon,* the backwardly directed current is enhanced by the alternate phase of suction and compression produced by each pair of gills moving slightly in advance of the pair next behind. This strengthening does not occur in *Leptophlebia* because of the differences in gill attitude. *Caenis horaria* differs markedly from the above species. It lives partly buried in mud and its second pair of gills is expanded into large plates which cover and protect the third to sixty pairs, forming a sort of branchial chamber.

During ventilation these plates lift so an angle of 30° to 40° to the body surface to allow the other gills to oscillate. The ventilatory current is completely lateral, either form left ot right or form right to left. The flow pulsates with weak reversals between each forward pulse. Each gill is a flat lamella with numerous marginal filaments. The effective stroke is downward and backward when the filaments are straight and close together to that water cannot pass between them. During the upward and forward return stroke the marginal filaments lag behind and spread out to allow water to pass between them. The gill also pivots and follws an elliptical path. To produce a flow from left to right, the underside of the gill faces the right on the effective downstroke and faces the left on the upward return stroke, thus producing a screw effect.

In both directions the gill presents a marked angle to its direction of motion, but the return stroke is ineffective in producing any current for the reasons outlined above. There is a delay of about one-third oscillation between successive gills (i.e. a phase lag of about 120°) and so the sixth gill is in the same position as the ipsilateral third gill. In this animal, the members of each pair of gills ar out of phase by about one-third oscillation also. When the current is from left to right, the left gill of each pair is in advance of the right gill. The lateral flow produced by the individual gills is enhanced by the lateral peristaltic wave caused by the phase lag between the members of a pair, and this in turn is reinforced by the longitudinal metachronal rhythm. The direction of the current is reversed by stopping the oscillatory movement and then starting it again, changing the direction of the elliptical path, the attitude of the gills, and the phase relationship between the two sides. When swimming, the gill movements cause the body to spin on its longitudinal axis, gill-less larvae swimming dorsal side up.

Respiration in Anisopteran (Odonata) Larvae

In anisopteran dragonfly larvae the gills are arranged in six longitudinal sets in a branchial chamber formed from the hindgut. The branchial chamber can be closed at its ends by the pre- and postbranchial valves. Behind it is a muscular vestibule, inserted onto which are six rows of dilator muscles. The vestibule communicates with the exterior via the anus and this opening is controlled by the anal valve. The first phase of normal ventilation is expiration and this can be recognized by the upward movements of the abdominal sterna, especially of the more posterior segments. Snodgrass (1954) states that there is also some compression of the terga, especially during strong expiratory movements. At the same time as the sterna are being raised, the anal valve is opened to about one-third of its maximum extent.

The rising sterna produce a pressure of 2 to 4 cm H_2O in the branchial chamber and this forces the contained water out through the anal valve. When they reach their uppermost position the sterna start to fall again, signifying the start of inspiration. At the same time the anal valve opens fully, this often being preceded by a brief closing movement. The net effect of these actions is to cause a negative pressure of about 0.5 cm H_2O in the branchial chamber and so water is drawn. It is though that this is aided by active dilation of the vestibule. The effect of the narrow anal valve opening, coupled with high branchial chamber pressure, during expiration, and the wide opening and low negative pressure during inspiration, is to prevent too much mixing of inspired with expired water. Apart from the musculature directly associated with the hindgut, other abdominal muscles are obviously involved.

In *Aeshna* and *Anax* each abdominal segment form the foruth to the ninth has three pairs of dorsoventral muscles. Wallengren (1914a) thought that all of these, and also the dorsoventral oblique

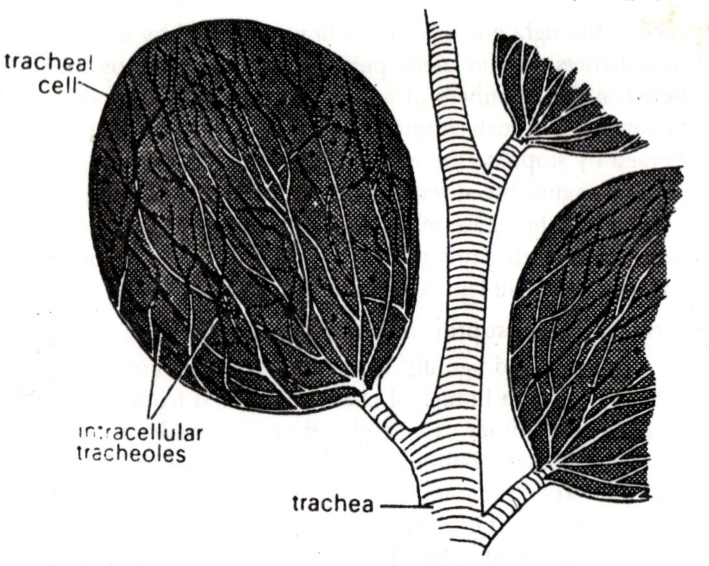

Fig. 4.15. Tracheal cells arising from a trachea in the larva of Gasterophilus.

muscles, are expiratory and help to lift the sterna, but Whedon (1918) that the so-called respiratory dorsoventral muscles are the middle or "respiratory" pair of dorsoventrals. It is confirmed by Mill and Hughes (1966) that the so-called respiratory dorsoventral muscles are the ones normally concerned with raising the sterna, although in certain circumstances other dorsoventral muscles may be involved. The respiratory dorsoventral ventral muscles differ from the other two pairs in that they are richly tracheaed and contain numerous mitochondria, regularly oriented on either side of the Z-lines. In *Libellula* two pairs of richly tracheated, segmental muscles are invariably concerned with lifting the sterna. Running transversely across the abdomen between segments 4 and 5 and segments 5 and 6 are a muscular diaphragm and a subintestinal muscle, respectively.

The diaphragm was first described by Amans (1881), who suggested that it was involved in ventilation, defecation, and extension of the labial mask; the subintestinal muscle was described by Wallengren (1914a). It was suggested by Snodgrass (1954) that the sterna return to their resting position solely as a result of the natural elasticity of the tergal plates. Tonner (1936) thought that contraction of the subintestinal muscles forces the sterna down, but that the diaphragm only starts to contract later and thus has little effect, except in widening the posterior segments. However, earlier Matula (1911) and Whedon (1918) were of the opinion that both muscles aid the downward movement of the sterna, and it has been shown that this is indeed the case, with both muscles contracting synchronously. However, tergal elasticity is still thought to play an imporant role. Normal ventilation is a repetitive event which may continue for sometime.

It is interrupted by other forms of ventilation and by periods of quiescence. Frequencies in the range 12-57/minute have been recorded (40-114/minute in *Libellula*), the frequency increasing with increase in temperature. Wallengren (1914b) has shown that a 4-centimeter long larva changes about 83% of the contents of its branchial chamber and vestibule with each oscillation. This amounts to about $0.05cm^2$. Thus, as a frequency of 20/minute, the larva pumps about 1 cm^3 of water/minute. When the environmental pO_2 falls below about 2.5 cm^3/liter at 17°C to 18°C the larva comes to the surface and pumps air instead of water. This has been called *Notatmung*, or *"emergency ventilation."* Tonner (1936) distinguished two other types of ventilation.

In *"gulping ventilation"* (*Schlukatmung*) the anal valve opens and the vestibule is filled with water. The anal valve then closes and the postbranchial valve opens. The vestibule contracts strongly forcing the water intot he branchial chamber. this procedure is repeated several times in rapid succession, thus *"pumping up"* the branchial chamber. The ensuring expiratory phase is rather slow. In many cases the water is retained in the banchial chamber for several seconds, when it is churned about by the action of the intrinsic musculature. This latter is "*chewing ventilation*" (*Kauatmung*). Hughes and Mill (1966) have recorded maintained positive pressure in the branchial chamber. They are of two types. In one, the pressure increases in a series of jerks similar to normal ventilatory cycles, but at a higher frequency which produces a temporal summation effect. Coincident with each pressure increase, the sterna show a slight expansion. When the maximum is reached (5 to 6 cm H_2O) the pressure starts to fall slowly and then drops abruptly, the sterna risign back to their resting level at the same time.

This is probably gulping ventilation. In the other type, the pressure increases smoothly and fairly rapidly up to as high as 35 cm H_2O, and is accompanied by lifting of the sterna. The pressure may fall immediately, but is usually maintained for several seconds. Finally, the jet-propulsive mechanism of locomotion also sub-serves respiration. This is basically a rapid ventilatory movement, but in addition some longitudinal contraction of the abdomen occurs and the legs are moved posteriorly to lie alongside the body. Pressure of up to 40 cm H_2O have been recorded, with a rate of increase of about 1000/cm/second as against 6-12 cm/second for normal ventilation.

The Control of Ventilation

The effect of various environmental parameters, such as pO_2, pCO_2, and temperature, on the initiation and frequency of ventilation has been mentioned in previous section. Here we are concerned primarily with the neural control of ventilation, for which information is only available for the pumping mechanism of aeshnid dragonfly larvae. Ventilation in aeshnid larvae consists of two phases, expiration and inspiration, both of which involve abdominal skeletal muscles. The segmental, expiratory "respiratory dorsoventral muscles" are innervated by the corresponding second segmental nerves, while the inspiratory "diaphragm" and "subintestinal transverse muscle" receive their nerve supply from the unpaired median nervous system. Rhythmic bursts of activity have been recorded in the nerves to all of these muscles.

The expiratory bursts in the second segmental nerves of *Aeshna* normally consist of a single unit, although a second may sometimes be present, whereas in *Anax* there are almost invariably three units. However, in both genera only one unit has every been shown to produce an electrical effect in the respirtory dorsoventral muscle. There is always a 1:1 relationship between nerve and muscle potentials and the firing pattern shows a characteristic increase in frequency which causes facilitation of the muscle potentials. The other segmental neves also now show activity synchronized with the expiratory bursts in the second nerves, but neither this nor the other active units in the second nerves has been seen to instigate any muscular contraction in dissected preparations. Some units could be inhibitory, others could serve to increase the tone in certain muscles, and indeed, in chronic preparations, synchronous activity has been recorded from the anterior dorsoventral muscles on a few occasions.

The expiratory bursts on opposite sides of the same segment are in phase and there may or may not be a 1:1 relationship between the action potentials. On the other hand, the activity spreads from behind forward. The expiratory burst first appear from the last (eighth) abdominal ganglion and then from successively more anterior ganglia (as far forward as the fourth) with an intersegmental delay of about 100 m seconds. Thus, there is a pacemaker in the eighth ganglion. This posterior-to-anterior rhythm is also seen in the control of swimmeret movements in decapod crustaceans, in contrast to the anterior-to posterior metachronal rhythm of the gills of ephemeropteran larvae.

The bursts cease simultaneously in all segments. However, his occurs before the sterna are fully raised, and records of hte tension developed by individual muscles show that tension starts to fall as soon as electrical activity ceases. The continued upward movement of the sterna could be due to the intertia effect of water leaving the branchial chamber or to the sterna "clicking" into a stable position after being lifted past a cirtical level. Inspiration is closely coupled to expiration and as soon as the dorsoventral msucles start to relax the diaphragm and subintestinal muscle contract. Recordings from these muscles show in-phase bursts of activity.The bursts are fairly uniform, with possibly a slight overall increase in frequency toward the end.

Alternate expiratory and inspiratory bursts have been recorded from the fifth segmental nerves of the last abdominal ganglion, which innervates all of the musculature of the tenth segment. Jet propulsion also involves the respiratory dorsoventral muscles and recordings form them show short bursts of activity in which the frequency is high and the muscle potentials large. Other abdominal muscles are also involved and synchronous activity has so far been recorded from the anterior and posterior dorsoventral muscles. Electrical stimulation of a first segmental nerve has no effect during an expiratory burst. However, if the stimulus nerve is applied between bursts, if often elicits a normal expiratory burst which then resets the rhythm.

Furthermore, it was shown by Mill (1970) that repetitive electrical stimuli, delivered at a frequency slightly greater than that of ventilation and starting in an interexpiraotory period, are capable of causing the ventilatory cycle to become phase-locked with this new frequency. A similar effect has been shown by Miller (1971) in a mantid (*Sphodromantis lineola*), while in *Periplaneta americana* only a decrease in frequency can be obtained. However, the effect of proprioceptive feedback on the control mechanism is probably very small. Matula (1911) demonstrated that removal of the head ganglia caused an increase in ventilatory frequency, whereas removal of the prothoracic ganglion caused a decrease (as also did removal of the legs). He revealed also that removal of the head ganglia had no effect on the relationship of ventilatory frequency wiht temperature. Furthermore, he showed the removal of either head or prothoracic ganglia negated the frequency increase associated with decrease in environmental pO_2. From this the concluded that, in addition to the abdominal ventilatory centers, there are centers in the head and prothoracic ganglia sensitive to lowering of the environmental pO_2.

However, Wallengren (1913) found that the frequency increase caused by removal of the head ganglia was only temporary and these animals were still sensitive to pO_2 changes, but he agreed that the prothoracic ganglion must contain a center which controls ventilation according to environmental oxygen tension. It has been suggested by Mill and Hughes (1966) that the expiratory center in the last abdominal ganglion is activated by a command interneuron or interneurons from a higher center in the head or thoracic ganglia. Although normally the center in the eighth ganglion fires first, with successively more anterior centers then firing in sequence, it has seen in some preparations that one of the centers is firing out of turn.

Thus, the centers as far forward as the sixth at least presumably receive a direct input from the command system. A possible neural arrangement which counts for the known parameters in a single ganglion, and this also demonstrates how the system may oscilalte between expiration and inspiration. Some of the details of the control of ventilation in these animals may well have parallels in the crayfish swimmeret control system.

RESPIRATORY PIGMENTS

Of the four respiratory pigments, only hemoglobin has been found in insects and this in only a very few species. Presumably this is a result of the tracheal system method of supplying hte tissues with oxygen, the vascular system thus having little respiratory significance.

Hemoglobin as a High Affinity Pigment

Hemoglobin was first recorded in larvae of *Chironomus* by Lankester (1867), but it was nto for several decades after this that any work was carried out on the function of this pigment. *Chironomus plumosus* (and probably the other chironomids as well) has the smallest known hemoglobin, consisting of only two unit molecultes (MW = 31,400) and with a sedimentation constant of 2.0 Fox (1945), working on *C. riparius,* demonstrated that it has a high affinity for oxygen, with a p_{50} of 0.6 mm Hg at 17°C. There is no Bohr effect and the temperature effects is very small indeed ($p50$ = 0.5 at 10°C). The oxygen uptake of normal larvae has been compared with that of larvae in which the hemoglobin was rendered functionless by treatment with carbon monoxide to produce carboxyhemoglobin.

Harnisch (1936) concluded that in *C.thummi* the hemoglobin is functional at all oxygen tensions and that its main function is to increase the recovery rate after anaerobis. In contrast, Ewer (1942), working on *C.plumosus,* found that the hemoglobin any transports oxygen at low environmental tension of this gas, a condition presumably experienced at times in the animal's natural habitat (they live in mud burrows in still water). He showed a that at 17°C the pigment only functions below about 40% air saturation (i.e., at 3 ml O_2/liter) and down to 15% air saturation oxygen uptake is independent of environmental pO_2. This is in agreement with the earlier work of Leitch (1916) who found that larvae of *Chironomus sp.* only appear to use their hemoglobin below an environmental pO_2 of about 7 mm Hg at 17°C. Walshe (1950) investigated the behaviour of *C. plumosus* larvae in glass U-tubes. She found that in well-aerated water they spent about 50% of their time ventilating the tube, with pauses for filter feeding or rest.

A decrease in environmental pO_2 caused an increase in the amount of time spent ventilating and, below 10% air saturation, feeding ceased and the animal ventilated continuously. Under completely anaerobic conditions the larvae became immobile. In contrast, larvae treated with carbon monoxide stopped feeding at a much higher environmental pO_2 (26% air saturation) and became immobile at about 9% saturation. Thus, hemoglobin helps to keep the larvae aerobic, and therefore active, at low oxygen tensions. Since both environmental oxygen and carbon dioxide have an effect Walshe (1950) suggested that are pH sensitive. If the larvae undergo a period of anaerobis they only build up a small oxgyen debt and this is repaid when the water is aerated.

Initially there is a large increase in oxygen uptake as the larvae become active, but this soon subsides, remaining slightly above normal for about 1 temperature, the initial excessive increase in oxygen consumption does not occur and the animals take twice as long to repay their debt. According to Walshe (1947a) the hemoglobin is only involved during the initial increase, at which time the water was not fully aerated in her experiments. Furthermore, Walshe (1950) has shown that *Chironomus* larvae can repay their oxygen debt in water which is only 7% saturated. Thus, *Chironomus* hemoglobin is used to transport oxygen at low tension of environmental oxygen. A number of authors have suggested a storage function for this pigment. Leitch (1916) showed that its storage capacity is limited to about 12 minutes and Walshe (1950) confirmed this with a figure of about 9 minutes for the storage capacity in a resting animal. Thus, a storage function is not significant during long periods of anaerobis. The larva of another chironomid which possesses hemoglobin, *Tanytarsus brunnipes,* unlike that of *Chironomus,* is very intolerant of low environmental pO_2.

In spite of this its hemoglobin does not transport oxygen about 25% air saturation at 17°C, and even below this level contributes little to the metabolic rate. In fact, at all tensions below air saturation, its oxygen uptake varies with environmental pO_2. Tanytarsus also lives in mud tubes but in well-aerated water, and nothing is known about its normal metabolic rate. This may be of considerable importance in reaching an understanding of the role of this pigment since the polychaete *Nereis* has a low metabolism when it is in its tube and Eriksen (1963a) has shown that burrowing ephmeropteran larvae have a much lower metabolic rate when they are provided with a normal substrate in which to burrow. The chironomids are unique in having a low molecular weight respiratory pigment in the plasma; in all other cases pigments of low molecular wieght are contained in cells. This may be possible here because insects do not use ultrafiltration for excretion, a method which has been shown to remove low molecular weight proteins from plasma.

Hemoglobin as a Low Affinity Pigment

Larvae of *Gastrophilus* and the adults and larvae of the notonectids *Anisops* and *Buenoa* have hemoglobin contained in richly tracheated groups of abdominal cells. The hemoglobin of *Anisops pellucens* differ from that of chrironomid larvae in that it has a very low affinity for oxygen (p_{50} = 28 mm Hg at 24°C). It shows no Bohr effect, but there is a considerable temperature effect, increase in the latter shifting the oxygen dissociation curve to the right. When *Anisops* and *Buenoa* dive they show a period of neutral buoyancy and this is possible almost entirely because of the storage function of the hemoglobin. The adults possess a compressible gas gill consisting of a ventral air film connected to a bubble lying between the bases of the legs and in the subelytral space.

Unlike *Notonecta,* however, there is no air film on the wings of the adults and so they are less buoyant and have a much smaller water-air interface. In adults of *A. pellucens* the average duration of a dive is about 5 minutes and this is reduced to about 1 minute in animals treated with carbon monoxide. Thus, the hemoglobin provides a storage of oxygen which can last for about 4 minutes. This compares with 1-2 minutes in *A. debilis* and 4 minutes in *Gastrophilus* larvae. The environmental pO_2 does not affect the duration of a dive and so the gas gill is probably acting mainly as an oxygen store also. This has led Miller (1966) to suggest the following sequence of events during a dive by *Anisops*. When the animal dives it is at first active and the gas gill starts to decrease in size, and so it become less buoyant.

At the onset of the phase of neutral buoyancy, activity is minimal and the hemoglobin starts to give up its oxygen. This passes into the abdominal tracheae and hence into the gas gill to maintain the neutral buoyancy. From here, most oxygen is porbably taken up via the thoracic spiracles. At a gill pO_2 of about mm Hg the hemoglobin will be unloaded and the animal will start to lose its gas gill again. This decreases its buoyancy, resulting in more activity and the animal soon has to return to the surface to replenish its stores.

Other Functions of the Respiratory System

Compressible gas gills are normally sufficiently large to make the animal positively buoyant and it is decrease in this buoyancy, caused by the shrinking glll, that provides the stimulus for the animal to return to the surface. The rather unusual case of *Anisops* and *Buenoa* has been dealt with in the previous section. In the plastron-bearing adults of *Aphelocheirus* there is a pair of sense organs, each consisting of an oval plate bearing hydrofuge hairs which are longer and less abundant than those of the plastron. Interspersed with these are sensory hairs. These organs are sensitive to uniform and differrential pressure changes. they about against the spiracular rosettes of the second abdominal segment and so are in close contact with the tracheal system. The trachea from these same spiracles each connect with a small air sac (one on each side and Thorpe and Crisp (1947c) suggests that these sacs damp out flutuations in the gas pressure in the tracheal system caused by body movements or changes in environmental gas tension. Three pairs of similar sense organs are present in *Nepa,* but Thorpe and Crisp (1947c) have shown that their response is not so finely graded as in *Aphelocheirus* and they only respond to differential, not abdolute, pressure. *Corethra* and *Mocylonyx* (Diptera-Nematocerca) larvae contain tracheal air sacs, whch have a hydrostatic function.

In *Corethra* there are two of these sacs and they are about all that isleft of the tracheal system. The gas inside the air sacs equilibrates with the gas mixture in the environment and their size can be varied. Thus, the buoyancy of the animal can be altered to enable it to float at different levels in the water. In the larvae of the mayfly *Hexagenia* the gills appear to have a balancing role as well as being important sites for gaseous exchange, and this probably applies to other larvae with lateral abdominal gills.

5

HAEMOLYMPH

In most of animals including all vertebrates the blood flows through special vessels such as arteries, capillaries and veins. This is known as a *closed circulatory system.* The circulatory system of insects, like that of all arthropods, is of the "open" type; this is, the fluid flows freely among the body organs. An open system results from the development, in evolution, of a haemocoel rather than a true coelom. A consequence of the open system is that insects have only one extracellular fluid, haemolymph, in contract to vertebrates which have two such fluids, blood and lymph. The occurrence of an open system does nto mean that haemolymph simply bathes the organs it surrounds because usually thin granular membranes separate the tissues from the haemolymph itself.

Insects generally possess pumping structures and various diaphragms to ensure that haemolymph flows through the body along a definite route. As the only extracellular fluid, it is perhaps not surprising that haemolymph, in general, serves the functions of both blood and lymph of vertebrates. Thus, plasma is important in providing the correct milieu for body cells and is transport system for nutrients, hormones, and metabolic wastes, while haemocytes provide the major defence mechanism against foreign organisms which enter the body and are important in would repair and in the metabolic of specific compounds. Blood cells may be involved in connective tissue formation, but in many cases connective tissue is formed by the cells of other tissues.

The functions of connective tissue include supporting and binding tissues together and in this the tracheal system may play an important role, but it is also possible that some connective tissue serves to conduct secretions from their origins to the target cells. The volume of blood may vary at different stages in the life cycle and according to the physiological condition of the insect. The plasma contains various inorganic ions, of which sodium and chloride may be the most important, but insects differ from other animals in that other ions may be present in higher concentrations than these two.

Organic substances are also present and in higher insects amino acids make a considerable contribution to the total osmolar concentration of the haemolymph. Proteins are also present and vary in concentration in the course of the life history. The plasma serves primarily as a means by which substances may be transported round by body, although it plays little part in respiration. It may also provide a store of substances such as sugars and proteins, while its water acts as a reservoir for the maintenance of the tissue fluids. The hydrostatic pressure of the haemolymph is important in the movements of soft-bodied larvae, in expansion after moulting and in other ways.

Haemocytes

The blood or haemolymph circulates round the body cavity between the various organs, bathing them directly. It consists of a fluid plasma in which are suspended the blood cells or haemocytes.

Types of Haemocyte

Many different types of haemocyte have been described, but a comprehensive classification is difficult because individual cells can have very different appearances under different conditions and a variety of techniques have been used in their study. J.C. Jones (1962, 1964) recognises four main types of cell, which are found in most of the insects studied.

1. *Cystocytes* (*coagulocytes*) which, when viewed with the phase contract microscope, have a small, sharply defined nucleus and a pale, hyaline cytoplasm containing scattered black granules, while other types of haemocytes have a larger, paler nucleus and darker cytoplasm. The cytocytes are probably specialised granular haemocytes.
2. *Granular haemocytes* are also phagocytic, but are characterised by the possession of acidophilic granules in the cytoplasm.

3. *Prohaemocytes* are small rounded cells with relatively large nuclei and intensely basophilic cytoplasm. The divide at frequent intervals and give rise to other types of cell.
4. *Plasmatocytes* are frequently the most abundant cell type. They are variable in form, phagocytic and with a basophilic cytoplasm.

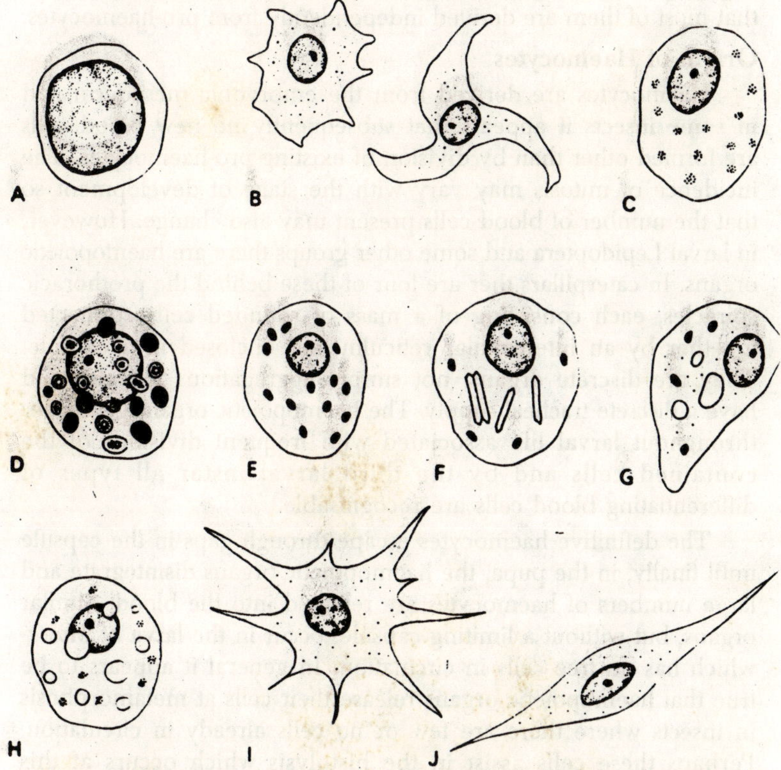

Fig. 5.1. Examples of haemocyte types. A–Prohemocyte. B–Plasmatocytes. C–Granular haemocyte. D–Spherule cell. E–Cystocyte. F and G–Oenocytoid cells. H–Adipohaemocyte. I–Podocyte. J–Vermiform cell.

In addition there are other types of cell only occur in certain insects. The commonest of these are oenocytoids, spherule cells and adipohaemocytes. Oenocytoids are found in Coleoptera, Lepidoptera and some Diptera and Heteroptera. They are usually large, thick, basophilic cells containing canaliculi, strands of granules or crystals. Spherule cells, found in Lepidoptera and diptera, are round or oval cells with large, non-refringent, usually acidophilic inclusions filling the whole cells. Adipohaemocytes (spheroidocytes) have been found in all the Lepidoptera and Diptera so far studied an in some representatives of various other groups. They are characterised by refringent fat droplets and other inclusions. It is not clear if these different types of haemocytes represent different stages in the development of individual cells, but it seems probable that most of them are derived independently from pro-haemocytes.

Origin of Haemocytes

Haemocytes are derived from the embryonic mesoderm and in some insects it appears that subsequently no new blood cells are formed other than by division of existing pro-haemocytes. The incidence of mitosis may vary with the stage of development so that the number of blood cells present may also change. However, in larval Lepidoptera and some other groups there are haemopoietic organs. In caterpillars ther are four of these behind the prothoracic spiracles, each consisting of a mass of rounded cells connected together by an intercellular reticulm and enclosed in a capsule. They are discrete organs, not simply aggregations of cells, and have a discrete tracheal supply. The haemopoietic organs get bigger throughout larval life associated with frequent divisions of the contained cells and by the third larval instar all types of differentiating blood cells are recognisable.

The definitive haemocytes escape through gaps in the capsule until finally, in the pupa, the haemopoietic organs disintegrate and large numbers of haemocytes are released into the blood. Similar organs, but without a limiting capsule, occur in the larva of *Musca*- which has no free cells in circulation. In general it appears to be true that haemopoietic organs release their cells at metamorphosis in insects where there are few or no cells already in circulation. Perhaps these cells assist in the histolysis which occurs at this time. *Calliphora,* however, is exceptional since in this species the blood cells are released from the haemopoietic organs in the second instar larva. Haemopoietic organs are not known to occur in adult insects.

Numbers of Haemocytes Present

The number of haemocytes present in the blood can fluctuate considerably over short periods because normally not all the cells are free in the circulation; many of them may adhere to the surfaces of tissues in the haemocoel, only appearing in the circulation at certain times. The number of cells in a unit volume of blood is also influenced by changes in the blood volume, but marked changes in the total numbers of blood cells are known to occur. Some insects, such as the larvae of *Musca* and Chironomus plumosus, normally have no haemocytes in circulation, while in others, like *Periplaneta,* there may be several million free blood cells. In general the number of circulating cells increases before a moult and decreases again after it.

In *Sarcophaga* the blood cell count rises from about 8000/mm^3 in the larva of 34,000 mm^3 just before pupation, possible following the release of the cells from haemopoietic organs. In the early pupa the number drops to 12,000 mm^3. probably as a result of many cells adhering to the tissues. These changes are largely due

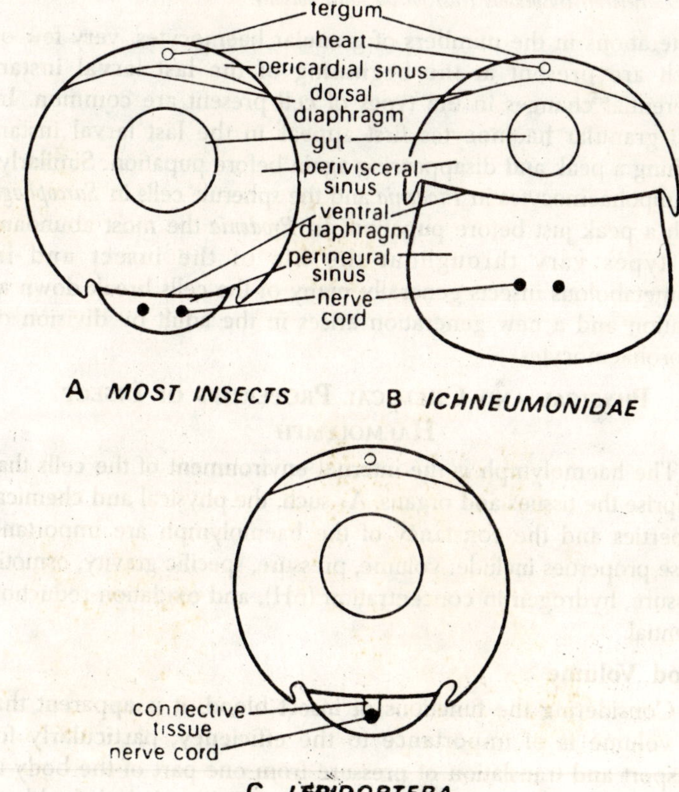

Fig. 5.2. Diagrammatic cross-sections of various insects showing the main sinuses of the haemocoel and the positions of heart, alimentary canal and nerve cord.

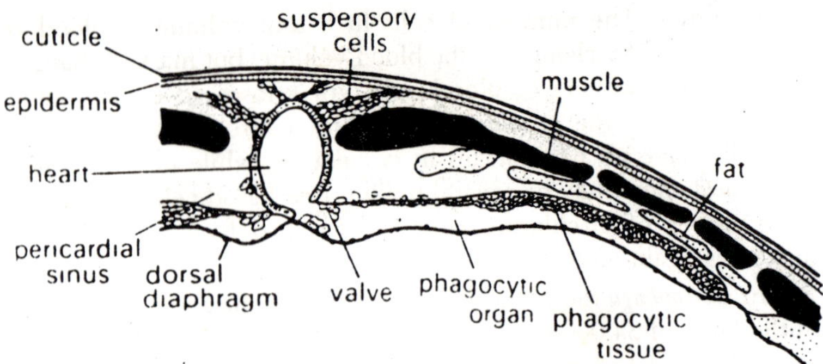

Fig. 5.3. Transverse section through the pericardial sinus in the abdomen of Gryllotalpa showing a phagocytic organ and phagocytic tissue. Pericardial cells form supporting elements (suspensory cells) for the heart dorsally.

to alterations in the numbers of granular haemocytes, very few of which are present at the beginning of the last larval instar. Differential changes in the types of cell present are common. In *Sialis* granular haemocytes first appear in the last larval instar, reaching a peak and disappearing again before pupation. Similarly, the adipohaemocytes in *Prodenia* and the spherule cells in *Sarcophaga* reach a peak just before pupation. In *Prodenia* the most abundant cell types vary throughout the life of the insect and in holometabolous insects generally many of the cells break down at pupation and a new generation arises in the adult by division of the prohaemocytes.

Physical and Chemical Properties of Insect Haemolymph

The haemolymph is the internal environment of the cells that comprise the tissues and organs. As such, the physical and chemical properties and the constancy of the haemolymph are important. These properties include: volume, pressure, specific gravity, osmotic pressure, hydrogen in concentration (pH), and oxidation-reduction potential.

Blood Volume

Considering the functions of insect blood, it is apparent that the volume is of importance to the efficiency, particularly for transport and translation of pressure from one part of the body to another. The body cavity is rarely filled with blood, and the blood volume varies with the physiological state. Tabular figures showing the blood volume relationships for a number of insect species are

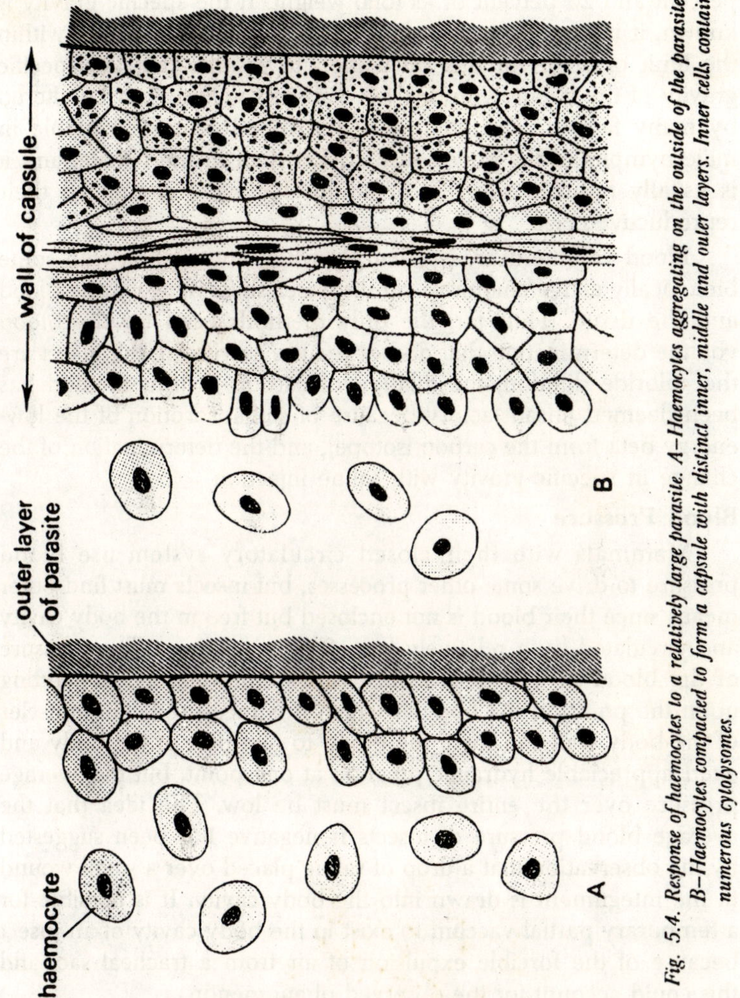

Fig. 5.4. Response of haemocytes to a relatively large parasite. A–Haemocytes aggregating on the outside of the parasite. B–Haemocytes compacted to form a capsule with distinct inner, middle and outer layers. Inner cells contain numerous cytolysomes.

found in Altman and Dittmer. These data were collected from several sources and represent measurements made by a variety of methods. These methods include; exsanguination, cell dilution, chloride dilution, dye (amaranth) dilution, Carbon-14 (inulin) dilution, and specific gravity changes with saline dilution. The results of these determinations are usually related to body weight following the convention used in mammalian physiology.

The correlation is poor, but there is a general agreement that the blood of an averrage normal insect accounts for between 16 percent and 20 percent of its total weight. If the specific gravity is known, it is possible to calculate the volume; and it is well within the limit of error for this calculation to assume that the specific gravity of the haemolymph is 1.000. Blood volume can be altered by many factors. The volume of the blood is more variable in male nymphs of the gray house cricket than in the female, and it is usually reduced severely in adults that have completed their reproductive cycle.

Blood volume decreases sharply when intoxication with some biologically active chemicals causes water shifts between the blood and the tissue. Thus far, the only methods used for the blood volume determination that do not require sacrificing the insect are the chloride dilution, the dilution of carbon-14 (this method has been deemed unsatisfactory because of self-absorption of the low-energy beta from the carbon isotope), and the determination of the change in specific gravity with saline injection.

Blood Pressure

Mammals with their closed circulatory system use blood pressure to drive some other processes, but insects must find other means since their blood is not enclosed but free in the body cavity and circulated by a relatively low-efficiency system. The pressure of the blood of insects probably varies considerably, depending upon the point on the body at which it is measured. The muscles of the body wall can force the blood to one part of the body and exert appreciable hydraulic pressure at one point, but the average pressure over the entire insect must be low. The idea that the average blood pressure of insects is negative has been suggested by the observation that a drop of saline placed over a small wound in the integument is drawn into the body cavity. It is possible for a temporary partial vaccum to exist in the body cavity of an insect because of the forcible expulsion of air from a tracheal sac, and this could account for the observed phenomenon.

Specific Gravity of the Haemolymph

Specific gravity is the mass per unit of volume compared to the mass of an equal volume of water at the same temperature. For practical purposes, specific gravity and density may be considered to be synonymous although they are equal only at 4°C. Specific gravity is very easy to measure precisely with small amounts of blood and serves as an index of the physiological state of individual insects. Many physiological conditions bring about shifts in the internal water balance, and all of these shifts are reflected in a change of blood specific gravity. Specific gravity measurements have been reported from data obtained by direct weighing (pycnometry), the application of Stoke's law of falling bodies, and the Linderstrom-Lang gradient column.

Specific gravities of the blood of several common laboratory species measured by the latter method ae shown in Table 5.1. The values are listed as a mean, plus or minus the strandard deviation. The range so described included two thirds of the individuals in a normal distribution included two thirds of the individuals in a normal distribution which can be considered to describe the normal range for the group. Specific gravity values outside this range are abnormal. For most insects, the haemolymph specific gravity lies between 1.015 and 1.060 (usually calculated in grams per milliliter). Considerable variations between species is evident, and the period during which an insect is molting is always marked by an increase in the specific gravity of the blood.

Table 5.1. Specific Gravity of the Haemolymph of Several Insect Species.

Insect	*Sex*	*State*	*Mean Specific Gravity*	*Standard Deviation*
Orthoptera				
Acheta domesticus	male	adult	1.0215	.0045
Acheta domesticus	female	adult	1.0195	.003
Acheta domesticus	male	nymph	1.0188	.0025
Acheta domesticus	female	nymph	1.0192	.0026
Periplaneta americana	male & female	nymph	1.0297	.0037
Leucophacd mediera	male & female	nymph	1.0293	.00064
Hemiptera				
Oncopeltus faciatus	male & female	nymph	1.0243	.0002
Diptera				
Musca domestica		larva	1.0479	.0041
Lepidoptera				
Galleria mellonella		larva	1.0546	.0037
Coleoptera				
Tenebrio molitor		larva	1.033	.004
Hymenoptera				
Apis mellifera	worker	larva	1.038	.006
Apis mellifera	drone	larva	1.050	.0054

Osmotic Pressure of the Haemolymph

Osmotic pressure is numerically equal to the hydrostatic pressure difference required to prevent penetration of a solvent through a semipermeable membrane into a solution. In biological systems the penetrating solvent is always water. All tissues and organs of insects are bathed in the blood, which according to the figures tabulated by Altman and Dittmer, ranges from 84 percent to 94 percent water. Since the blood is the internal environment of the insect, osmotic pressure is of great importance. Osmotic pressure is related empirically to the specific gravity since both of these values depend upon the amount of water present.

Osmotic pressure is directly related thermodynamically to the freezing point depression, the boiling point elevation, and any one of these properties that can be measured can be translated into another. Data describing osmotic pressure are expressed in several ways. Depression of the freezing point is the most common method of describing osmotic pressure, because it is the most popular method of measurement. When testing insect blood, the depression of the freezing point must be determined with caution. The proteins of the blood are denatured by freezing so that a second freezing point obtained from a single sample differs from the value of the first reading. The coagulation of a sample also alters the freezing point depression significantly. Another method of expressing data describing osmotic pressure is in terms of millimeters of mercury (pressure) or atmospheres. None of these methods is of direct value in experimental work, in which the application most needed is a guide to the concentration of a salt solution prepared sot that it is isomotic with the blood and tissues. The most useful value for this is the equivalent concentration of sodium chloride. Data for the osmotic pressures of the haemolymph of a number of insects were tabulated by Buck. Table 5.2 is derived from these data with the values recalculated to the sodium chloride equivalent.

The pH of Insect Haemolymph

pH is the negative logarithm of the hydrogen ion concentration of the blood, or a measure of the degree of acidity or alkalinity. Following the invention of measuring methods, determinations of the pH of almost every measurable substances became popular, and the blood of insects was no exception. A summary of the data of insect blood pH is found in the work of Altman and Dittmer. These values were derived by a number of methods, and many of the variations in results undoubtedly come from the technology. Repetition of this work is unnecessary because the pH range for almost all insects lies between 6.0 and 8.0. A safe generality is that the pH of insect blood is near neutrality.

Table 5.2. Osmotic Pressure of Insect Blood (Expressed as NaCl Equivalent)

Species	*NaCl Equivalent (percent Wt./vol.)*	*Stage*
Orthoptera		
Aeridia nasuta	1.48	Adult?
Blatta orientalis	1.12	Adult
Carausius morosus	0.83	Adult
Decticus albifrons	1.06	Adult
Gryllotalpa gryllotapa	1.25	Adult
Locusta vuridissima	1.12	Adult
Hemiptera		
Napa cinerea	1.17	Adult?
Notonecta glauca	0.88	Adult
Pyrrhocoris apterus	1.18	Adult
Ranatra linearis	1.12	Adult
Lepidoptera		
Bombyx mori	0.72	Larva
Ephestia elutella	1.68	Larva
Galleria mellonella	1.71	Larva
Galleria mellonella	1.58	Pupa
Prodenia eridania	1.26	Larva
Saturnia pyri	1.15	Larva
Diptera		
Aedes aegypti	0.6-0.75	Larva
Aedes detritus	0.6-1.2	Larva
Anopheles maculipennis	0.85-1.11	Larva, pupa, adult
Culex pipiens	0.6-0.75	Larva
Gastrophilus intestinalis	1.30	Larva
Coleoptera		
Carabus intricatus	1.42	Adult?
Dytiscus circumcinctus	0.83	Adult
Hydrophilus pistaceus	1.05	Adult
Melolantha vulgaris	1.22	Larva
Oryctes nasicornis	1.14	Larva
Popillia japonica	1.53	Larva
Silpha carinata	1.32	Adult
Tenebrio molitor	1.7-2.2	Larva
Tenebrio molitor	1.46	Adult
Temarcha tenebrrecosa	1.11	Adult
Hymenoptera		
Apis mellifera	1.29	Larva
Vespa crabro	1.30	Adult
Odonata		
Aeschna sp.	0.83	Larva

Shifts in pH occur with some physiological disturbances as a reflection of the failure of a vital system. These shifts are usually small and are rarely a limiting factor in the normal function. If the excretory system is intact, the concentrations of the various metabolites that might alter the pH of the blood are regulated. Of more importance than pH is the buffer capacity or buffer index of the blood. The buffer capacity is the ability of hte blood to resist changes in PH that might come from the production of acidic or basic metabolites. Experimental measurements show that the buffer capacity of insect blood is only slightly greater than that of water.

The maximum buffer index usually lies at a pH somewhat acid to the normal pH. This is explained by noting that the range of greatest buffer capacity is coincident with the isoelectric points of the amino acids that are present in high concentration. For the most part, pH of these amino acids ranges between 3.0 and 5.0. The buffering capacity at neutrality, such as it is, comes from the small amounts of carbonates and phosphates in solution in the blood.

The Oxidation-reduction Potential of Insect Haemolymph

The oxidation-reduction potential is a function of the electron potential (free negative charge) present by virtue of the constituents of the solution. The oxidation-reduction potential has received very little attention, but it is a property with considerably importance in the mechanics of action of many biologically active chemicals, because it exerts a controlling force on the valence state of various elements in the system. Many toxicants are sensitive to oxidation of reduction and may be rendered either toxic or nontoxic by changes. Systems that oxidize or reduce other things depend upon the oxidation potential, which is symbolized by Eo, the normal or modal electrode potential for the system, with the modal hydrogen electrode referred to as a standard.

In physical chemistry Eo is determined at pH = 0.0, but this is impossible in biological systems so that, instead, it is determined at pH = 7.0. Any system with an Eo value more positive than that of a second system will oxidize that second system, becoming reduced itself in the process. There are two apposed conventions used in this designation; however, the above usually applies to biochemical measurements. Oxidation-reduction potentials can be measured electrometrically with an appropriate electrode potentiometer system, but in biological application it is easier to obtain on estimate by bracketing the value with oxidation-reduction-sensitive dyes. These dyes change colour from the oxidized to the reduced state, and with a series of these compouns the oxidation-reduction potential of the blood can be determined. Most attention to oxidation-reduction potentials of the blood of insects has been given to the determination of the total reducing substances.

The results of this moderately popular study have been reported in terms of fermentable reducing substances (primarily sugars). That other reducing substances exist seems apparent, but they have not been positively identified. A tabulation by Buck summarizes the information on the reducing substances in insect blood. The variations in the values and in the methods of determination, and lack of information on the stage of development as well as the nutritional history of the insects from which the samples were taken make the data difficult to use.

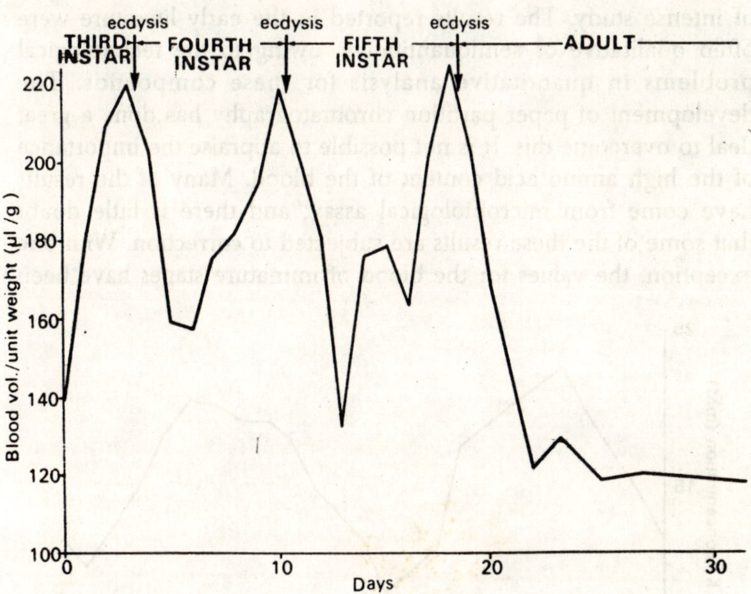

Fig. 5.5. Changes in the haemolymph volume during the life cycle of Schistocerca.

Composition of the Serum Fraction

The blood of insects has been the subject of many analyses, and like studies of other tissues, the results show great variation among taxonomic groups. A typical analysis complied by combining data from various sources is shown in Table. Comparable values for human serum ae used for reference. Many of the figures quoted are subjected to correction according to the method of analysis, the stage of the insect sampled, the sex, or the food upon which the insect was fed. Most of the values for insect serum are from analyses of whole blood, which introduces an obvious error when these values are compared with those for mammalian serum. There are several noteworthy differences between insect blood and mammalian serum. The first is the relatively high amino nitrogen content of insect blood, the second is the low or inverted sodium-potassium ratio, and the third is the relatively high magnesium content. Insect blood nearly always has a high uric acid content, which may reach the supersaturation point.

Amino Acids in Insect Haemolymph

The amino acid content of insect haemolymph has been subject of intense study. The results reported in the early literature were often qualitative or semiquantitative, owing to the technological problems in quantitative analysis for these compounds. The development of paper partition chromatography has done a great deal to overcome this. It is not possible to appraise the importance of the high amino acid content of the blood. Many of the results have come from microbiological assay, and there is little doubt that some of the these results are subjected to correction. With one exception, the values for the blood of immature stages have been

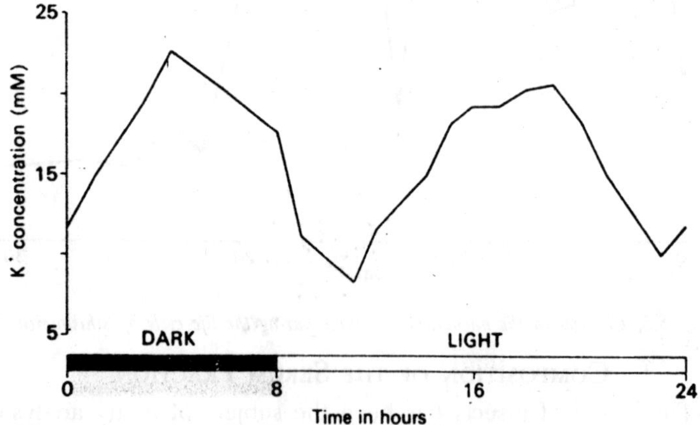

Fig. 5.6. Diurnal changes in potassium concentration in the haemolymph of Leucophaea.

selected, the reason being that these values are more likely to remain constant than those of adult blood, which show significant variation with age.

Table 5.3. The Chemical Composition of Insect Haemolymph (in mg/100 ml.)

Compound	*Haemolymph*	*Human Blood*
Total nitrogen	1400-1700	3000-3700
Protein nitrogen	700-900	1000-1265
Total protein	4375-6525	6500-8200
Nonprotein nitrogen	300-500	25-35
Amino nitrogen	200-300	5-8
Urea nitrogen	1-10	10-15
Uric acid	12-14	2-35
Potassium	20-40 (omnivorous)	178
	160-180 (foliage)	
Sodium	20-60	330
	300 (aquatics, adults, and some parasites)	
Calcium	35-40	9-11.5
	150 in some species	
Magnesium	10-25	1-3
Phosphorus (total)	64-245	34.9
Chloride	50-100	450-500
Reducing substances	0-25	70-100
(as glucose)	1000 (honeybee)	
Glycogen	24	5.5
Trehalose	700-800	
Lipids	398	652

Free amino acids are probably derived to a great extent directly from the proteins in the food. Some free amino acids, however, are produced as a result of metabolic synthesis in the individual, since they appear in the blood of insects that have been fed on diets deficient in free amino acids.

One characteristics of amino-acidemia in the blood of insects is its constancy, a subject discussed by Florkin. When concentrations of free amino acids are plotted against the concentrations of amino acids resulting from eimer hydrolyzed or nonhydrolyzed blood samples of a number of insects, the resulting curve is relatively uniform and destinctive. The free amino acids found in the blood in the largest qualities are not these amino acids essential to the nutrition of the insect. From this, it can be suggested that the high amino acid concentration in the blood of insects represents the storage of nitrogenous materials that can be drawn on, according to the needs of the tissues; or it may indicate that an excess of amino acids ae produced from the diet and the that the amino acids are stored in the blood until they can be eliminated by normal excretion. A great deal of information is available on what takes place in the amino acid metabolism of the insect, but relatively little is known about why it takes place.

The Sodium-potassium Ratio

That this ratio is inverted in some insects has led to speculation on its importance, but most of this has ben due to the popularity of emphasizing the exotic as it is observed in insects. Some insects that feed on plants with an unusually high potassium content show this inversion, but insects that feed on diets with less potassium do not. Bone showed that the potassium content of haemolymph is a function of the amount of potassium ingested with the food. Tobias reported further information on this factor for the American cockroach. His interest in the subject came from the study of water interchange of tissues, which interchange shouldbe altered by the amount of potassium in the blood. From these data, it can be stated that the serum of the American cockroach has more sodium than potassium but that the ratio is (6.2) is low compared to that of man (29.2). Tobias did find, however, that irritable tissues of the cockroach can function at a relatively high serum potassium level; but it can also be shown that a saline solution with an inverted sodium-potassium ratio is quite toxic to dissected preparations, particularly those made using insects in which the inversion is not normal. Duchateau arrived at the following conclusions in discussing the mineral salt content of the blood of insects.

The results of determinations of the sodium, potassium, calcium and magnesium concentration of the blood of a series of insects indicate the existence of two types of blood, which are extremes. the first blood type is probably the primitive type, which may be characterized by the Odonata. It is like that of other animals in which the concentration of sodium is somewhat greater than the concentration of potassium and magnesium. The second blood type is interpreted as a bio-chemical specialization, which is characterized by the Lepidoptera, Coleoptera and some other groups. In these groups, there is a low concentration of sodium and a high concentration of potassium and magnesium. The other insects fall more or less into one or the other of these types as a function of their evolutionary development and their nutritional habits. The content of inorganic bases in the food causes a departure from this normal condition, and insects deviate in varying degrees dependent upon the concentration or dilution of these inorganic bases in the blood but end up with a specific composition of the base as one might predict from the taxonomic position. The high content of magnesium in the blood of insects as compared to that in some other invertebrates is worthy of attention since this element has an anesthetic effect upon most invertebrates. This anesthetic effect is not important in insects.

Uric Acid in Insect Haemolymph

Most insects are uricotelic. Uric acid is produced princiaplly in the cells of the fat body and released into the blood, which transports it to the Malpighian tubes tobe excreted. In a normally functioning insect, the ability of hte Malpighian tubes to remove uric acid is balanced by uric acid production, and the concentration of uric acid remains at about the saturation point. When a disturbance causes the blood volume to decrease because of water loss, if the excretory function is impaired or if the metabolic rate is unusually high, crystals of uric acid may appear in the blood, indicating a supersaturation. The high uric acid content of the blood probably has no significance beyond indicating that the serum fraction functions as a transport mechanism.

Other Differences in Insect Blood

Two more differences between the chemical composition of the blood of insects and the serum of humans are apparent from the data in table. One is the low chloride content of insect blood, which is much too low to account for combination of the base elements to form chlorides. The other difference is te presence in relatively large quantities of the glucoside, trehalose. It is usually assumed that the basic elements in animal fluids are present in the form of chlorides, and the basic elements are furnished in the form of chlorides in most physiological salines and in diets. The analyses reported of insect fluids do not show enough chlorine to nearly account for the amount of sodium, potassium, calcium, and magnesium if these were to exist as chlorides. Thus, it is logical to assume that these elements are present, at least in part, as carbonates, bicarbonates, or phosphates.

This idea is of use later in explaining other phenomena that take place in insects, and it might be borne in mind when preparing experimental salines for in vitro work with insects. Trachalose is a glucoside, (α-D-glucosido)-α-D-glucoside, which has been found in cocoons of certain parasitic beetles for some years, and it is also found in nature in certain fungi. Trehalose is soluble in water and does not reduce copper but it fermentable by yeast. As the technology for analysis improves, trehalose likely with be reported from the blood of more and more insect species, in which it may serve as a storage form of carbohydrate, possibly replacing glycogen. Trehalose is not found in the blood of mammals.

The Proteins of Insect Blood

The total blood protein has been recorded for a number of species. These have been tabulated by Buck. Wide discrepancies exist in the values for different species, and many of them show a low protein level. A reason for these differences lies in the analytical prodedure and the method of collecting blood samples. These procedure may exaggerate differences that exist, and in some cases the low values may be artifacts. Patton demonstrated that heat

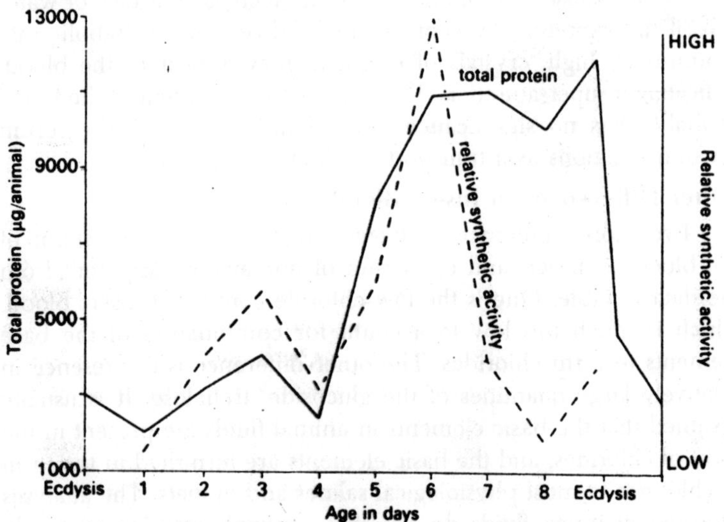

Fig. 5.7. Changes in total haemolymph protein and relative rate of synthesis of protein from amino acids in the haemolymph during the fifth instar of Locusta.

fixation before sampling reduces the blood protein titer of the gray house cricket (*Acheta domesticus*) by a factor of six. The protein in the blood are usually conjugated. They are intimately bound to or associated with triglycerides, phospholipids, sterols, or other non-protein compounds.

The properties of insect blood proteins have been reported in a number of miscellaneous papers. Blood proteins have been classified on the basis of their solubility in salt solutions, by serologic reactions, by ultracentrifugation, by free electrophoresis, by histochemistry, and by paper and starch block electrophoresis. The results of a series of studies using these methods were reported by Siakotos. Five factions of proteins were separated from the blood of the American cockroach. By various staining procedures these proteins were shown to be conjugated.

Two fractions (II and IV) were readily identified by their phospholipid, carbohydrate and protein composition. Fraction III, characterized by a high neutral lipid and sterol content, was found to be present in greater quantity in males than females. Fraction I also contains neutral lipid and sterol but has a greater mobility than III. Fraction V is the least mobile and has properties similar to those of human fibrinogen when subjected to electrophoresis. Siakotos was quite specific in pointing out that the electrophoretic similarities observed between insect and human proteins do not necessarily indicate functional similarities. By its electrophoretic properties, fraction I is comparable to albumin, and fraction II to α-globulin, and reactions III, IV and V are like β-globulins. From these observations it was concluded that the β- globulin-like proteins may be important in the transport of lipid to and from the fat reserves. The plasma glycoproteins may be important in the transport of carbohydrate particularly during the molting process. Except for osmotic control and transport of unsterified fatty acids, which are controlled by the albumin fraction of vertebrate blood, the fractions of insect blood and vertebrate blood appear to have comparable functions when the adjustments necessary to compensate for the requirements of the insect system are considered.

Haemocyte Functions

Although quite a variety of functions have been attributed to haemocytes, only four are well documented and generally accepted. They are (1) phagocytosis of small particles, (2) encapsulation of large, foreign objects (3) coagulation of the blood by cellular agglutination and/or by contributing to plasma precipiration, and (4) storage and distribution of nutritive materials. These are basic functions through which the haemocytes may become involved in a number of different activities such as the healing of wounds, nodule formation about clumps of foreign particles, the removal of lysed tissues, and so forth, ad some less well-demonstrated actions such as tissue formation, control of hormone action, detoxication of chemicals, and melanization of cuticle.

Phagocytosis

Over the years it has been demonstrated numerous times that the haemocytes of many insects can engulf a wide variety of naturally occurring and injected foreign particles within the size range of microorganisms. On the list of engulfed materials are erythrocytes, polystyrene beads, yeast cell walls, a variety of microorganisms and virus particles, and some autolyzing tissues. In most insects the plasmatocytes are the most active phagocytes, but in some the granular haemocytes (mainly adiphaemocytes) are quite active. Prohaemocytes are rarely phagocytic, and there are few references to phagocytosis spherule cells and oenocytoids. Strangely enough, the evidence for phagocytosis is largely circumstantial. The activity has not been oberved in its entirely in *vivo* or *in vitro*, yet when suitable particles are injected into the organism they are founder later in the cytoplasm of the haemocytes.

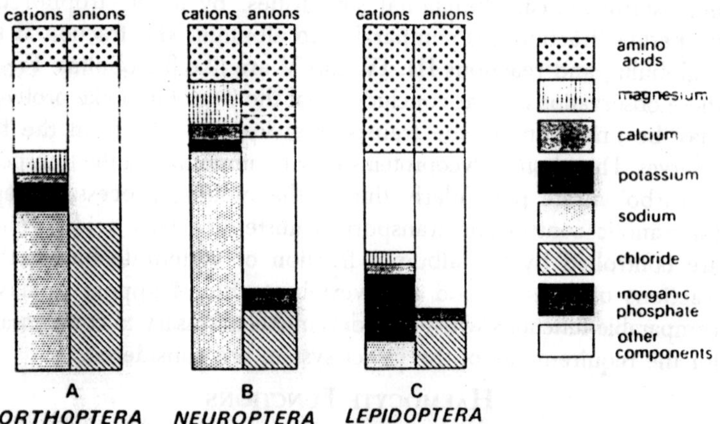

Fig. 5.8. Osmotic components of the haemolymph in different groups of insects expressed as percentage of the total osmolar concentration. Each vertical column represents 50% of the total concentration.

In the case of some virus particles, entrance into the haemocytes may not be by phagocytosis. The act of phagocytosis is assumed to be the same for haemocytes as for mammalian leukocytes or for *Amoeba*. Salt (1970), in his excellent monograph on cellular defense reactions of insects, infers that haemocytes ingest particles in at least three ways: (1) by formation of pinocytotic vesicles to engulf fluid that contains small particles; (2) by encircling particles with pseudopodia; (3) by close contact and spreading of the plasma membrane.

There is no evidence that haemocytes direct their movements toward particles, but there is some evidence that they can discriminate between particles, at least between normal and modified erythrocytes. Although phagocytosis is often though to be an efficient defense mechanism against invading microorganisms, its effectiveness is quite variable. It may be inactive or only feebly active against some organisms, or it may be very efficient. It is said to be more efficient in immunized than in normal insects, but Stephens (1963) found no such effect. She considered immunity to be entirely humoral. Phagocytosis may be important also in the removal of autolyzed tissues, or it may play no part. Adipohaemocytes are said to be involved in the destruction of neurilamella. Against some microorganisms, the phagocytic cells are themselves destroyed. Toxic proteinase from *Pseudomonus aeruginosa* causes death of haemocytes in larvae of *Galleria*. Nevertheless, the potential efficiency of phagocytes seems great. Single cells can ingest a number of particles. The mean number of cells that ingest injected particles (Phagocytic index) is usually very high, and the blood is often cleared of injected organisms within a short period.

Furthermore, insects are reported to be more suceptible to infection when the activity of phagocytes has been blocked. On the other hand, the phagocytic defense mechanism is to no vail against a number of bacilli, including *B. thuringiensis*. *B. cereus*, some microsporidia a such as *Nosema sp.*, some viruses, and certain fungi, *Sorosporella uvella*: *Coelomomyces sp.* However, no change was observed in the hemocyte picture during progression of nucleopolyhedrosis. Of course it is difficult to determine the effectiveness of the mechanism in nature.

Encapsulation

Encapsulation of large objects by haemocytes has long been considered a primary defense reaction of insects to internal metazoan parasites. The effectiveness of the reaction has been questioned by a number of authors, but there is little doubt that healthy insects, especially Lepidoptera, encapsulate any parasite that has not developed some means of avoiding the cells. Hemiptera seem to lack an efficient system of encapsulation, possibly due to the paucity of haemocytes. The entire subject has been throughly discussed by Salt (1970), whose work is a constant source of reference in this short presentation. Despite some suggestions to the contrary, encapsulation of foreign bodies is carried out exclusively by haemocytes. Plasmatocytes are the cells chiefly concerned in most insects, but lamellocytes (modified plasmatocytes) and crystals cells (oenocytoids) are involved in some Diptera.

An association with spherule cells has been inferred in the case of one Lepidoptera. The reaction appears to be general among insects, but there are obvious differences in this vigor at different stages of the insect's development, in different species, and in population from, separate localities. Some of the differences are likely due to variations in the number of haemocytes available at different times and in different species. This is indicated by the breakdown in immunity in superparasitized larvae. The reaction is normally evoked by a wide variety of endophagous Metazoa, and is essentially a reaction of the cells against all foreign bodies that are too large to be phagocytized. It can be evoked also by a wide variety of implanted bodies, including dead parasites, glass rods, nylon thread, transplanted organs, and by injured tissues of the host. Surface properties of the objects seem to be the critical factor in inciting encapsulation, but in some cases seem not to be involved. Some natural invaders are not encapsulated because of specific adaptations which enable them to avoid or suppress the reaction. Some habitual parasites ae encapsulated without challenge and without effect.

Sometimes immunity to a parasite does not involve a hemocyte reaction. The possibility that the haemocytes exhibit chemotaxis in response to parasitization is raised by Nappi and Stoffolano (1972a). The most common reaction, the development of cellular capsules, is evident where haemocytes cover the foreign body in a consolidated mass, 50 cells or more thick. Such capsules are usually made up of three layers. Cells in the inner and outer layers are mostly rounded or lightly flattened. The middle layer comprises about half the total thickness, and is composed of excessively flattened cells, but there is no evidence of loss of cellular integrity in any layer. Intercellular substances, mainly mucopoly-saccharides, seem involved in, the initial cohesion and eventual cementing of the cells together. Melanization occurs at different intervals after encapsulation and close to the surface of the object. There is strong evidence that the chemical involved is indeed melanin, and that it derives from the blood cells, perhaps especially from oenocytoids. There is considerable variations in the incidence of melanization and in its contribution to the death of parasites. In some it forms so long after the capsule that it seems to contribute little. In a few it is essential for the death of parasites.

In some it forms so long after the capsule that it seems to contribute little. In a few it is essential for the death of the parasite, and in *Drosophila* it quickly stops development of the hymenotperous parasite *Pseudeucoila bochei.* Melanization is said to be involved also in the formation of tumbours in *Drosophila.* Another reaction, the formation of seheath capsules, has features of cellulr encapsulation and melanization. It is characterized by rapid development of a tough, thin, pigmented layer at the surface of the object, and overlaid by relatively few layers of cells. Capsules that answer this description has been reported by a number of authors for several hosts and parasites, but mainly where the objects are eggs or nematodes.

Habitual parasites have evolved a variety of devices to avoid reactions of their hosts, and there is recent evidence that some may cause deterioration of the host haemocytes. Brief mention should be made also of "nodule formation" by haemocytes, a reaction not always distinguished from capsule formation. This too has been examined by Salt (1970), and appears not to be an especially significant defense reaction. It combines features of phagocytosis and encapsulation to isolate clumps of bacteria. It seems to be the type of structure found also in tumorlike lesions.

Coagulation

This important subject is dealt with separately in this text, but bears mention here for the sake of continuity. Some years ago, Yeager and Knight (1933) arranged quite a number of insects into three groups according to the manner in which their blood coagulated. In one group coagulation was a feature of the plasma alone. In another it was largely by cellular agglutination. In the third both phenomena occurred. Since then, there has been no cause to change their view, and a somewhat similar separation was made by Gregoire (1961) from an examination of several hundred species. Gregoire and his colleagues have found three main patterns of clotting among insects as well as some insects where no clotting occurs. For each pattern, special blood cells (coagulocytes = cystocytes?) are involved. They act either by bursting and releasing materials (probably from mitochondria, Frake, 1960) which form islets of coagulation, or by extruding threads of cytoplasm which form a filopodial network where other cells adhere and where the plasma gelates in the form of thin plastic veils. Early attempts to explain coagulation in insects in terms of the mechanism found in mammalian blood have failed.

The process seems to involve cell-serum interactions and special forms of cellular adhesions which can occur independetly or together in the same or in different species. This seems to be general among arthropods. In *Calliphora,* coagulation can occur in the absence of haemocytes. There seems to be some question yet on the individuality of coagulocytes. The question is simply whether the cells are separate entities with a very specific function, or hyaline cells which exhibit short-lived physiological responses. At the fine structure level. Lai-Fook (1970) found that plasmatocytes of *Rhodnius* resemble coagulocytes, granulocytes, and early oenocytoids of *Locuste* as described by Hoffmann et.al. (1968a,c) and suggested that the degree of density and homogeneity of granules in haemocytes depends on the physiological state of the cell. She suggests too that the changes in granules of plasmatocytes in *Rhodnius* after wounding resemble those which occur in the clotting of blood cells in *Limulus* (Merostomata). Taylor (1969) also found a similarity between the granules of haemocytes in lesions in the cockroach *Leucophaea* and those of haemocytes involved in clotting in *Limulus.*

Storage and Distribution of Nutritive Materials

Although the haemocytes are not commonly considered considered as a major storage tissue, there is no doubt that in some insect they play significant role in this capacity. Their numbers in some species are large, and their inclusions occupy a large portion of the cell volume. Viewed in this way, they are seen to be capable of storing a considerable bulk of material. The have been reported to contain a variety of nutritive substances, such as glycogen, lipids, glycoproteins, and mucopolysaccharides. All types of haemocytes are involved, including the prohaemocytes, but the various kinds of granular cells are especially notable. Spherule cells have obvious potential as storage vehicles. Few enzymes have been identified, but they can likely be revealed by suitable techniques.

Evidence is fairly strong that the haemocytes contribute some or all of these materials on occasion to various developing tissues, but there is a real need to determine the manner, extent, and result of the exchanges. Zeller's description (1938) of hemocyte involvement in the development of wings in *Ephestia* gives fairly strong evidence for a nutritive function. It showed the haemocytes adhering to the middle membrane of the wing, their distinegration, and the reappearance of their contents in the membrane. Evidence is derived too from the work of Yeager and Munson (1941, 1944) where a direct relationship is shown between food intake by the insect and glycogen content of the haemocytes. The view is supported also by Wigglesworth's suggestion (1959) that changes in *Rhodinus* plasmatocytes after feeding may signal the beginning of synthesis of mucopolysaccharide inclusions, and Whitten's observation (1964, 1969) that a direct transfer of osmiophilic hemocyte inclusions to hypodermal cells takes place at metamorphosis is *Sarcophaga.*

Haemocyte Activities

One or more of the four acknowledge functions of haemocytes may be involved in a number of hemocyte activities, i.e., roles that the haemocytes play in conjunction with other tissues. Functions that have received particular notice are the healing of wounds, formation of membranes, regulations of growth and development, detoxication of poisons, and the tanning of cuticle.

Healing of Wounds

Clearly, the haemocytes in some insects play an important part in initiating the healing of internal and external wounds. Their actions in this capacity include accumulation at the site of the wound and plugging of the opening, increase in numbers by mitosis, phagocytosis of cellular debris and formation of a layer of cells. Plasmatocytes are the main cells involved, but Gupta (1970) suggests that spherule cells too are engaged in phagocytizing the disintegrating tissues at the wound site. The formation of tissue over the wound is through to come about in somewhat the same way that membranes form *in vivo* and *in vitro.* Plasmatocytes ae released quickly from depots within the insect and aid in would closure. Their numbers increase further by mitosis. At first they move about the wound site, then gradually arrange themselves to form a nearly continuous sheet.

A fairly stable configration of the cells is achieved later when the membrane becomes multilayered. The epithelioid character-of haemocytes in reaction to injury has been noted by Dawe *et.al.* (1967). The same principle may be involved in the aggregation of haemocytes at sites of nerve transaction in *Periplaneta.* There the cell aggregate forms no tissue, but provide substance upon which the regenerating nerve fibers advance.

Membrane Formation

Membrane formation at a would site is not unlike the formation of connective tissue by haemocytes. Wigglesworth noted also the migration of large numbers of plasmatocytes to sites of membrane formation e.g., beneath the epidiermis just before the new cuticle forms, around the fat-body, and around developing muscles. Scharrer too (1965b) suggests that haemocytes in the prothoracic gland of *Leucophaea* form and maintain the connective-tissue matrix Lai-Fook (1970) has shown that haemocytes can form a continuous layer, and that the cells have a normal form of cellular junction as seen at the ultrastructural level. On the other hand, haemocytes were not involved in the formation of connective tissue around the adult nervous system in *Galleria,* or in the embryo of *Schistocerca.*

Regulation of Growth

The involvement of haemocytes in regulating growth and molting has been claimed mainly for *Rhodinus.* The suggestion is based on observations of the increase in number and size of oenocytoids during molting, and on experiments where hemocyte activity was blocked by injection of various substances. Blockage caused a long delay in molting it it occurred within three days after feeding. Apparently the action involved prevention of full activation of the thoracic gland by the hormone from the brain. It was suggested that the haemocytes might normally produce a factor that either activates the gland or aids some way in the transport of brain hormone to the gland. This concept is supported somewhat by the finding of Hodgson and Geldiay (1959) that haemocytes occur in the brain of hyperactive cockroaches and may be involved in transporting neurosecretory materials. Similarly, the note by Scharrer (1965) that haemocytes occur within prothoracic glands of the cockroach *Leucophaea* suggests that the cells may facilitate the transport of specific substances to and from endocrine cells.

Detoxication of Poisons

The reaction of haemocytes to various poisons has studied by a number of authors. In general, these studies show that the haemocytes are damaged in various ways and that there are changes in the hemograms, but seldon indicate a direct mechanism of defense against the toxicants. This seems to be the conclusion reached by Wittig (1962). Arnold (1952b) indicated a mechanism of recovery that supposedly involved metabolism of lipids, but only a few authors have suggested that the cells themselves are active in detoxication. Patton's work indicated that nonspecific esterases in the haemocytes might be involved in detoxifying parathion. Whitten (1969) too suggests an active role in detoxication of contact insecticides that penetrate the tarsal cuticle.

Phenol Metabolism

The darkening of blood on exposure to air is presumably due to the action of polyphenoloxidase on a phenolic substance in the haemolymph. The involvement of haemocytes in phenol metabolism and thus in the process of hardening and darkening of the cuticle and in the melanization has been indicated by the presence of tyrosinase in some of them. Tyrosinase has been reported from spherule cells in Diptera and in oenocytoids from other insects. In *Drosophila,* Rizki and Rizki (1959) observed that trysone occurs in the crystalline inclusions of crystals cells (oenocytoids) while tyrosinase is in the surrounding cytoplasm, possibly associated with mitochondrial or microsomal structures. They suggest that any factor that causes returning of the cells allows the enzyme and substrate to come together. A somewhat similar observation is made in the beetle *Diabrotica.* There the haemocytes lyse on contact with the cuticle of a parasitic nematode and their cytoplasm transforms into the melanin layer around it. Steps in the tanning of egg capsule in *Periplaneta* are suggested by the presence of tryosine decarboxylase in haemocytes and the implied biosynthesis of protocatechuic acid from tyrosine in the collateral system. Precisely how the process occurs is not established. On the other hand, Stephens (1963) noted that blood of actively immunized *Galleria* does not darken in air, even though the number of cells containing tyrosinase is unchanged.

Inorganic Constituents

Chloride is the most abundant inorganic anion in insect blood. The concentration of chloride is high in Apterygota and hemimetabolous insects, but is characteristically low, usually amounting to less than 10% of the total asmolar concentration, in holometabolous forms. Other inorganic anion; present are carbonate and phosphate, but these are rarely found in any quantity. Phosphates are important in *Carausius* and *Anabrus* (Orthoptera). The most abundant cation is usually sodium, but in Lepidoptera, Hymenoptera and some Coleoptera there is relatively little and it contributes only about 10% of the total osmolar concentration.

The absolute concentration of potassium is usually lower than that of sodium, 2-10% of the total osomolar concentrations, but the Na/K ratio varies considerably. In *Petrobius* Na/K -40 and relatively high values occur in Odonata, Orthoptera, Diptera and some Coleoptera, but in phytophagous Coleoptera, Hymenoptera and Lepidoptera Na/K is much lower and may be less than one. It is suggested that these last groups evolved with the angiosperms which contained relatively little sodium and the adaptation of the insects to a lower Na/K is much lower and may be less than one. It is suggested that these last groups evolved with the angiosperms which contained relatively little sodium and the adaptation of the insects to a lower Na/K ratio reduced the amount of regulation necessary to maintain a more or less constant level in the blood. Hoyle (1954) has shown that the level of potassium in the blood influences the activity of locusts. A low level of patassium raises the muscle resting potential.

As a result the action potential arising from stimulation is higher, the twitch tension of the muscle is increased and the insect will jump farther or much more readily than when the potassium concentration is high. Ellis and Hoyle suggested that feeding raised the potassium level in the blood and hence produced the characteristic period of quiescence after a meal, but field work suggests that the potassium concentration in the blood is influenced more by changes in the blood volume than by the potassium concentration in the food. The potassium concentration increases markedly before moulting and this is probably responsible for much of the inactivity occurring at the time. The concentration of magnesium in the blood is often relatively high. In phytophagous insects this to some extent reflects the high level of magnesium in the diet, since it is a constituent of chlorophyll, and in Lepidoptera the level in the blood falls when the larvae stop feeding. A good deal of magnesium still occurs, however, and all insects appear to concentrate this metal.

In Phasmida it almost completely replaces sodium. calcium is usually less important than the previous metallic elements but is essential for the development of the end plate potential at the muscle.

Table. 5.4. Concentrations, in Milli-equivalents/litre, of the Major Inorganic Cations in the Haemolymph of Two Insects in Different Stages of Development.

Insect	*Stage*	*Cation*			
		Na	K	Ca	Mg
	larva	15	46	24	101
Bombyx	pupa	11	41	24	69
		22	55	29	87
	adult	14	36	14	47
	larva	26	56	19	24
		48	41	–	–
Vespula	pupa	23	61	11	19
	adult	93	18	2	3
		153	22	2	1

In many insects the ionic concentrations in the blood are roughly similar in larval and adult stages, although adult Lepidoptera have less magnesium than the larvae. There is relatively little information, however, on different stages of holometabolous insects and in *Vespula* the concentrations change markedly between the pupal and adult stages, the sodium concentration increasing, while all the other cations decrease. Various metallic trace elements are also found in the blood. The most frequent are copper, which is a constituent of tyrosine, iron, present in the cytochromes, zinc and manganese. It is possible that the metallic elements do not exist wholly as free ions, but that a proportions is bound in organic complexes. In *Anteraea* there is no evidence for any binding of potassium, but 15-20% of calcium and magnesium is bound to macro-molecules.

Organic Constituents

Insect blood is characterised by the very high level of amino acids present in the plasma. Most of the known amino acids have been recognised in various insects, but they vary considerably both qualitatively and quantitatively from one species to another and in different stages of the same species. To some extent the amino acids present depend on those available in the food. In general, members of the more primitive inect orders possess fewer amino acids. Most insects have high concentrations of glutamine, which may be concerned in uric acid synthesis, proline and one or more of arginine, lysine and histidine. Amino acids constitute 35-65% of the non-protein nitrogen present in the blood. The concentrations of amino acids may change at different stages in the life cycle.

Tyrosine, for instance, commonly accumulates before each moult and then decreases shrply as it is used in tanning and melanisation of the new cuticle. Similarly in the larva of *Bombyx* the amino acids, such as glutamic acid and aspartic acid, which are concerned with silk production fall to a very low level while the larva is spinning its cocoon, increasing again when this is complete. Others such as methionine, are not concerned with silk production, but increases at the time of the pupal moult as a result of the histolysis of the larval tissues. In *Rhodnius,* however, the amino acid concentration in the haemolymph rises after feeding, but then remains constant right through the period of moulting. It appears that in this insect the utilisation of amino acids is offset by the slow, continuous digestion of the stored blood meal. Other non-protein nitrogen in the plasma is mainly in the form of the end products of nitrogen metabolism.

Uric acid is always present and in addition there may be allantoin, urea and ammonia. Apart from sometimes constituting half the total carbohydrate in the blood. Numerous proteins are present in the haemolymph; 19 are recorded from *Drosophila* and 21 from *Locusta.* They are not all present at the same time, but there are progressive changes through the life cycle of the insects. For instance, the first instar larva of *Locusta* emerges from the egg with seven different haemolymph protein fractions. Within a day or two more fractions (16 and 17) have developed and these increase in concentration, but before the end of the instar four of the original fractions have disappeared (1, 4, 6 and 11). These probably represent the remains of yolk protein which are being utilised. Another protein band regularly reaches a maximum at the time of moulting and then disappears again (14). The significance of the haemolymph proteins is not always clear.

In the early larva of *Malacosoma* none of these proteins is detectable in the tissues, but at the onset of pupation they are selectively and differentially absorbed by the fat body, the wall of the midgut and the muscles of the heart. Similarly in *Hyalophora* a haemolymph protein which develops in the female pupa is selectively absorbed by the oocytes. The uptake of thse proteins at metamorphosis and in oogenesis may be under hormonal control. Other proteins, which do not appear in the tissues, may be broken down to provide a source of amino acids which are then resynthesised *in situ* in the tissues, while others may be enzymes. Just before puparium formation in *Sarcophaga* the enzyme tyrosinase is released into the plasma from the blood cells. Trehalase and other carbohydrases are also present as well as various other enzymes, some of which probably leak out from the surrounding tissues. Non-amino organic acids may be present in some quantity in the plasma.

Citrate is usually present in high concentration, although this varies considerably from one species to another, and there is consistenly more in the larva than in the adult. In *Prodenia* the level of citrate is not markedly affected by diet so that it must be endogenous in origin. The organic phosphate concentration in insect blood is also usually high and in *Hyalophora* a-glycerophosphate and phosphocholine, in particular, contribute to this. There is characteristically a high concentration of trehalose, a non-reducing disaccharide, in insect blood. Trachalose is a source of energy so that its level in the blood is reduced by starvation and also by activity, such as flight, but it is increased after feeding because other sugars are converted to it. *Apis* is exceptional in having high concentrations of glucose and fructose in the blood. Glycerol, or some equivalent, is probably always present, sometimes in very high concentrations, in insects which are able to tolerate freezing. It is well-known that glycerol helps to preserve living tissues in the forzen state, although the way in which it acts is uncertain.

Variations in Plasma in Different Insects Orders

Sutcliffe (1963) groups the plasma of pterygote insects into three broad categories:

1. Sodium and chloride account for most of the osmolar concentration. This is probably the basic type of insect blood and is similar to that in most other arthropods. This type occurs in Ephemeroptera. Odonata, Plecoptera, Orthoptera and Homoptera.
2. Chloride is low relative to sodium which constitutes 21-48% of the total osmolar concentration. Amino acids are also present

in high concentration. This type is found in Trichoptera, Diptera, Megaloptera, Neuroptera, Mecoptera and most Coeloptera.

3. Amino acids account for about 40% of the total osmolar concentration. There is a large unknown factor, but one of the other substances accounts for more than 10% of the total. Lepidoptera and Hymenoptera have this type of blood.

Pigments

Apart from *Chironomus* larvae in which hemoglobin is present in solution in the plasma, respiratory pigments do not occur in insect blood. The blood may, nevertheless, assume a variety of colour due to various pigments. Commonly the blood is green due to the presence of insectoverdin, a mixture of carotenoid and bile pigment. Insectoverdin is present in the blood of Lepidoptera and is solitary locusts, but in locusts it becomes altered with starvation as the pigments are metabolised and if the insects are crowded together the insectoverdin is replaced by blue mesobiliverdin. In the blood of *Bombyx* larva carotene, xanthophyll and flavines are present; in the pupa of *Hyalophora* are a-carotene, taraxanthin, riboflavin and chlorophyll. Aphids frequently have a purplish-red pigment in the blood in large quantities, up to 2% of the wet weight. This pigment is protoaphin belonging to a class of pigments, the aphins, so far found only in members of the Aphididae and the genus *Adelges*.

The Blood during Metamorphosis

During the period of transformation from the larva to the imago in holometabolous insects, the blood becomes so charged with the debris of disintegrating tissues that the pupa appears to be filled with a thick, creamy pulp. At the beginning of metamorphosis the cells of the fat body detach themselves from one another and float free in the body liquid. Very soon they undergo a dissolution by which their contents, consisting now principally of fat globules an great numbers of small protein granules, are set free in the blood. Among this mass of liberated material there may still be seen a few normal haemocytes. With many insects, during the time of metamorphosis, the prevailing type of blood cells are the phagocytes, which are often distended with particles of degenerating tissues, especially muscle fragments. In the pupa of Diptera spherical bodies filled with muscle fragments (*sarcolytes*) are particularly abundant and are known as *spherules of granules,* or *Kornchenkugeln.*

Most investigators have regarded these sperules as blood phagocytes gorged with ingested sarcolytes. It cannot always be demonstrated, however, that such bodies are of a cellular nature, and the writer (1924) has found that sarcolyte spherules, occurring in great abundace in the larva of the apple fruit fly *Rhagoletis pomonella* are almost certainly mere masses of sarcolytes given off directly from muscles undergoing histolysis. These sarcolyte spherules eventually break up and add their contents to the accumulation of material already in the blood. Included in the later, besides the bodies already mentioned, are also numerous small nucleated masses of protoplasm (*caryolytes*) probably derived from the muscles, and great numbers of minute grains and other unidentifiable fragments of disintegrating tissues. Toward the end of the pupal period these temporary inclusions of the blood begin to disappear, probably going into solution by complete histolysis, and, shortly after the transformation to the adult, the blood again becomes clear and regains its normal condition.

Clotting of Blood

With most insects when the blood issues from a small wound it thickens and forms a viscous plug closing the aperture in the integument. This "clotting" of the insect blood has well been studied. Yeager and Knight after investigating the blood of 47 species, found that insects can be divided into three groups according to the clotting properties of the blood, as follows: (1) species in which clotting does not take place, (2) species in which clotting is produced mainly by an agglutination of the haemocytes, and (3) species in which clotting is principally a coagulation of the plasma. The grouping of species according to blood clotting has no relation to taxonomic classification. Insects having no clotting of the blood occur among Homoptera, Coleoptera, Lepidoptera and Hymenoptera. The honey bee larva furnishes a typical example in this class, the nonclotting properties of its blood having been shown by Bishop, Briggs and Rosoni (1925).

Insects forming a cellular coagulum include orthoptera, Homoptera, Coleoptera, Lepidoptera, Hymenoptera, and Diptera; those in which the blood clot is principally a plasmatic coagulum occur in Heteroptera, Orthoptera, Coleoptera and Lepidoptera. Where clotting consists essentially of a coagulum formed by the blood cells, as in the cockroach, described by Yeager, Shull, and Farrar, the haemocytes throw out fine threadlike pseudopodia and become agglutinated into clumps, which spread out and appear to disintegrate. The plasma undergoes little change visible under the microscope and exhibits no fibrin formation, but there appears in it a granular precipitate. When the haemocytes take on part in the clotting, a fibrous coagulum is formed in the plasma. An intermediate type of clotting involving both the plasma and the haemocytes is described by Muttkowski (1924a). In this case, the gelatin of the blood coagulates and the haemocytes become agglutinated, while those of the amoebiform type throw out pseudopodia and form a meshwork which contracts tighly and draws the lips of the wound together. In addition, however, a fibrous net is formed among the haemocytes, which, by contraction, brings about a still better closure of the wound and a complete stoppage of the blood flow. Muttkowski was able to demonstrate the effectiveness of the haemolymph alone to produce a clot by killing the haemocytes with cyanide fumes.

6

Circulatory System

The Organs of Circulation

The primary blood cavity of an animal is the embryonic haemocoele, or body space between the ectoderm and endoderm, which is the remnant of the earlier blastocoele left after the invasion of the latter by the gastric endoderm. The haemocoele in triploblastic animals is then again invaded by the mesoderm and in most cases is finally reduced to tracts enclosed in mesodermal walls, whch constitute the blood vessels. The cardiac sinus of the young insect embryo is the dorsal part of the haemeocle lying above the yolk or the alimentary canal between the upper ends of lateral sheets of mesoderm. As development proceeds, the mesoderm layers approach eachother from oppostie sides in the cardiac sinus; their opposing surfaces become hollowed lengthwise, and, when the two layers meet, the lips of their furrows unite to form a median dorsal tube. This mesodermal tube is the *dorsal blood vessel.* Its lumen is merely a restricted part of the cardiac sinus and is filled with blood liquid containing blood cells.

The dorsal vessle is the only part of the haemocoele of insects that is closed by definite walls; elsewhere the blood is contained in the body cavity because of the embryonic union of the ventral part of the haemocoele with the cavities of the coelomic sacs. In some of the other arthropods, however, as in the larger Crustacea and Arachnida, there may be present an elaborate system of blood vesses branching from the dorsal organ. The parts of the dorsal mesoderm not included in the cardiac rudiments extend lateral in the upper part of the body from the walls of the dorsal vessel and are drawn out into thin sheets of cells. These cells give rise to the transverse dorsal muscles of the body and to connective tissue membranes more or less uniting the muscle fibers. The two wings thus formed extending lateral from the dorsal vessel constitute the *dorsal diaphragm.* The muscles become attached to the body wall along the lateral parts of the dorsum; the membranes become continuous with the delicate peritoneal covering of the somatic muscles.

The dorsal diaphragm shuts off a space in the upper part of the body cavity containing the dorsal blood vessel, which is known as the *pericardial cavity,* or *dorsal sinus.* In some insects the ventral transverse body muscles, stretched between the lateral margins of the sternal region of the body wall, form a continuous series of

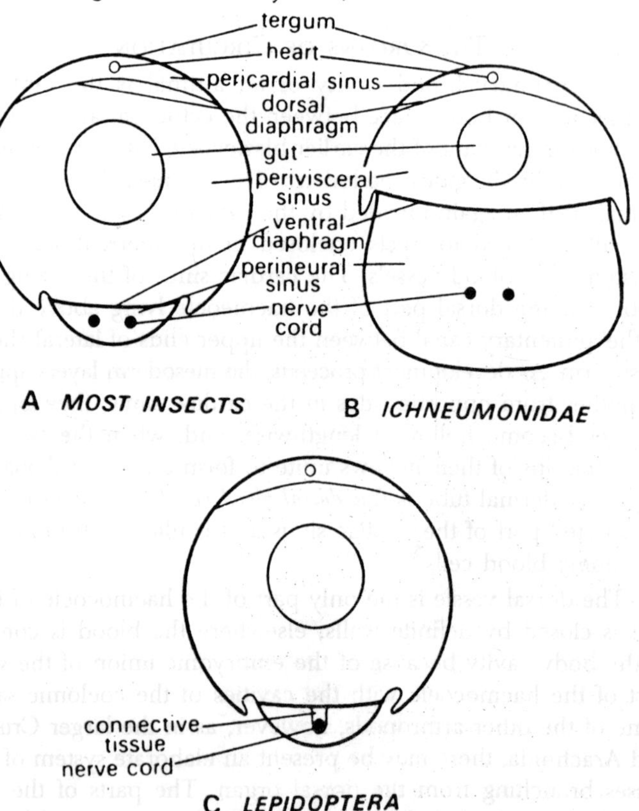

Fig. 6.1. Diagrammatic cross-sections of various insects showing the main sinuses of the haemocoel and the positions of heart, alimentary canal and nerve cord.

the fine fibers ofen held in a delicate membrane. This structure, when present, constitutes a *ventral diaphragm.* Beneath the ventral diaphragm is a free space enclosing the nerve cord and the ventral longitudinal muscles; it represents a part of the embryonic epineural sinus and is known as the *ventral sinus* in the adult anatomy. The body cavity between the dorsal diaphragm and the ventral diaphragm, when the latter is present, containing the principal visceral organs in the *perivisceral sinus.* The dorsal vessel is regularly pulsatile and is the principal organ by which the blood is kept in motion. The dorsal and ventral diaphragms, however, are rhythmically contractile and functions as important adjuncts to the dorsal vessel in the circulation of the blood. Finally, there are present in many insects other pulsating organs located in various parts of the body and in the appendages, which apparently serve to drive the blood into the extremites.

The Dorsal Vessel

The dorsal blood vessel when fully developed extends from the posterior end of the abdomen into the head. The organ is differentiated into an anterior part known as the *aorta,* and a posterior part called the *heart,* but, while the two sections are anatomically different, their distingusihing features are difficult to define. The heart is in general the pulsating part of the tube, though the aorta frequently is provided with pulsating vesicular diverticula, and, according to Wettinger (1927), the dorsal vessel of the larva of *Tipula selene* pulsates throughout its length. The cardiac section is usually confined to the abdomen, but it may extend into the thorax. The heart is characteristically chambered, that is, slightly dilated in each body segment, but the pulsating vesicles of the aorta also are segmental dilations that may be regarded as of the aorta also are segmental dilations that may be regarded as of the same nature as the aradiac chambers. The heart is provided with opening (ostia) for the admission of the blood, and the aorta likewise may have apertures for the same purpose.

The Heart

The cardiac part of the dorsal vessel is usually marked by the presence of more or less distinct symmetrical segmental swellings of the tube. These dilations are known as the *chambers* of the heart. Typically, each heart chamber has a pair of vertical or oblique slitlike openings, the *ostia,* in its lateral walls, one ostium on each side, placed generally behind the middle of the chamber, and

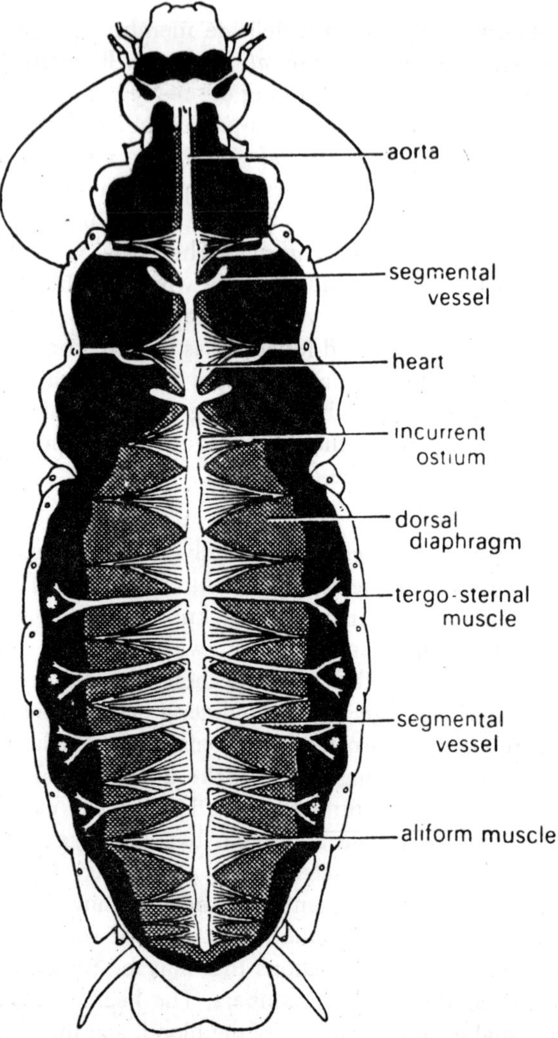

Fig. 6.2. Ventral dissection of Blaberus to show the dorsal and segmental vessels. The dorsal and segmental vessel.

sometimes close to the posterior end. In some insects cardiac chambers are separated by well-marked constrictions; in others they are apparent only as slightly widened parts of the vessel. In genral there are no internal valves between the heart chambers except the ostial flaps, to be described presently, but Wettinger (1927) reports the presence of well developed valvular structure in the heart walls of the larva of *Tipula selene* in addition to the sotial flaps, and Popovici-Baznosanu (1905) describes a pair of valves in the heart of certain chironomid larvae lying just before the last cardiac chamber, which evidently prevent the backward flow of the blood into the latter.

In some other chironomid larvae the same writer finds swellings of the heart wall between the consecutive pairs of ostia, but these structures he says are protruding heart cells and not valves. Valve-like folds of the heart wall in some case may be the lips of closed and otherwise functionless ostia. The chambered part of the dorsal vessel is usually limited to the abdominal region of the body and commonly begins in the second abdominal segment. In Blattidae, however, according to Brocher (1922), and *Japyx,* as shown by Grassi (1887), there is a typical cardiac chamber not only in the first abdominal segment but also in the third and the second segments of the thorax, each provided with pair of lateral ostia. In the larvae of Ephemerida, according to Popovici-Baznosanu (1905), a pair of ostia is present in the metathorax. Ampulla-like enlargements of the aorta occur in the two wing-bearing segments and in the first abdominal segment of various insects, sometimes provided with ostial openings, but these aortic swellings or diverticula do not have the usual structure of the cardiac chambers. The number of chambers in the abdominal section of the dorsal vessel may coincide with the number of segments occupied here by the tube, the maximum being nine.

But generally there is not a distinct chamber in the first segment, and the number of chambers may be variously reduced to a minimum of one, as in Mallophaga and Anoplura, though the single large terminal swelling of the heart in these insects in probably a compound chamber, since it contains two or three pairs of ostia (Fulmek, 1906). The heart of larval Odonata has a single large posterior chamber (Zawarzin, 1911; Brocher, 1917a) provided with two pairs of ostia in Aeschnidae and one pair in Agrionidae, and having two pairs of alary muscles. In the hemipteron *Nezara* the heart, as described by Malouf (1933), consists likewise of a single large swelling of the dorsal vessel, provided with three pairs of ostia. Though reduction in the number of chambers in the heart usually results form a suppression of the cardiac swelling in the anterior abdominal segments, the heart may be shortened also at the posterior end.

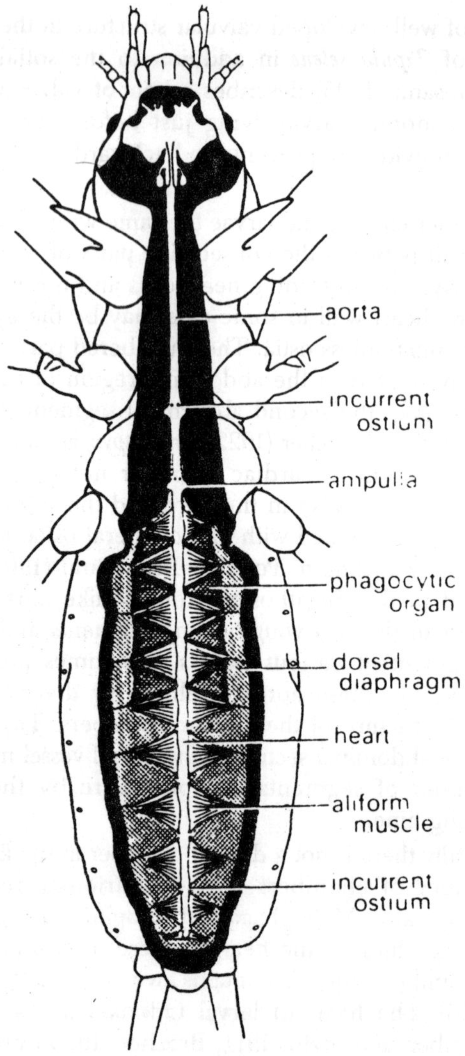

Fig. 6.3. Ventral dissection of Gryllotalpa to show the dorsal vessel and phagocytic organs.

The last chamber of the heart commonly ends either abruptly or in a narrowed tapering process that may extend posteriorly through the tenth abdomen segment. The end of the vessel in either case is generally closed; certain dipterous larvae (*Tipula, Chironomus*) the heart is aid to have a dorsal terminal opening, but it is possible that this median ostia represents the Ist pair of ostia united dorsally in a common aperture. In the larvae of the Ephemerida the heart terminates in three slender branches which open into the bases of the cerci and the median caudal filament. The typical cardiac ostia have usually the form of vertical or oblique slits in the sides of the segmental chambers of the heart.

Frequently the openings lie in deep lateral inflections of the heart walls, and these ostial pouches, projecting inward and forward within the lumen of the heart, form a series of paired valve-like flaps having the true ostia at their free inner ends. The flaps virtually mark the division of the cardiac tube into chambers. The ostial valves evidently are so contructed as to admit the blood into the heart when the wave of the diastole runs forward through the tube, and to prevent its escape during systole; they possibly serve also to obstruct a rearward flow of blood within the vessel; and yet the heart of certain insects is known at times to reverse the direction of its beat, and to cause a backward flow of blood when the systolic waves run caudad. The walls of the heart consists of muscle tissue immediately derived from the cardioblasts, or heart-forming cells of the embryonic mesoderm, which are converted into semicircular or circular straited muscle fibres that compose the heart tube. In larval insects the striations of the cardiac fibers are sometimes indistinct or not evident.

The ostial valves are also muscular, and therefore probably contract. The lining of the heart is simply the normal muscle surface, constant usually of a thin layer of hyaline sarcoplasm beneath a delicate membrane. Outside the heart there may be a sheath of adventition nective tissue. The vessle is usually suspended from the dorsal the abdomen by fine radiating strands attached to the epiderminal which appear to be filamentous processes of the heart wall. The mechanism of the heart is not fully understood. The concentrations of the organ are produced undoubtedly by the muscles heart walls.

Diastole, it is often assumed, is effected by the muscles of the diaphragm, but these muscles are attached to the ventral wall of the heart, and it has been observed in some cases that the heart continues to beat when the diaphragm muscles are cut. Since the systolic waves run through the tube, it would seem that successive chambers may be expanded by the blood forced into them. While ordinarily the succession of contractions in the dorsal vessel is in a forward direction, a temporary reversal of the heart beat has been observed in various insects. This appears to be a normal process and recently has been carefully studied by Gerould. A periodic reversal of the heart beat, Gerould finds, occurs particularly in Lepidoptera at the end of the larval period and may continue through the pupal and imaginal periods.

In the silkworm, phases of antiperistalsis begins about 48 hours before pupation, alternating with phases of forward beating. Pupation is accompanied by a vigorous backward pulsation, and during the pupal and imaginal periods the heart beats regularly with alternating forward and backward series of pulsations, the time of each phase being greatly shortened in the adult. A reversed flow of the blood during antiperistalsis, Gerould says, can be shown by injections of india ink. The dorsal vessl in innervated both from the occipital ganglion of the stomodeal nervous system and from ganglia of the ventral nerve cord.

The Aorta

The aorta is the slender part of the dorsal vessel continued forward from the first chamber of the heart into the head. Usually it begins in the first segment of the abdomen; but if the cardiac region of the vessel extends into the first abdominal segment or into the thorax, the aorta is correspondingly shortened. The aorta is not always a simple tube; in the honey bee, as it enters the thorax, it is thrown into a series of short loops closely bound together in a sheath of connective tissue; in the Lepidoptera it makes a large bend upoward in the mesothorax. Dorsal diverticula

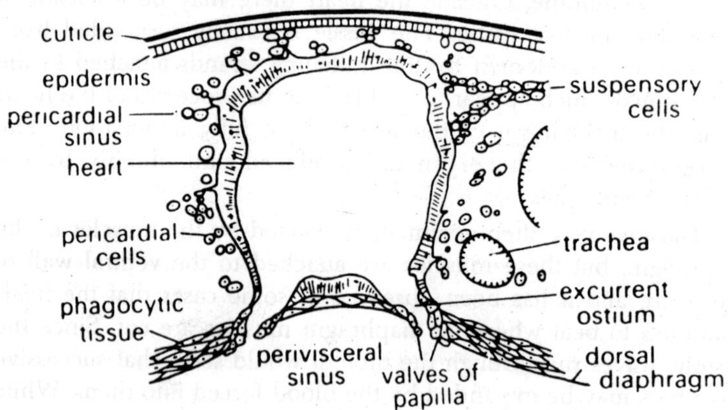

Fig. 6.4. Transverse section of the heart of Taeniopoda showing the excurrent ostia opening directly to the perivisceral sinus.

of the aorta occur in Odonata, Orthoptera, Coleoptera, and Lepidoptera. Within the head the aorta opens into the head cavity behind or beneath the brain, and its lateral and ventral walls, continued before the brain, end in delicate attachments of various tissues of the head. The ampullalike swellings or diverticula of the aorta are of particular interest since by their structure they suggest that they are modified segmental chambers of the dorsal vessel anterior to the true cardiac region.

In the acridid *Dissosteira* the aorta begins in the first abdominal segment, but it has there a large dorsal dilation (not visible from below), which lies in the posterior part of the concavity of the strongly elevated tergal plate of this segment. Again, in the metathorax and the mesothorax the aorta gives off dorsal diverticula forming two sacs lodged in the scutellar cavities of the metatergum and mesotergum, respectively. A weakly developed extension of the diaphragm is present in each of these segments attached to the dorsal vessel in the usual manner. The aortic sacs of *Dissosteira* are attached to the walls of the sacs, and throughout its length the aorta is accompanied by strands of similar cells, which are probably pericardial nephrocytes. The larvae of Odonata, as described by Brocher (1917a), have a dorsal tubular diverticulum of the aorta in the mesothorax in the metathorax.

In each segment a delicate membrane containing transverse muscle fibers is stretched above the diverticulum between the bases of the wings. Each membrane is perforated by two apertures having valvular lips opening into the aortic diverticulum beneath it. These membranes, Brocher claims, are pulsating organs that draw the blood into the dorsal sinuses above them and discharge it into the aorta. In *Dytiscus* likewise the aorta gives off dorsal diverticula in the mesothorax and the metathorax, and each diverticulum is covered by a sheet of muscle tissue attached laterally on the wall of the tergum, above which is a blood-filled sinus.

The aortic diverticula have the form of stalked ampullae windened above into flattened chambers. The first is lodged in the cavity of the mesotergal scutellum, the second lies beneath the median tongue of the metascutellum. The aortic ampullae and associated structures of *Dytiscus* are described by Korschelt (1924) from the work of Oberle and Kuhl. These authors believe that the muscular membranes serve to compress the ampullae, and that the latter thus act as accessory pulsating organs to drive the blood

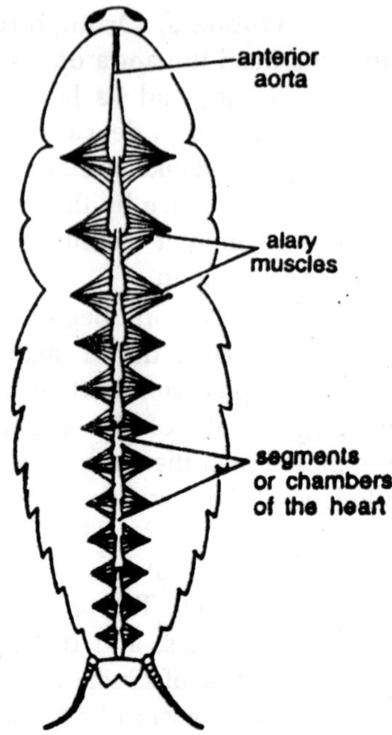

Fig. 6.5. Periplaneta. Heart in dorsal view.

forward through the aorta. Brocher (1916), however, claims that there is a pair of openings into each aortic ampulla from the sinus above it, and he explains the structures as organs for creating a circulation of the blood through the wings, the blood returning to the sinus being then sent into the aorta. A pulsating organ connected with the aorta occurs also, according to Brocher (1919), in the mesothorax of Lepidoptera.

A muscular membrane is stretched across the anterior half of the cavity of the scutellum above a tracheal air sac lodged in the median scutellar lobe and encloses a small sinus in the dorsal part of the thorax. The aorta entering the thorax dips ventrally beneath the second phragma and then makes a large loop dorsally in the mesothorax. From the dorsal part of the loop a diverticulum goes straight upward to the back and terminates in a small bilobed vesicle in the anterior end of the sinus above the pulsating membrane. The posterior wall of the vesicle is perforated by two small ostia, each having its ventral lip prolonged into the lumen of the vesicle in the form of a valvular flap. Brocher gives experimental evidence showing that the pulsations of the membrane draw the blood into the sinus above it from the thorax and from the wings and discharge it through the ostia into the aortic vesicle, whence it is carried forward in the aorta to the head.

The Dorsal Diaphragm

The dorsal diaphragm, when typically developed, consists of two delicate connective tissue membranes enclosing between them the dorsal transverse body muscles, which are inserted medially on the ventral wall of the heart. The degree of development in both the membranous and the muscular elements of the diaphragm, however, varies much in different insects. Usually the dorsal diaphragm is well-developed only in the abdomen, though it may extend in a much reduced condition into the thorax. The diaphragm muscles are commonly known as the "wing muscles" of the heart (alary muscles) because typically they occur in fan-shaped groups of fibers spreading from their points of origin on the tergal plates to their insertions on the heart.

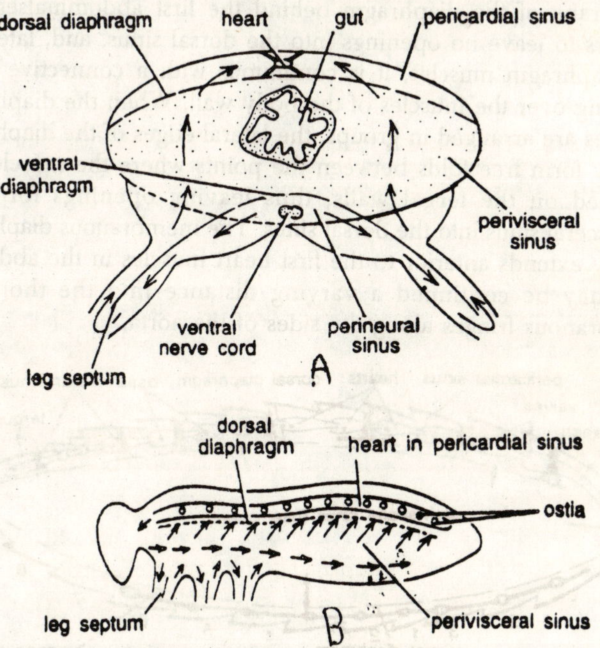

Fig. 6.6. Periplaneta. A–Course of circulation of blood in T.S. of thoracic segment; B–Course of circulation of blood in L.S. of body.

In some insects, however, the diaphragm fibers are all approximately transversely parallel and arise serially along the laterodoral parts of the body wall. The median ends of the muscles terminate in fine branching, tendonlike fibrils either attached to the lateroventral parts of the heart wall or continuous across the ventral wall with fibrils from the opposite side. In general the diaphragm muscles are present only in the body segments containing a chamebr of the heart and are therefore usually limited to the abdomen. In the Blattidae, however, in which the heart and dorsal diaphragm are continued into the metathorax and mesothorax, there are, according to Brocher (1922), groups of muscle fibers in each of these thoracic segments. The diaphragm membranes in some cases are almost entirely absent or form but a scant binding between the muscle fibers, consisting of a weblike tissue full of large and small fenestrae.

On the other hand, the membranes may form a continuous and fairly tough septum entirely separating the dorsal sinus from the perivisceral space of the body cavity. In *Dissostheria* the membrane of the diaphragm behind the first abdominalsegment appears to leave no openings into the dorsal sinus, and, lateral of the diaphragm muscles, it is continuous with a connective tissue covering over the muscles of the body wall. When the diaphragm muscles are arranged in groups, the lateral edges of the diaphragm usually form free folds between the points where the muscles are attached on the tergal walls, thus leaving openings form the perivisceral sinus into the dorsal sinus. The membranous diaphragm usually extends anterior to the first heart muscles in the abdomen and may be continued a varying distance into the thorax as membranous fringes along the sides of the aorta.

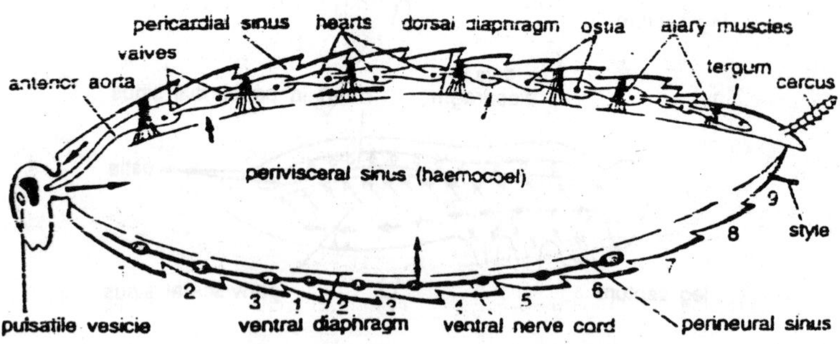

Fig. 6.7. Periplaneta. Blood vascular system.

The Dorsal Sinus

The dorsal, or pericardial, sinus is coincident in extent with the doral diaphragm, and, according to the development of the diaphragm, it is more or less shut off by the latter from the perivisceral sinus. The dorsal sinus contains, besides the doral vessel, some of the median longitudinal muscles of the body wall, the dorsal longitudinal tracheal trunks when the latter are present, masses of fat cells, and usually sheets or masses of spical "*pericardial cells*" resting on the diaphragm along each side of the heart. The pecicardial cells are in most cases nephrocytes. Segmental tracheal trunks enter the dorsal sinus laterally, either above the lateral margins of the diaphragm or between the groups of muscle fibers, and unite here with the dorsal longitudinal tracheal trunks if the latter are present; otherwise the transverse trunks from opposite sides may become continuous in dorsal commissures above the heart. The dorsal diaphragm is sometimes penetrated by parts of the Malpighian tubules, which make convoluted loops within the sinus.

The Ventral Diphragm

A ventral diaphragm is not a constant feature of insect anatomy. When well-developed, as in Acrididae and Hymenoptera, the ventral diaphragm forms a continuous ventral sheet of tissue composed mostly of the ventral transverse muscles of the abdomen. In *Dissosteira* the ventral diaphragm extends through the length of the body from the head into the seventh abdominal segment. In the anterior part of the thorax it is a very delicate membrane without muscles and appears to be attached laterally to the salivary glands and sheets of fat tissue. Between the spreading bases of the sternal apophyses in the metathorax, however, there appears in the diaphragm a series of fine transverse muscle fibers attached laterally on the apophyses. The fibers continue throughout the length of the abdominal part of the diaphragm as it sprincipal tissue.

In the abdomen the muscles have their attachments on the sterna the bases of the anterior and lateral sternal apodemes. The anterior and posterior fibers in each segment spread somewhat forward and rearward to bridge the spaces intervening between the consecutive segments. By this arrangement there is left a series of intersegmental notches along the lateral margins of the diaphragm where the latter has no connection with the body wall, and the openings thus formed appear to be the only connection between the ventral sinus and the perivisceral cavity of the abdomen, except for the wide space below the free posterior margin of the diaphragm. In some insects the ventral transverse muscles consist of compact bundles of fibers in each segment, and in such cases there is no ventral diaphragm.

Accessory Pulsatile Organs

In addition to the dorsal vessel there are often other pulsating structures connected with the haemocoel which are concerned with maintaining a circulation through the appendates. In the mesothorax and sometimes also in the metathorax there is a pulsatile organ concerned with the circulation through the wings. The veins of the posterior part of the wing connect with a blood space beneath the tergum via the axillary cord. In Odonata the blood space, or reservoir, opens through a terminal ostium into an ampulla at the end of a dorsal diverticulum of the aorta. Contraction of this ampulla drives blood into the dorsal vessel; when it relaxes the ostium opens, drawing blood in from the reservoir beneath the tergum and hence, indirectly, from the wings.

In many Lepidoptera the dorsal vessel itself loops up to the dorsal surface of the thorax and forms the so-called *pulsatile organ.*

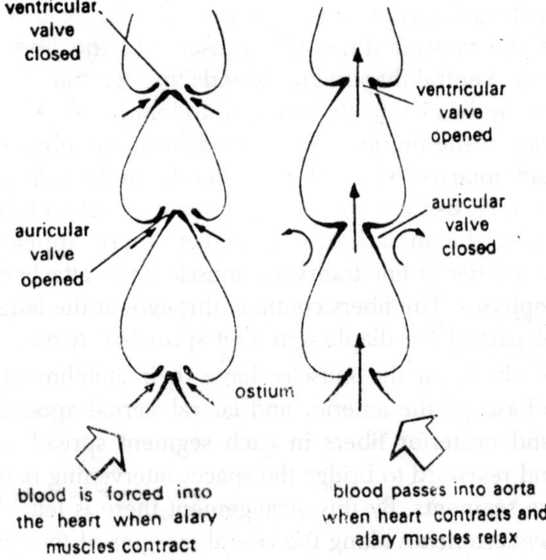

Fig. 6.8. Periplaneta. Diagrammatic representation of the working of valves in the heart.

A reservoir is cut off beneath the tergum by a muscular membrane and this connects with the heart by a pair of ostia at the top of the loop. At diastole blood is drawn into the heart from the reservoir, while at systole it is pumped forwards in the normal way and at the same time the muscular diaphragm falls, drawing in a fresh supply of blood from the wings and thorax. Gerould (1938) maintains that the muscular membrane moves passively, but Brocher (1919) regards the membrane as actively forcing blood through the ostia. The circulation of blood through the wings is also often aided by pulsatile membranes in the veins. For instance, in each wing of *Drosophila* there are four such membranes in the veins which conduct centripetally and one in a vein conducting centrifugally.

The structure and mode of action of these membranes is not understood, but their activity is probably dependent on the activity of the thoracic pulsatile organs. Orthoptera and probably many other insects have a small ampulla at the base of each antenna. This communicates with the haemocoel by a valved opening and extends as a vessel into the antenna. When the ampulla expands blood is drawn into it from the haemocoel; when it contracts blood is forced into the antenna. Other pulsating organs occur in the legs of Heteroptera.

Innervation of the Heart

In some insects, such as *Anopheles*, the heart is entirely without any nerve supply although there are segmental nerves to the aliform muscles. On the other hand the heart of *Periplaneta* is innervated from three sources. Nerves from the corpora cardiaca and from the segmental ganglia combine to form a longitudinal nerve on either side of the heart from which nerve endings remify in the wall of he heart and the aliform muscles. In addition, supposedly sensory fibres arise from the heart and join the sensory nerves in the dorsal body wall. Between these two extremes are various intermediate degrees of innervation and *Prodenia*, for instance, has only segmental nerves. In the cockroach, and probably in most orthopteroids, scattered nerve cells known as ganglion cells, occur along the lateral heart nerves, but these are not always present in other insects.

Circulation

The Course of Circulation

In normal circulation the blood is pumped forwards through the heart at systole, passing out of the heart via the excurrent ostia and, anteriorly, from the aorta. The valves on the incurrent ostia prevent the escape of blood through these openings. The blood driven forwards by the heart increases the blood pressure anteriorly in the perivisceral sinus so that in this sinus blood tends to pass backwards along a pressure gradient. Blood percolates down to the perineural sinus where it is agitated by movements of the ventral diaphragm which assist the blood supply to the nervous system and possibly produce a backward flow of blood. The dorsal diphgragm is usually convex above so that contraction of the alary muscles tends to flatten it. This flattening increases the volume of the pericardial sinus at the expense of the perivisceral sinus and then at diastole is drawn into the heart through the incurrent ostia. Many insects have a well-defined, but variable, circulation through the wings although in some, apparently, circulation only occurs in the young adult.

Normally blood passes out along the anterior veins, back to the posterior veins via cross veins and smaller tissues spaces and then back to the body via the posterior veins and the axillary cord. Changes in pressure modify the wing circulation by pumping blood into spaces which were previously empty or stagnant, but the general course of the circulation remains the same. If, however, the pressure changes in the thorax are very marked the direction of blood flow along the veins may be reversed, particularly in the anterior veins. The flow at the base of the wings is directed by fusion of the dorsal and ventral articular membranes so that the space between them is largely occluded. Anteriorly the membranes are held apart by the axillary sclerites so that they contain a space, the anterior sinus, which is continuous anteriorly with the perivisceral sinus.

Behind the axillary sclerites the membranes are fused except for a few small, irregular channels so that blood from the anterior sinus in mostly directed back to the perivisceral sinus or out along the anterior veins. Posteriorly the anal veins connect with the axillary cord via channels between the two fused membranes and from here it is aspirated by the pulsatile organ in the thorax. In this way the normal circulation in the wings is maintained although in *Anopheles* the contractions of the pulsatile organs are very irregular. The circulation is reduced when the wings are folded because of the occlusion of the channels in the articular membrane. In the absence of the wing circulation the tracheae in the wings of *Blattela* collapse, and the wing structure becomes dry and brittle. Pulsatile organs also pump blood into the antennae and in Heteroptera serve to aspirate blood from the legs.

In most insects the cavity of the legs is divided into anterior and posterior channels by a longitudinal septum. Blood passes down the posterior channel from the perineural sinus and up the anterior channel to the spaces between the wing muscles in the perivisceral sinus. It is thought that pressure differences between these two sinuses maintain the direction of flow. The circulation is affected in an irregular manner by movements of the alimentary canal and by respiratory movements. Any activity which tends to induce pressure differences in different parts of the body must affect the circulation.

Heartbeat

Systole, the contraction phase of the heartbeat, results from the contractions of the muscles in the heart wall which start posteriorly and spread forwards as a wave. Diastole, the relaxation phase, results from relaxation of the muscles assisted by the elastic filaments supporting the heart and, in some cases, by the contraction of the aliform muscles, whether these are inserted directly into the heart wall or are only indirectly connected to it by connective tissue. The contractions of the aliform muscles are in antiphase with the contractions of the heart although they may not coincide exactly with diastole.

After diastole there is a third phase in the heart cycle, known as diastasis, in which the heart rests in the expanded condition. Increases in the frequency of the heartbeat result from reductions in the period of diastasis. In a mechanical recording of heart activity there is often a slight dip in the trace immediately before systole indicating a slight expansion before the contraction. This dip is known as the presystolic notch and probably results from an increase in hydrostatic pressure within the heart due to the start of systole in the more posterior segments.

Rate of heartbeat

The frequency with which the heart contracts varies considerably. Beard (1953) gives a range of from 14 beats per minute in the larva of *Lucanus* to 150 beats per minute in *Campodea.* In general the frequency of beating is higher in early than in later instar larvae and also depends on the age within an instar, becoming very low just before moulting. In the larva of *Sphinx* the rate drops from over 80 beats per minute in the first instar to less than 50 in the fifth; just before moulting it drops to 30 minute and in the pupa is only about 22 beats per minute. The heart of the young pupa of *Anopheles* sometimes stops beating altogether and in old pupae on beating is observed. The heart beats faster in the adult than in immature stages; 150 beats per minute compared with 100-130 in *Anopheles*. Other factors also affect the rate of heartbeat; high temperature and activity increase it, strong movements of the gut may slow it or even stop it for short periods.

In general, the heart stops beating at temperatures below 1-5^{0}C. or above 45-50^0, but it is interesting that the heart of *Periplaneta* continues to beat when the insect is in a state of cold stupor. The heart may undergo periodic reversals in which contractions start at the front and move backwards. This occurs particularly in late larval instars, pupae and adult insects, and in female *Anopheles* 31% of the beats start at the front of the heart. Often the frequency of the heartbeat is lower than normal during these periods of reversal. With reversed beat blood is forced out of the excurrent ostia and Nutting (1951) records powerful current passing out of the subterminal incurrent ostia in the heart of *Gryllotapa*.

Control of heartbeat

Since the heart sometimes has no nerve supply, in these instances the contractions of the heart muscle must be myogenic. Where a nerve supply to the heart is present it is not clear whether the beat is myogenic or neurogenic. Since, in a number of insects, the heart continues to beat after removal of connections to all parts of the nervous system a myogenic beat is suggested, but it is possible in these cases that there are intrinsic ganglion cells on or near the heart as in *Periplaneta*. Pharmacological evidence is conflicting, but possibly favours a neurogenic origin for the beat. Krijgaman (1952) concludes that the heart muscle contracts myogenically, but often has neurogenic pacemakers.

The pacemaker cells has cholinergic properties and acts via motor fibres with adregergic properties. In turn the pacemaker cell may be stimulated and its output varied by sensory input from extrinsic fibres and possibly also by sensory fibres from the heart itself. When contraction is purely myogenic activity may be initiated by contraction of the aliform muscles, as in *Chironomus* larva, but this is not always so because the heart will beat with all the aliform muscles cut and de Wilde (1947) suggests that in Lepidoptera it is the heartbeat which initiates contraction of the aliform muscles. The heart is caused to beat faster by a substance from the corpora cardiaca which is presumably normally released into the blood. This substance promotes the activity of the pericardial cells which enlarge and become vacuolated, producing a second substance which acts on the muscles of the heart. Release of the first substance from the corpora cardiaca is known to be induced in *Periplaneta* by feeding on glucose. As this is ingested sensilla on the labrum are stimulated and impulses pass from them to the corpora cardiaca via the brain and the frontal ganglion.

Other activities probably affect the corpora cardiaca through different sensilla. There is no evidence that this role of the pericardial cells is a general phenomenon, while there is some evidence that hormones may act directly on the heart. A beat can be initiated in any part of the heart, but normally contraction starts posteriorly so that it move forwards. The direction of beat may be related to the distribution of blood pressures. If pressure at the front of the heart becomes so high that a back pressure is set up the heartbeat is reversed. The direction of beat after transaction of the heart adds support to this suggestion and possibly the prevalence of a reversed beat in pupal insects results from the blockage of excurrent ostia by the abundant fat and histolysed tissue present at this time.

In *Anopheles* the direction of heartbeat is sometimes correlated with abdominal ventilation. If ventilation starts posteriorly the heart beats forwards; if ventilatin starts anteriorly the heart beats backwards. These changes might well be due to differences in pressure. Alternatively or additionally the direction of heartbeat might be related to the availability of oxygen. In the absence of a good oxygen supply the rate of heartbeat is strongly reduced indicating the need for an adequate supply for normal working.

The larva of *Bombyx* has a better tracheal supply to the posterior end of the heart than to the anterior end. Thus, the posterior end has a better oxygen supply and so might be dominant, producing the normal forward heartbeat. If however, the posterior spiracles are occluded so that the oxygen supply is reduced the direction of heartbeat is reversed since the anterior part now has the better oxygen supply. In the pupa the tracheal system of the whole heart is poor and the rate of beating is low with reversals, while in the adult the tracheal system is dense both anteriorly and posteriorly and the heartbeat is rapid, again with reversals.

7

OSMOREGULATION

An organism is living till it carries metabolism, which is the sum of all chemical reactions going on inside the organism. Excretion is a regulatory process by which the internal environment of the animal is kept constant. In this process of regulation, excess materials are removed from the body fluids and eliminated with the solid excreta. The term *elimination* in an unrestricted sense, refers to the physiological discharged of any useless or waste substances from the body tissues; by a more limited definition, *excretion* is the elimination of waste products of metabolism. Many substances discharged through the excretory tissues or organs, and therefore classed as excreta, are simply such as salts, unavoidably taken into the body in excess of the need for such matter. In insects, the internal medium is the haemolymph. The physical and chemical properties of this fluid govern the environment of the cells and tissues that make up the organs, and the uniformity of the haemolymph is controlled by excretion. On the other hand, waste products of metabolism that are truly excretory substances are not necessarily removed from the body; they may be merely separated from certain tissues where their presence would be detrimental and stored in others where they become harmless.

Finally, excretory substances, whether stored or eliminated, may serve some useful purpose in the economy of the animal. Unused or indigestible parts of the food material ejected from the alimentary canal are not exereta, though much excretory matter may be voided with the food refuse in the faeces. Eliminated substances are solid, liquid, or gaseous in form. The principal gas expelled as a waste is carbon dioxide, and the principal liquid is water. Solid excreta are given off mostly in crystalline masses or in aqueous solution. They include nitrogenous and nonnitrogenous organic compounds and inorganic salts. Almost any epithelial tissue of the body may assume an excretory function, but special organs of excretion are developed generally from the ectoderm or the mesoderm or from both these germ layers.

Most of the arthropods have no excretory organs corresponding morphologically to the nephridia of Annelida and Onyehophora, their excretory functions being accomplished by the integument and the walls of the alimentary canal. Exceptions are found perhaps in the head glands of Crustacea, which are generally regarded as modified nephridial organs. The organs or tissues of insects that serve to eliminate substances that cannot be physiologically utilized in the body include (1) the general body integument; (2) certain specialized parts of the integument, such as the surfaces of gills and of integumentary glands; (3) the capillary end tubes of the tracheal system; (4) the walls of the alimentary canal; (5) specific excretory tubes opening into the proctodaeum, known as the Malpighian tubules.

In addition, there are the masses of pericardial cells, and groups of similar cells in toher parts of the body, generally called "nephrocytes," which have been supposed to eliminate foreign and waste substances from the blood and to store tham in their cytoplasm, but which perhaps function in a manner preliminary to excretion by changing colloidal substances to crystalloids that can be eliminated through the Malpighian tubules and excreting tissues in other parts of the body. Though we define excretory substances as the end products of metabolism and excess matter that cannot be put to any physiological use, this is not to say that such substances may not be made to serve some practical purpose in the biological economy of the animal. Numerous instances might be cited in which insects make use of their excreted matter. Excretory substances discharged from glands often have offensive odors which become a part of the insect's means of defense. The products of the Malpighian tubules may be employed in the construction of larval cases or injected into the fabric of the cocoon.

Other excretory substances held within the body may be deposited within the integument, where they serve as pigments giving surface colour markings, and it is possible even that nitrogenous component of the cuitcula, a most important part of the physical organization of insects, is to be regarded as an excretory product eliminated with the moults.

Body Wall

The body wall of many animals serves as an excretory organ in the elimination of waste substances from the body, including inorganic salts and nitrogenous compounds. The arthropod integument has, as its outer layer, a cuticula in which the nitorgenous substance chitin is a prevailing if not the predominant element, and in which there usually occur calcium salts or other incrustations in varying amounts. The cuitcula is periodically cast off and renewed. In this way the arthropod loses with each moult a large amount of nitrogen and whatever calcium or other substances may be contained in the exuviae. It has been suggested, therefore, that moulting in the arthropods is primarily a process of excretion, particularly of nitrogen excretion.

Chitin constitutes probably 30 to 40 percent of the insect cuticula, but, as Uvarov (1928) points out, there are other substances associated with it which are also nitrogenous. These substances include pigments, such as those in the scales of Lepidoptera, which Uvarov says, "are clearly the final products of metabolism transferred to certain parts of the integument, and deposited there instead of being excreted." In this connection it is interesting to recall the suggestion of N. Holmgren (1902) that the cuticular covering of the integument has originated as a hardening of secretions thrown off from the epidermis between filaments on the outer surface of the latter. The products of most of the integumentary glands of insects serve some specific purpose in the biology of each species, but, as already mentioned, it is probable that some of them are primarily excretory substances. Respiration in insects, no matter what mechanical devices are developed to faciliate it, always takes place through some part of the ectoderm, and, in this way, the ectoderm serves as an excretory tisssue for the elimination of carbon dioxide and water.

Insects, such as most of the Collembola, lacking tracheae or certain larvae having closed or rudimentary tracheal systems respire directly through the integument, and it is known that many insects having well-developed tracheae discharge at least a part of their carbon dioxide directly through the body wall. But the tracheal tubes themselves are merely invaginations of the body wall, and the ultimate tracheoles are developed in cells of the ectodermal tracheal epithelium. Both external gills and internal rectal gills are also parts of the ecoderm specially modified for respiratory purposes.

Alimentary Canal

Many investigators have observed the accumulation of crystalline bodies in the walls of the mesenteron of various insects. Most of these bodies are salts of calcium, but some also are said to react to tests for uric acid salts. In either case the bodies are probably excretory products or excess substances that cannot be utilized. The *insect, making little use of calcium in its body structures,* must necessarily eliminate most of the calcium *absorbed from its food,* and analyses of the faeces always show a high percentage of calcium in the latter. It seems probable that excretory matter may be eliminated also directly through the walls of the proctodaeum, but most of the intestinal excretion takes place through the Malpighian tubules. The function of the rectal glands, as we have seen, has not been exactly determined.

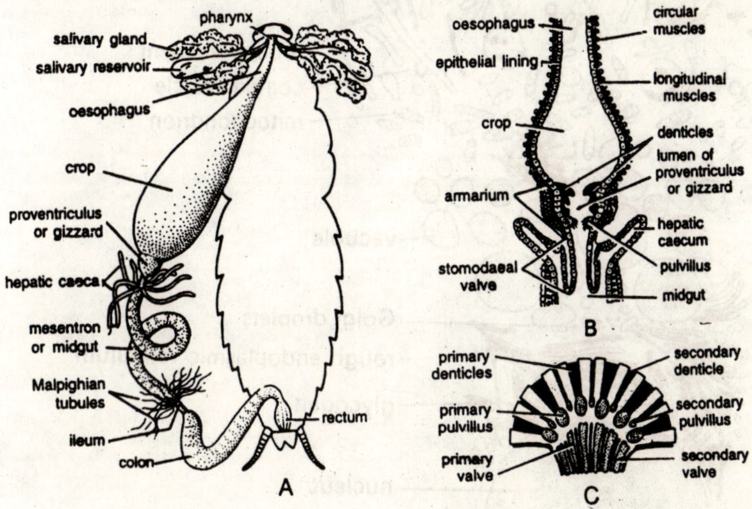

Fig. 7.1. P. americana. A–Alimentary canal and salivary apparatus. B–Crop and gizzard in L.S. C–Proventriculus split longitudinal and laid open.

The Nephrocytes

In nearly all the Arthropoda and in the Onchophora, variously distributed throughout the body and even in the appendages, there are groups of special cells having the common property of absorbing ammonia carmine injected into the blood, and of retaining a precipitate of carmine in their cytoplasm. Because of their action on carmine, these cells have been supposed to have a similar action on other substances naturally present in the blood, such as waste products of metabolism or other injurious bodies, and for this reason they have been termed "storage kidneys" (reins d' accumulation), or nephrocytes. The carmine-absorbing cells of insect occur particularly in the pericardial sinus, where they form masses or long strands of cells on each side of the heart, known as the *pericardial cells. Nephrocytes,* or *pericardial cells,* are cells occuring

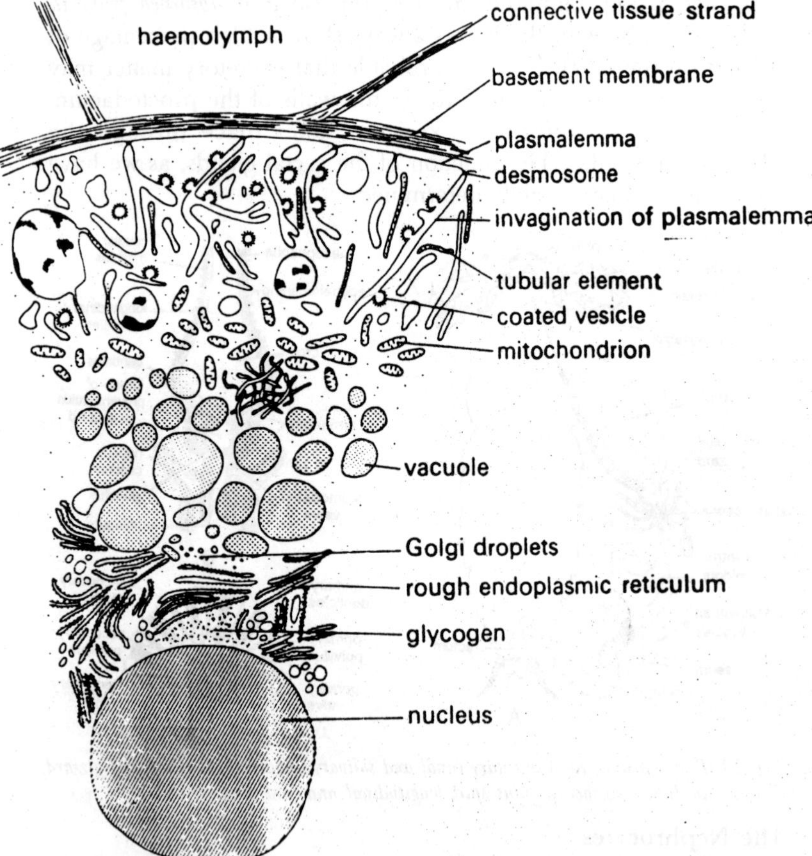

Fig. 7.2. Diagram of a part of a nephrocyte of the larva of Call:phora showing the distinct zonation of the organelles.

singly or in groups in various parts of the body. They may be very large, as in dipterous larvae, or small and numerous and usually they contain more than one nucleus.

In larval *Galleria* they are syncytial. They are usually present on the surface of the heart, or lie on the pericardial septum or the alary muscles. In larval Odonata they are scattered throughout the fat body and in *Pediculus* (Siphunculata), in addition, form a group on either side of the oesophagus. In larval Cyclorrhapha they form a conspicuous chain running between the salivary glands. The nephrocytes undergo cycles of development. In *Drosophila* they are seen to bud off pinosomes internally from the deeply invaginated plasma membrane and it is suggested that in this way materials too complex for immediate excretion are removed from the haemolymph. Within the cell, pinosomes are believed to coalesce and their contents crystallise. The crystals are then degraded and the products held in a large vacuole which is ultimately discharged into the haemolymph.

The effect of the suggested activity would be to transform the original waste materials into a form which could be dealt with by the normal metabolic pathways. Other authors believe the nephrocytes to play a part in protein and lipoprotein metabolism. Nephrocytes also take up dyes and probably colloidal particles from the haemolymph, and they play a part in the control of the heartbeat. The pericardial cells of insects are generally large and are often binucleate. They are always found to have an acid reaction. They are of mesodermal origin and are derived, according to Heymons (1895), from the same parts of the mesodermal layers that form the heart and the dorsal diaphragm. Hollande (1922) says that the cells of the larva generally persist during metamorphosis and become the pericardial cells of the imago.

Functionall, Hollande finds that the pericardial cells have preponderantly the power of absorbing colloids, such as albumins and globulins, or their derivatives, and he claims that their well-known property of absorbing certain pigments results from their affinity for colloids in general. For the most part the cells take up only colouring matter of a colloidal nature. Rarely they contain crystalloids. At the approach of the pupal metamorphosis, Hollande observes, the cytoplasm of the pericardial cells in some insects, particularly in Coleoptera, Trichoptera, and Lepidoptera, becomes charged with albuminoid inclusions, but these inclusions usually disappear by the end of the transformation period. The studies of Hollande (1922) on the physiology of the pericardial cells of hte insects sustains the view that these vells play an important role in excretion, but they discredit the idea that the cells have a storage function.

Hollande claims that, when the cells absorb ammonia carmine, they precipitate the carmine, and that the latter remains in the cytoplasm because it is but little soluble in the cell juices. Normally, he finds, the pericardial cells are agents for breaking down complex colloids, which are transformed by ferments produced in the cells into crystalloids. The latter are then given off into the blood, from which they are removed by the Malpighian tubules. This view is in accord with that earlier expressed by Cuenot (1896). Hollande summarizes his findings on the function of the pericardial cells as follows: Each cell is a ductless glandular body with a merocrine type of secretion, possessing the property of neutralizing alkaline substances present in excess in the blood, and also that of absorbing certain colloid substances; by means of its diastases, functioning in an acid medium produced by the cell itself, it splits the complex colloidal molecules, transforming them into simpler crystalloid compounds that are reduced into the blood, from which they are finally eliminated by the Malpighian tubules.

The pericardial cells, therefore, Hollande points out, may be likened in some respects of the liver of a vertebrate animal and might be called *hepatic cells* more appropriately than "renal" cells, or "nephrocytes," though they differ physiologically from a vertebrate liver in that no glycogen function has been detected in connection with them. In any case, however, the pericardial cells and presumably also the similar cells in other parts of the body play an important part in the physiology of the insect.

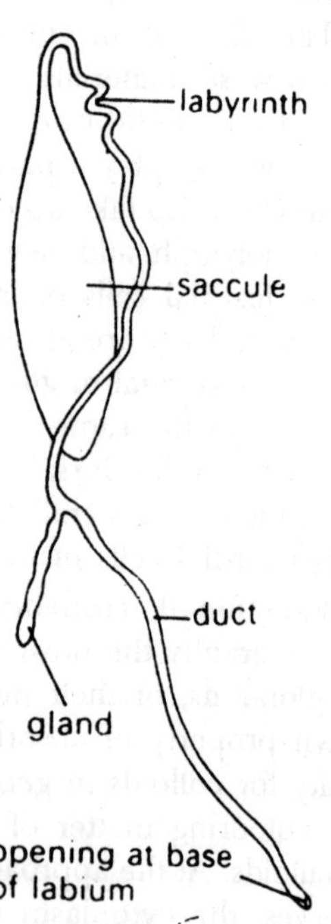

Fig. 7.3. Labial glands of a collembolan.

Malpighian Tubules

The Malpighian tubules are long, thin, blindly ending tubes arising from the gut the junction of midgut and hindgut and lying freely in the body cavity. The Malpighian tubules were described in 1669 by Italian scientist, Marcello Malpighi, who called them *vasa varicosa* and thought that they served a biliary function. In 1816, Herold observed that their function was excretory, and in 1820 Meckel gave them the name *Malpighian tubes*. Malpighian tubes are found in all insects except the Collembola, and the family Aphididae (Homoptera). These structures are long slender tubes, which are often convoluted. Malpighian tubes attach to the digestive tract at a point that is by definition the junction of the midgut and the hindgut.

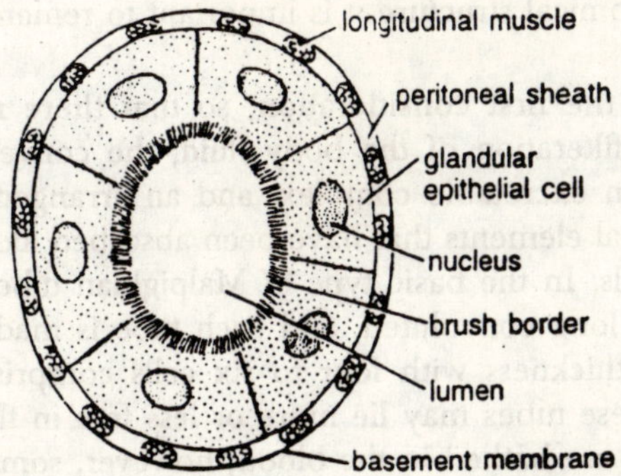

Fig. 7.4. P. americana. T.S. of Malpighian tubele.

Embryologically it is thought the tubes arise from evaginations of the hindgut tissue and that they must be ecodermal. However, according to Savage, the Malpighian tubes of *schistocerca gregaria* have an endodermal origin. In this species, six primary tubes develop, and from these tubes more arise until the average number is as great as 250. The primitive number is usually considered to be six, and they almost always exist in multiples of two. Mosquitoes are an exception to this in that they have five Malpighian tubes. Regardless of the number, direct insertion of the tubes into the digestive tract is unusual. Instead, the tubes anastomose at the proximal end to form an ampulla structure that is usually bilaterally symmetrical and forms the connection between the Malpighian tubes and the digestive tract. From the descriptive literature it is possible to contruct schematic drawings that represent the common types

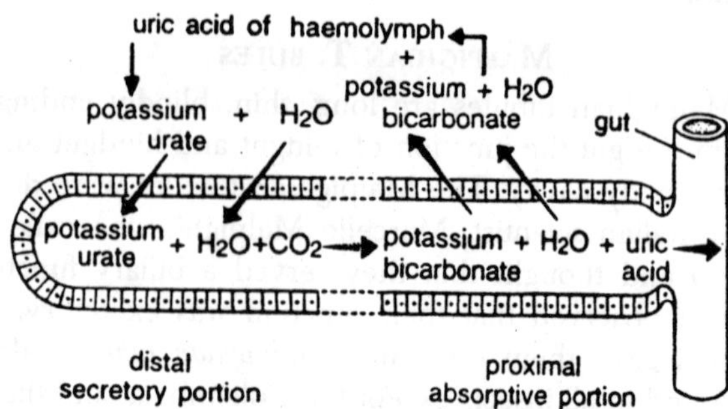

Fig. 7.5. P. americana. Diagrammatic representation of Physiology of Malpighian tubule.

of Malpighian tube systems. These are shown in Fig. and consist of a basic type (A) with three variants (B), (C) and (D). While visualizing the antomical structure it is important to remember the function.

Regulation is the first consideration, so that there must be provision for the filteration of the body fluid, the conversion of the filterate into an excretable complex, and an arrangement for recovery of essential elements that have been absorbed along with the excess materials. In the basic type of Malpighian tube system (A), the tubes are long convoluted, and each tube is made up of walls one cell in thickness with four to six cells comprising the circumference. These tubes may lie more or less free in the body cavity, where they are bathed in the blood; however, some of the tubes are always closely associated with the fat body and digestive tract. Each tube is supplied with a fine trachea, and the cells of the Malpighian system are richly supplied with tracheoles, indicating that the tissues have high metabolic activity. In the Orthoptera, there are a large number of tubes, and the arrangement may appear to be irregular; however, careful observation will show a repetition of the pattern. In these insects (type A), no visible difference exists in the appearance of the tubes over their entire length, but functional differences can be demonstrated.

The distal ends of the tubes may be simply closed with a terminal cell, or the ends may be attached to and imbedded in the tissues that surround the rectum on the hindgut. Tubes with the distal attachment are described as *cryptosolenic* or *cryptonephridic.* According to Lison the latter term is the more descriptive. Tubes of this type are typial of the Coleoptera, in which there is no visible differentiation over the entire length, and they are

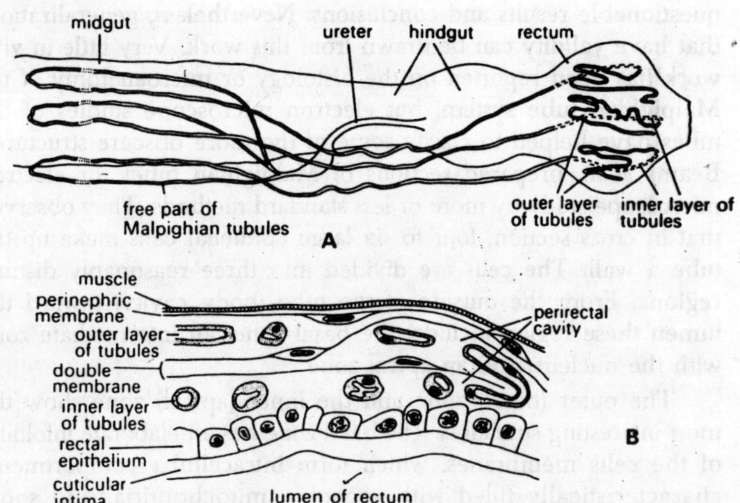

Fig. 7.6. Cryptonephridial arrangement of the Malpighian tubules of the larva of Aglais urticae (Lepidoptera). A–General arrangement showing the close association of the distal ends with the rectum. B–Section of rectum and associated tubules.

represented schematically by type B. The Lepidoptera (type D) also have cryptonephridic tubes, but thse tubes are distinctly differentiated into two sections. The distal part is filled with a clear fluid, and the proxiaml part has solid particles that make it appear opaque.

It is easy to demonstrate that the solid contents include purine base compounds by applying as simple murexide test. Type C is typical of the Hemiptera, and through the extensive work of Wigglesworth with *Rhodnius*, this type has been studied quite thoroughly. This type of tube is free at the distal end but shows a visible differentiation near the midpoint with the proximal half opaque and the distal half clear. Wigglesworth also described two other tube types in which the granular material filled the entire length of the tube in one, and only the distal part in the other. There is not logical way to explain these two types in terms of function as it is observed in the other types of that these must be considered to be either exotic or pathological.

Histology

The histological structure of the Malpighian tubes of various insect species has been reported. The delicate nature of the tissues involved and the drastic treatment of fixation have produce some questionable results and conclusions. Nevertheless, generalizations that have validity can be drawn from this work. Very little in vivo work has been reported on the histology or microanatomy of the Malpighian tube system, but electron microscope studies of the tubes have helped to clarify some of the more obscure structures. Beams *et.al.*, prepared sections of Malpighian tubes for electron microscope study by more or less standard methods. They observed that in cross section, four to six large epithelial cells make up the tube a wall. The cells are divided into three reasonably distinct regions. From the outside of the tube (body cavity) toward the lumen these region include; the basal zone, an intermediate zone with the nucleus, and an apical zone.

The outer (basal) zone and the inner (apical) zone show the most interesting structures. The basal zone has an elaborate infolding of the cells membranes, which form intracellular compartments characteristically filled with vacuoles, mitochondria, and some sharply defined bodies that are otherwise unidentified. The apical zone has a brush border, which has been observed from normal histological study, and the filaments are made up of vertically

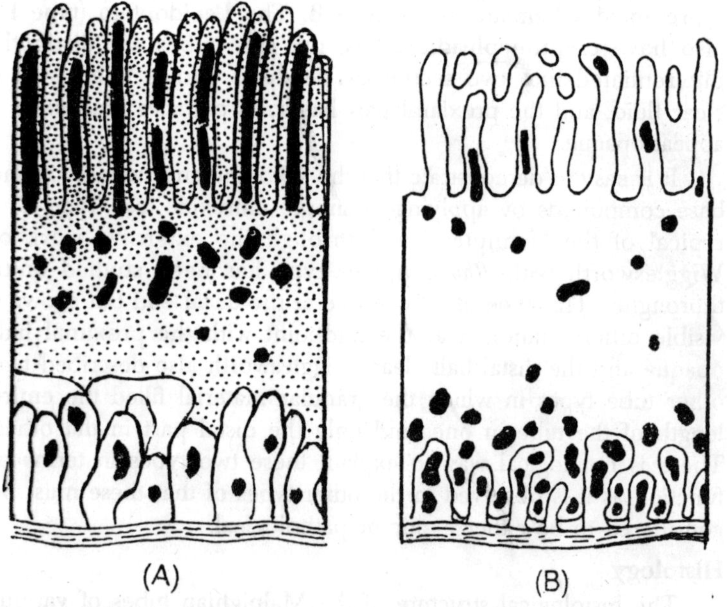

Fig. 7.7. Ultrastructure of (A) honey-comb and (B) brush-border epithelium of Malpighian tubules.

arranged protoplasmic process. Section taken from tubes in various stages of activity show that the mitochondria originate in the basal zone and migrate into the filamentous processes of the apical zone. In some cases, the tips of these processes become swollen and bulbous from the contents. Thus, it has been concluded that mitochondria actively transport materials from the basal part across the cytoplasm of the tube wall and into the processes where there material can be discharged into the lumen by a pinching off of the end of the process. It is presumed that the mitochondria are replaced continuously in the basal zone. Tubes of types C and D, in which there are obvious functional differences between the proximal and distal parts, also show differences in histological structure.

The distal part, which normally contains a clear fluid, has a structure described as a honeycomb, made up of parallel filaments of protoplasm. The proximal part, which contains the granular material, has elongate extrusions of protoplasm that give the impression of cilia. There has been some speculation that these cilia-like extrusions may function to sweep the crystals in the lumen toward the junction with the digestive tract, but there are authentic obsrvations in which cilia-like movement has been proved. The lateral waving motion may be passive, and evidence that the processes can be extended or retracted into the cells of the tube wall is moderately good. At the point of the junction in the proximal-distal differentiated tubes a sharp change in the cell types is displayed. The cells of the distal part contain considerably more mitochondria and other cellular inclusions that are characteristic of actively secreting cells.

It is quite common to observe cells of this region discharging materials into the lumen of the tube by either a merocrine or holocrine secretory process. Unfortunately, in vivo observation are rare or lacking, and most observations have come from the study of fixed tissue, which has introduced the possibility that the bursting cells are artifacts of fixation. To counter this criticism, however, cells of this type are not observed in the proximal part of tubes subjected to the same treatment. Histological evidence indicates that the cells of the distal section of the differentiated tubes and the entire length of the non-differentiated type are assigned a secretory function, by which they actively take up materials from outside the tube, pass them across the cell membrane; and discharge them into the lumen.

The cells of the proximal part in those types where differentiation occurs are likewise given the function of reabsorption. In types A and B, the reabsorption of essential must take place entirely in some other part of the insect, the most likely site being the rectum of the hindgut. In cryptonephridic tubes reabsorption must be a function of the tubes associated with the wall of the rectum. The histological structure of the tubes attached to the rectum has been given special attention. Ishimori suggested using their configuration as a taxonomic character to separate the immature forms of various Lepidoptera, and his work has a good description of the histological structures. From the inside of the gut toward the body cavity there are five layers of tissue. These include: a chitinous intima; an epithelial layer; a double, thin membrane; a single thin membrane; and a layer of the muscles.

The Malpighian tubes lie in two levels with one level of tubes lying between the single and the double membrane and the other level lying between the double membrane and the epithelial cells. In this latter level, the cells that make up the tubes are large and thick. Usually these cells have a single nucleus, but binucleate cells have been observed. Most tubes are lined with an intima that is presumed to be chitin. No basement membrane is discernible. The functional mechanism of the Malpighian tubes surrounding the rectum has not been entirely worked out, but it apparently is associated with the reabsorption of water and probably other essential materials from the hindgut. That substances other than water are reabsorbed and that these substances are liberated into the blood are strongly suggested by immersing the freshly dissected rectal structure from *Teneb rio molitor* larvae in dilute silver nitrate. Immediately, small round spots of white silver chloride appear at the apies of the convolutions of the tubes.

Histological examination shows that at each of these locations is situated a specialized cell that has a form suggestive of a poppet valve in a gasoline engine, but it cannot function in this manner. Evidence of the circulation of essential salts, specifically sodium chloride, which appears in the excreta of most insects, was obtained by Patton and Craig, who traced radioactive sodium -24 through the Malpighian tubes and into a saline surrounding the isolate restum. Historically, this experiment may be of some interest, because it is believed to have been the first application of a man-made radioactive isotope as a tracer in insect physiology.

Intrinsic Muscles and Movements of the Malpighian Tubes

In vivo observation often reveal that the Malpighian tubes are in motion. The movement may be a springlike contraction or a peristaltic wave, and in each case the movement has been traced to the activity of intrinsic muscle intimately associated with the Malpighian tubes. Palm studied these structures extensively. During the course of his work he examined over 3000 insects and described four types of contraction. The Malpighian tubes of species of Orthoptera, Odonata, Hemiptera, Neuroptera, Tricoptera, Lepidoptera, Coleoptera, Diptera and Hymenoptera were included in the survey, and all were found capable of movement.

In Thyanura, Dermaptera, and Thysanoptera, there are no muscles, and the tubes are quite. The muscular elements exist only in the proximal part of the tubes of Lepidoptera, Diptera and Hemiptera, but they are present as strong bands along the entire length of the tubes of Odonata and Hymenoptera. In the Neuroptera and Coleoptera, and in the genus *Gryllotalpa* (Orthoptera), a network of fine muscle fibers are formed. In one genus Palm observed a well defined median band of muscles; this was the only part of the tube capable of contraction. In the cryptonephridic forms, the tubes surrounding the rectum are capable of contractions that are independent of those of the rest of the system. In Orthoptera, the muscle system offers a special case.

Careful observation of the tubes, as they lie in what appears superficially to be tangled mass, reveals that they tend to form helical coils with a single strand of muscle fiber running through the center of the helix. As the muscle contracts, it causes the tube to compress in a motion resembling the action of a coil spring. Various parts of the tubes are supplied with separate fibers, which contract individually so that only sections of an individual tube may be active while the remainder may remain motionless.

Control of the Malpighian Tubes

All of Palm's observation indicate that the muscle fibers involved in the contraction of the Malpighian tubes are striated, and temperature contraction response data indicates that the rates of contraction are too great to be attributed to smooth muscle. Further studies with the effects of combinations of temperature and humidity indicate the contractions are myogenic and probably are initiated by an intrinsic stimulation center somewhat like that already postulated for the heart. A change in the partial pressure of oxygen in the air surrounding the insect does not alter the rate of contractions, nor does HCN gas until its concentration reaches the point of asphyxiation. An increase in the carbon dioxide tension in the air decreases the amplitude of the contractions. Osmotic pressure and salt balance of the perfusion salines cause changes in the rates of contraction of dissected preparations, and the persistence of movement can be prolonged by the addition of glucose to the perfusion saline. The rate of contractions seems to be increased slightly by acid solutions, and slow changes in pH are tolerated. The addition of various common muscle stimulants has no appreciable effect upon the contractile process.

The Function of tube movement

It seems logical that any anatomical structure as well-developed as the Malpighian tube muscles must serve a physiological function, but what the tube movement accomplishes is a matter of speculation. It has been suggested that these movements causes the contents of the tube to be propelled in a forward direction. It has also been suggested that these movements keep the contents mixed, which might enhance the penetration of materials through the walls of the tube and increase the efficiency of reclaiming essentials. Other possible functions remain obscure.

Mechanics of Malpighian tube function

Regardless of the histological structure, the physiological function of the Malpighian system is to regulate the concentrations of the various constitutents and metabolites in the blood. In mammals, the kidney filters blood through a structure called the *glomerulus,* with blood pressure as the driving force. After the blood filters through the glomerulus, a tubular structure reabsorbs the essentials from the ultrafiltrate and returns them to the blood. In insects, the end result of the process is similar, but the mechanism must be different since the blood pressure of insects is negligible. Perfusion experiments by Patton and Craig showed that the rate of penetration of salt solution through the Malpighian tubes of *Tenebrio molitor* could not be increased by adding hydrostatic pressure to the system, nor does a change in osmotic pressure of the perfusion saline alter the rate of absorption as related to a hyper- or hypo-osmotic system. Instead, these changes or forces tend to decrease the rate of absorption in all instances.

The time during which decrease the rate of absorption in all instances. The time during which filtration from a partially dissected system takes place can be increased by saturating the perfusion saline with oxygen. These observation seems to rule out osmotic pressure as a significant force in the filtering process. Wigglesworth studied the excretory function of *Rhodnius,* and the results of his work laid an excellent foundation for that of Ramsey, which came later. Ramsey's experimented those of Wigglesworth and added data that show the existence of a physiological differentiation in the function of the proximal and distal sections of the tube.

Since *Rhodnius* is a blood-sucking insect that engorges once between molts, it lends itself well to the type of experiments performed by Wigglesworth and Ramsey. It would be difficult to duplicate these experiments with other insects. Among the early experiments of Wigglesworth was one in which he dissected the insects so as to expose the malpighian tubes. Using a soft was, he placed ligatures at various levels along the tube and observed the results. If two ligatures were placed so that they were both in the porximal (opaque) part of the tube shortly after the insect had fed, the part of the tube between the ligatures would remain clear, while on each side the tube would fill with white granular material. If the ligatures were placed with one at the junction of the differentiated (proximal) section and the other proximal to this, the part of the tube between the ligatures remained clear but that part between the second ligature and the entrance into the digestive tract filled with crystalline material. In this experiment, the distal part of the tube become distended. These observations indicated that there is a one-way penetration of the wall of the Malpighian tube at the distal part, and that there is a force developed sufficient to cause the distension of the tube. That the section of the tube between the ligatures in the proximal region is clear is interpreted to indicate that no penetration of the filtrate takes place in this region, but that the proximal part of the tube functions in the reclamation of water and other essential materials.

It is a little difficult to rationalize the presence of crystalline material in the extreme proximal section, but until this material is otherwise explained, it may be considered the result of a back-up of the excretory effluent from an adjacent tube, probably by way of the ampulla. From his observations with *Rhodnius,* Wigglesworth proposed a scheme that could explain how the highly insoluble uric acid in solution in low concentration in the blood can be transported across the membranes into the tubes, converted to uric acid in an impure crystalline form, and eliminated. Wigglesworth's hypothesis lacks proof by demonstration, but it remains as a logical explanation of the observed phenomena. In this scheme, uric acid in the blood is converted to a sodium or potassium utrate by the reaction between uric acid and bicarbonate. The urate salts are relatively more soluble than uric acid itself and can cross the cell membrane in solution. Within the tube the urates are broken down to regenerate the bicarbonate, which is reabsorbed, and the utric acid is precipitated. It is necessary for the tissues of the Malpighian tubes to have carbonic anhydrase for this mechanism to work. This enzyme system has been demonstrated from insect tissue by Anderson *et.al.*, and it is likely that it exists in the Malpighian tissues; however, the same cycle could take place by substituting phosphate for bicarbonate, and the necessary enzymes for this reaction have been demonstrated from insect tissues.

Ramsey had contributed a number of details to elucidate the more obscure points regarding the mechanics of Malpighian tube function. His work included studies with mosquito larvae, *Rhodnius,* and the isolated Malpighian tubes of the walking stick (*Dixippus morosus*). In mosquito larvae, osmoregulation is aided and partly controlled by the excretory system. When the external medium (water) is low in salts, the fluid excreta contains less sodium ion than the blood. It is not possible for Malpighian tubes to form urine higher in sodium content than the blood. As a follow-up to this observation, Ramsey injected potassium (as KCI) into the blood of *Rhodnius* and checked the formation of urine and the concentrations of ions as the fluid passed along the length of the tube. He was able to show that this insect can remove potassium ion against a concentration gradient. Urine in the distal portion of the tube contains more potassium and less sodium than the blood.

In the proximal region, the urine becomes more nearly like blood in its concentrations of sodium and potassium. From these data, Ramsey concluded that sodium and water are reabsorbed from the region of the rectum. By comparing studies of *Rhodnius* with studies of mosquito, it is possible to substantiate the observations, using an intact insect. In this situation, concentrations of sodium and potassium, concentrations of sodium and potassium in the blood can be altered physiologically. By increasing the concentration of salts in water surrounding the mosquito larvae, the concentration of salts in the blood can be altered by the absorption of salts through the anal gills. Concentrations of both sodium and potassium increase in the urine with increased concentration in the blood, and absorption in the rectum decreases. Circulation of potassium from the blood to the tube, to the gut, and back to the blood takes place. The concentration of potassium in the blood of the larvae is controlled by the rate of voiding, which also affects the amount of reabsorption; this rate is controlled by peristaltic movement of the gut.

Further studies with eight other insect species led Ramsey to conclude that the concentration of potassium is always greater in the urine than in the blood, it was concluded that while potassium is actively transported into the tube, sodium enters by simple diffusion. With isolated Malpighian tubes of the walking stick, Ramsey observed that the urine (of the insect) is either isosmotic or slightly hypotonic to the blood. He measured the rate of urine fromation and found that it would amount to about six microliters per hour, which in 24 hours would be equivalent to the total blood volume. The tubes of the walking stick actively secrete sodium, potassium, and water in all sections; but the sodium-to-potassium ratio of the contents is greater in the proximal part. Potassium acts as a diuretic, but sodium has little effect.

The rate of secretion of potassium is more than 10 times that of sodium. The same observation also has been made in the larvae of *Sialus* (Sialidae, Neuroptera). In addition to the behaviour of sodium and potassium, Ramsey studied several divalent ions and some organic compounds. The results indicate that the concentration of a substance in the urine is nearly independent of the concentration of the same substance in the blood. There does not appear to be a maximum or minimum concentration for the removal of amino acids (for example) from the saline, although rates of penetration may be different amino acids. The sum of these observations leads to the following conclusions:

1. Substances required by the insect are absorbed by the rectum and returned to the blood.
2. This stream of saline passes rapidly down the tube and flushes out the diffusion substances.
3. According to Ramsey, it is simpler for an organism to develop ways of absorbing needed substance than to develop ways of excreting all possible undesirable substances.

4. The Malpighian tube is freely permeable, so that most metabolites of small molecules may pass through under the influence of a concentration gradient, if such exists.
5. Some salts and water are actively secreted, and it is not clear whether interaction between water and salts takes place.

In considering these conclusions, the only question that arises is that of maintaining the favourable concentration gradient to cause continuous filtration of the protein-free fluid part of the hemolymph. Experiments in which the activity of the Malpighian tubes in excreting a dye (indigo cramine) is blocked by a chemical (ethylene glycol) show that the tubes also lose their ability to reduce blue tetrazolium, an indicator of oxidase activity. The tubes also their fluorescence. These combined observations indicate existence of an active secretory mechanism that is driven by an enzyme reaction (or enzyme reactions). This conclusion is supported by the existence of a definite temperature maximum for excretor function (in the American roach), with a very sharp cut-off when this maximum is exceeded. A great deal is still to be learned about the mechanics of excretory function in insects.

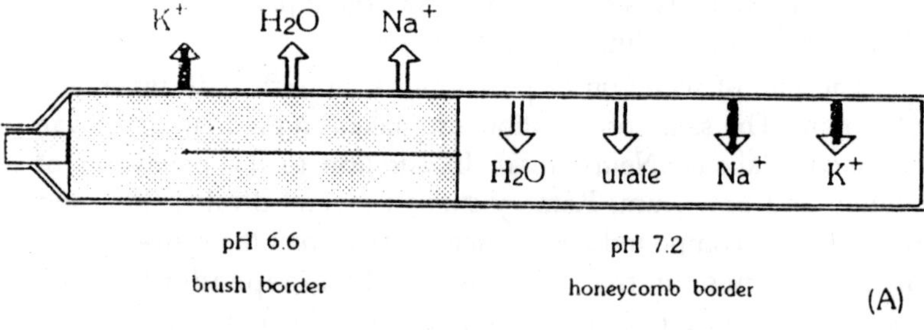

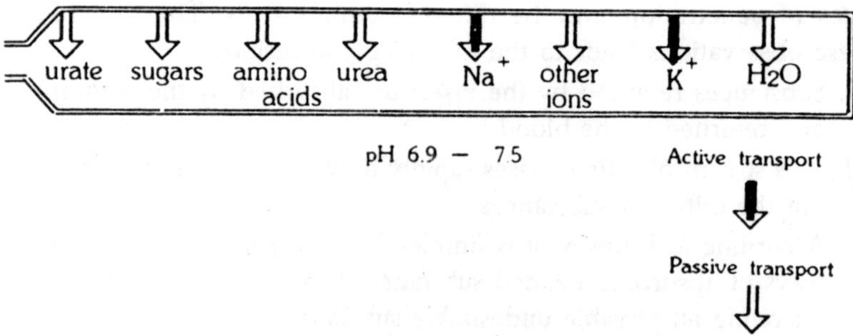

Fig. 7.8. Mechanism of Excretion–(A) Rhodnius and (B) Carausius.

The whole process may be described as a filtration of the fluid portion of the hemolymph, excepting the protein that has molecules too large to pass through the membranes. The filtrate penetrates the wall of the tube either by diffusion or by active secretion, with the possibility that both processes are important. The essential substances are reabsorbed up to a threshold value, either in the proximal part of the tubes or in the rectal structures of the hindgut.

Enzymes of the Malpighian Tubes

The tissues of the Malpighian tubes have been described as active in several metabolic processes, and like the mammalian kidney, these tissues should be expected to contain enzymes with relatively high activity. A summary of some of the enzymes from Malpighian tubes was compiled by Craig. In addition to the enzymes shown in Table, which is only a partial list of those enzymes that must exist in these tissues, several water-soluble vitamins associated with metabolic processes are found in high concentrations.

Metcalf reported the concentrations of various of these water-soluble vitamins from bioassay of Malpighian tube tissue from the American cockroach. The values are as follows: riboflavin, 840 to 1000 µg. per gram of fresh tissues, a value that is one of the highest ever recorded from an animal tissue; thiamin, 33 to 50 mg. per gram; niacin, 200 to 460 mg. per gram; pantothenic acid, 80 mg. per gram; and ascorbic acid, 600 to 1000 mg. per gram. Metcalf was unable to demonstrate the oxidative enzymes normally associated with riboflavin in highly active tissues, but the probability remains that riboflavin is associated with and probably essential to an unidentified enzyme system that produces energy to drive the secretory processes necessary for the function of the Malpighian tubes.

Table 7.1. Distribution of Enzymes in the Malpighian Tubes.

	Insect			
Enzyme	*Acrida bicolor*	*Mantis religiosa*	*Balaps gibba*	*Apis mellifera*
Alkaline phosphatase	+++	++	+++	+++
Acid phosphatase	++++	++	+++	++++++
Lipase	++	++	++++	+++++
Succinic dehydrogenase	++++++		++++	++++

Role of Digestive Tract in Excretion

Collembola and members of the family Aphididae are without Malpighian tubes, so that the regulatory function must be assumed for other organs. According to Waterhouse and Day, the midgut of the Collembola takes on the function, and it has been suggested that loss of the midgut epithelia, which in this insect takes place just before a molt, is part of the process. In aphids, the process was studied by Gersch, who used fluorescein dye to trace the process of elimination. He found that this dye was picked up readily by parts of the digestive tract and eliminated, but it must be remembered that this group has a very highly specialized digestive system.

Other examples cited by Waterhouse and Day include records of various dyes and iron salts that are absorbed by cells of various parts of the digestive tract and discharged into the lumen of the gut. They also point out that uric acid is present in the midgut of larval Hymenoptera, which has no through connection with the hindgut. Undoubtedly the digestive tract, in addition to the rectal structure, functions as a regulatory organs in the excretory process. Regulation of water is probably the principal contribution of the digestive tract to the overall process; however, no good evidence exists that the digestive tract rivals the Malpighian tubes in the general excretory functions.

NITROGENOUS WASTES

In nitrogenous wastes structural complexity, toxicity, and solubility go hand in hand. The simplest form of waste (ammonia) is highly toxic and very water-soluble. It contains a high proportion of hydrogen which can be used in production of water. It is generally found as an excretory product, therefore, only in those insects which have available large amounts of water, for example, freshwater forms and the larvae of meat-eating flies. Generally, however, in insects, as in other terrestiral organisms, water must be conserved, and more complex nitrogenous wastes must be produced, which are both less toxic and less soluble. In the egg and pupal stage the problem is accentuated because water lost cannot be replaced, and nitrogenous wastes must remain in the body in the absence of a functional excretory system.

Most insect, then, excrete their waste nitrogen as uric acid. This is only slightly water-soluble relatively nontoxic, and contains a smaller proportion of hydrogen compared with ammonia. It should be realized, however, that uric acid is not the only form of nitrogenous waste. Usually, traces of other materials (especially the related compounds allantoin and allotoic acid) can be detected, and in many species one of these has become the predominant excretory product. Urea is only rarely a major constituent of insect urine. Usually, it represents less than 10% of the nitrogen excreted. Traces of amino acids can be found in the excreta of many insects, but their presence should be regarded as accidental loss rather than deliberate excretion by an insect. Only rarely has the excretion of particular amino acids been authenticated; for example, the clothes moth *Tineola* and the carpet beetle *Attagenus* excrete large amounts of the sulfur-containing amino acid cystine. Although in tsetse flies uric acid is the primary excretory product, two amino acids, arginine and histidine, are important components of the urine. These make up about 10% of the protein amino acids in human blood; because their nitrogen content is high, it is probably uneconomical to degrade them, and they are therefore excreted unchanged.

Table 7.2. Nitrogenous Excretory Products of Various Insects

	Uric acid	*Allantoin*	*Allantoin acid*	*Urea*	*Ammonia*	*Amino acid*
Odonata						
Aeshna cyanea (larva)	0.08	–	0.00	–	1.00	–
Dictyoptera Phasmida						
Periplaneta americana	1.00	0.00	0.00	–	–	–
Blatta orientalis	0.64	0.64	1.00	–	–	–
Dixippus morosus	0.69	1.00	0.44	–	–	–
Hemiptera						
Dysdercus fasciatus	0.00	1.00	0.00	0.26	–	0.24
Rhodnius prolixus	1.00	–	–	0.33	–	trace
Coleoptera						
Melolontha vulgaris	1.00	0.00	0.00	–	–	–
Attagenus piceus	0.72	–	–	1.00	0.57	0.50
Diptera						
Lucilia sericata	1.00	0.30	–	–	0.30	–
Lucilia sericata (pupa)	1.00	0.00	–	–	0.15	–
Lucilia sercata (larva)	0.05	0.02	–	–	1.00	–
Lepidoptera						
Pieris brassicae	1.00	0.04	0.01	–	–	–
Pieris brassicae (pupa)	1.00	0.03	0.05	–	–	–
Pieris brassicae (larva)	0.28	0.16	1.00	–	–	–

The amino acids excreted in vast quantity by plant-sucking Hemiptera must, of course, be considered as fecal and not metabolic waste products. Because of the large amount of water taken in by aphids, it has been suggested that they might produce ammonia as their nitrogenous waste. Indeed, uric acid, allantoin and allantoin acid cannot be detected in the excreta. However ammonia makes only 0.5% of the total nitrogen excreted, which has led to the suggestion that it is used (and detoxified) by the symbiotic micro-organisms residing in the mycetome. Table contains selected examples to show the variety of nitrogenous wastes producted by insects.

Uric acid and the other nitrogenous waste products are derived from the sources, nucleic acids and proteins. Degradation of nucleic acids is of minor importance; most nitrogenous waste comes from protein breakdown followed by synthesis of hypoxanthine from amino acids. The biochemical reactions which lead to syntehsis of this purine appear to be similar to those found in other uric acid-excreting organisms. In addition to the enzymes for uric acid synthesis there are also uricolytic enzymes which catalyze the degradation of this molecule in many insects. Uricase has a wide distribution within the insect class.

Active preparations of allontoinase have been obtained from many species, but the distribution of this enzyme appears to be rather restricted compared with uricase. Although there are reports that indicate the occurrence of allantoicase and urease in tissue extracts from a few insects, their presence should not be regarded as having been established unequivocally. In other words, when urea and ammonia are produced in significant amounts, they are probably derived in a manner other than by the degradation of uric acid. The existence of an ornithine cycle for urea production, such as is found in vertebrates, has not been proved conclusively, even though the constituent molecules of the cycle (arginine, ornithine, and citrulline) and the enzyme organize have been identified in several species.

Similarly, the way in which ammonia is produced (especially in those insects in which it is a major excretory molecule) is poorly understood. It is generally assumed to result from deamination of amino acids, but the precise way in which this occurs in unclear. It has been suggested that the most primitive state was that in which the complete series of uricolytic enzymes was present, and ammonia was the excretory material. As insects became more independent of water, selection pressures led to loss of the terminal enzymes and production of more appropriate excretory molecules. This simple view should be regarded with caution. Thus, in some caterpillars diet can affect the nature of the nitrogenous waste. In certain insects substantial quantities of a particular nitrogenous waste molecule are produced, yet the appropriate enzyme in the uricolytic pathway has not been demonstrated, and *vice versa;* that is, the effects of other metabolic pathways may override the uricolytic system.

In many insects (especially endopterygotes) the predominant nitrogenous excretory product changes during development. For example, in *Pieris brassicae* (Lepidoptera) the major excretory product in the pupa and adult is uric acid; in the larva this compound constitutes only about 20% of the nitrogenous waste, allantoic acid being the predominant end product. Indeed, in some Lepidoptera, the ratio of uric acid to allantoin may fluctuate widely from day to day. Of great interest will be determination of factors which stimulate inhibition or activation (degradation or synthesis?) of uricolytic enzymes so that the most suitable form of nitrogenous waste is produced under a given set of conditions.

Physiology of Nitrogenous Excretion

Uric acid is produced in the fat body and/or Malpighian tubules (occasionally the midgut) and released into the haemolymph. It is secreted into the lumen of the tubules as the sodium or potassium salt, along with other ions, water, and various low molecular weight organic molecules. In *Dixippus* secretion occurs along the entire length of the tubule. No resorption of materials takes place across the tubule wall, and urate leaves the tubule in solution. In the rectum resorption of water and sodium and potassium ions occurs, and the pH of the fluid decreases from 6.8-7.5 to 3.5-4.5. The combined effect of water resorption and pH change in to cause massive precipitation of uric acid. Useful organic molecules such as amino acids and sugar are also resorbed through the rectal wall.

The Malpighian tuble-rectal wall excretory system thus shows certain functional analogies with the vertebrate nephron. In *Rhodnius*, whose tubules show structural differentiation along their length, the process of excretion is basically the same as in *Dixippus*. However, in *Rhodnius* only the distal portion of the tubule is secretory and resorption of water and cations begins in the proximal part. Slight change in pH occurs (from 7.2 to 6.6) as the fluid

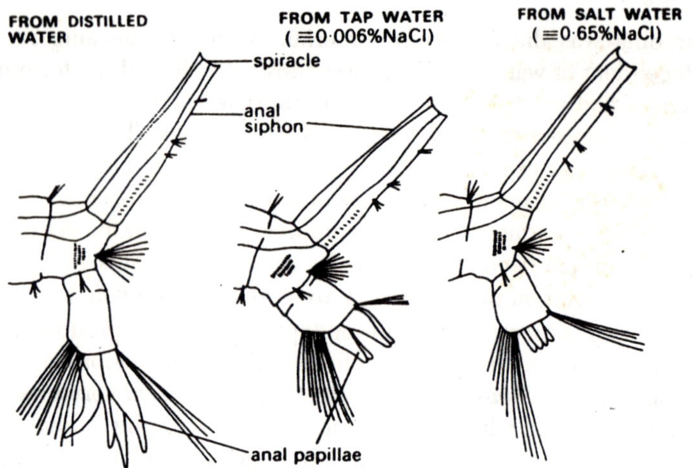

Fig. 7.9. Posterior end of larval Culex pipiens reared in different media showing the variation in size of the anal papillae.

passes along the tubule and this is sufficient to initiate uric acid precipitation. Further water and slat resorption occurs in the rectum (pH 6.0), causing precipitaion of the remaining waste. Although allantoin is the major nitrogenous waste in many insects, it made of excretion appears to have been studied in only one species, *Dysdercus fasciatus* (Hemiptera). This insect is required because of its diet to excrete large quantities of unwanted ions (magnesium, potassium and phosphate). This, combined with the insect's inability to actively resorb water from the rectum, results in the production of a large volume of urine.

Because no resorption or acidification occurs which could cause precipitation of uric acid, this molecule is not longer used as an excretory product. Thus, allantoin, which is 10 times more soluble than uric acid, this molecule is not longer used as an excretory product. Thus, allantoin, which is 10 times more soluble than uric acid (yet of equally low toxicity), is preferred. However, the insect does not posess a mechanism for actively transporting this molecule from the hemolymph to tubule lumen; that is, allantoin only moves passively across the wall of the tubule. It is therefore maintained in high concentration in the hemolymph to achieve a sufficient rate of diffusion into the tubule. Whether a similar mechanism occurs in other allantoin-excreting insects remains to be seen. It may be significant that many other allantoin producers are herbivorous and have the problem of removing large quantities of unwanted ions. The physiological mechanisms for excretion of other nitrogenous wastes are not known.

Storage Excretion

Storage excretion is the retention of waste material within the body, rather than its removal through the excretory system. Thus, in Collembola and other apterygotes which lack Malpighian tubules, in those larval Hymenotera whose tubules are not fully developed, in many endoparasites, and in *Periplaneta* where the tubules appear not to function in nitrogenous waste removal, there are specialized urate cells in the fat body in which uric acid concretions develop. These cells replace, functionally speaking the Malpighian tubule system. At other time storage of urate occurs even when the tubules are working normally and may be regarded as a supplementary excretory mechanism for occasions when the tubules cannot cope with all the waste that is being produced.

In the larval stags of many species uric acid crytallizes out in ordinary fat body cells and epidermis, even though the Malpighian tubules are functional. It appears that thisis caused by the metabolic activity of the cells themselves (that is, they are not accumulating uric acid from the hemolymph), and crystallization occurs by virtue of the particular conditions (pH, ionic content, etc.) existing in the cells. During the later stages of pupation the crystals disappear, the uric acid apparently having been transferred to the meconium via the excretory system. It is worth noting that in many species the malpighian tubules are entirely reconstituted during the pupal stage. Thus, storage of uric acid in fat body and epidermal cells is of great importance at this time. Often the uric acid is stored in specific regions of the body.

In *Dysdercus* for example, the malpighian tubules cease to function toward the end of each instar. At this time uric acid is deposited in the epidermal cells of the abdomen forming distinct, white transverse bands. The uric acid is stored permanently in these cells and thus serves an important pigmentary function. Storage excretion of material other than uric acid takes place. Calcium salts (especially carbonate and oxalate) are found in the fat body of many plant eating insect larvae. During metamorphosis they are released and dissolved, to be excreted via the Malpighian tubules in the adult. Dyes present in food are often acculumated in fat body cells where they appear to become associated with particular proteins. These proteins ae then transferred to the egg during vitellogenesis and the dyes subsequently "*excreted*" during oviposition. Nephrocytes accumulate a variety of substances, especially pigments, and it is thought by many workers that storage excretion is one of their major functions. According to Wigglesworth (1965), this view is mistaken; if they are involved in excretion it is probably in the intermediary metabolism of waste materials.

Salt and Water Balance

Salt and water balance involves more than simply the control of hemolymph osmotic pressure; the relative proportions of the ions that contribute to this pressure must be maintained within narrow limits. The osmotic pressure of the hemolymph is generally within the same limits as that of the blood of other organisms, but it can be increased considerably under specific conditions (by the addition, for example, of glycerol which serves as an antifreeze during hibernation). Regulation of the salt and water content is obviously related to the nature of he external environment. Different osmotic problems are faced by insects in different habitats. Nevertheless, they have been solved using the same basic mechanism, namely, the production of a *"primary excretory fluid"* in the Malpighian tubules followed by differential resorption from or

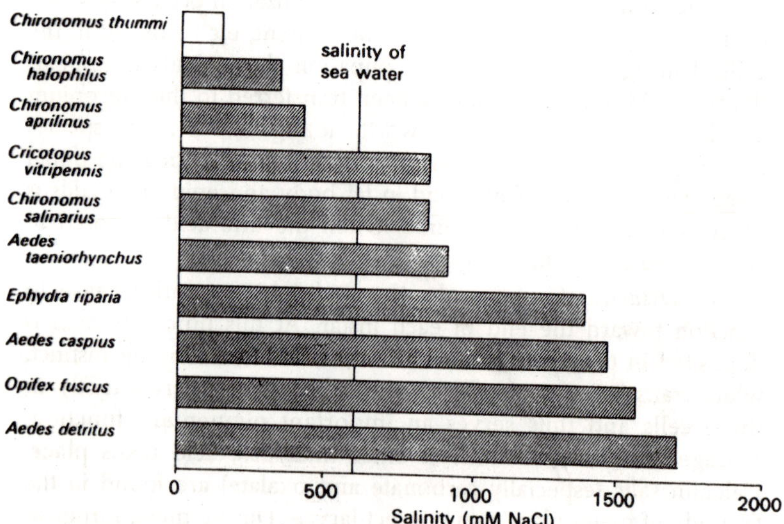

Fig. 7.10. The ranges of salinity tolerated by some salt-water dipterous larvae. Chironomus thummi is a freshwater species included for comparison.

secretion into this fluid when it reaches the rectum. For clearity we shall consider separately the problems of insects living on land, in brackish or saline water, or in fresh water; considerable similarity in the solution of these problems will, however, be seen.

Terrestrial Insects

Although only a limited amount of data is available, it appears that terrestrial insects can regulate their hemolymph osmotic pressure over a wide range of conditions. For example, in *Tenebrio* the hemolymph osmotic pressure varies only from 223 to 365 mM/liter (measured as the equivalent of a sodium chloride solution) over a range of relative humidity from 0 to 100%. In starving *Schistocerca* there is only a 30% difference in hemolymph osmotic pressure between animals kept in air at 100% relative humidity and given only tap water and those kept in air at 70% relative humidity and given only tap water and those kept in air at 70% relative humidity and given saline (osmotic pressure equivanent to 500 mM/litre sodium chloride) to drink. In terrestrial insects water is lost (1) by evaporation across the integument, although this is considerably reduced by the presence of the wax layer in the epicuticle; (2) during respiration through the spiracle (many insects possess devices both physiological and structural for reducing the loss; and (3) during excretion. The major source of water for most terrestrial insects is obviously food and drink. Some insects eat excessively solely for the water content of the food. Where sufficient water cannot be obtained by drinking or in food, the insect must obtain it by other means.

One source is the water produced during metabolism. Absorption of water vapor from the atmosphere is a method employed by a few insects (for example, *Thermobia* and *Tenebrio*) which are normally found in very dry conditions. Interestingly, the site of absorption is the rectum, which, as is noted below, is the site of uptake of liquid water in the rectum, which, as is noted below, is the site of uptake of liquid water in other terrestrial and brackish-water insects. Small amounts of ions are lost from the body via the excretory system, and these are readily made up by absorption across the midgut wall. Indeed, in terrestrial insects the usual problem is removal of unwanted ions present in the diet. The food often contains ions in concentrations which are widely different from those of the hemolymph. It is probable that these ions enter the hemolymph passively in the same proportions as

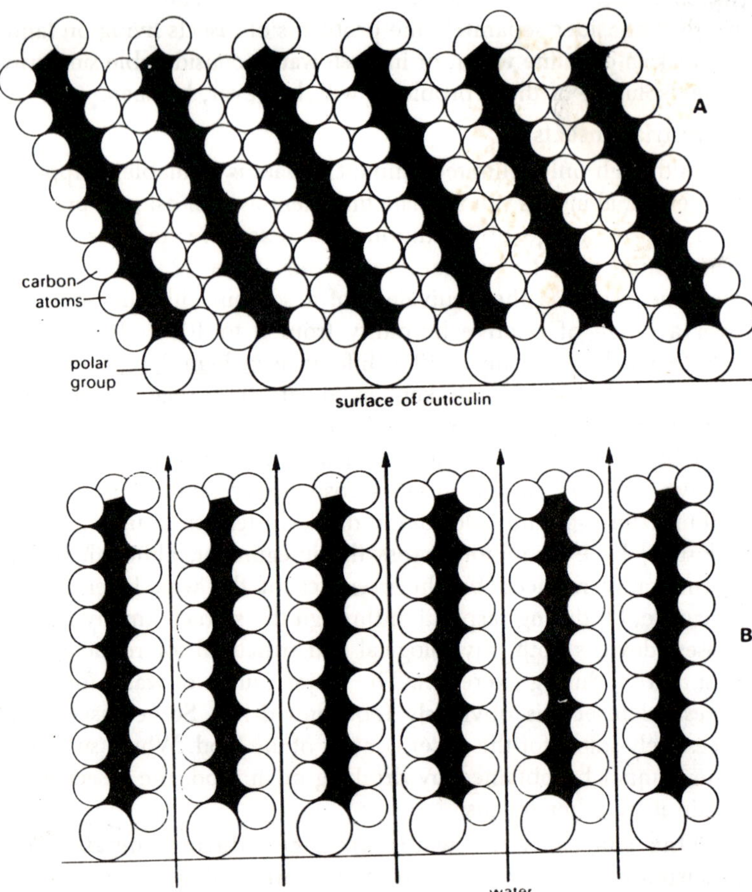

Fig. 7.11. Diagrammatic representation of the molecules of the wax monolayer provided the waterproofing layer of the insect cuticle. A–Molecules oriented at about 25° to the vertical with carbon atoms of adjacent chains interfitting so that there are no gaps between them. B–Molecules in the vertical position leaving gaps through which water can escape.

they occur in diet, and excesses are subsequently expelled via the excretory system. In other words, the midgut does not act as a selectively permeable barrier to the entry of ions.

The role of the Malpighian tubules and rectum has been investigated by examination of the ionic composition of the fluids within them and, more recently, by the use of radioisotopes to measure the direction and rate of movement of individual ions.

The studies of Ramsay in the 1950s revealed that the fluid in the tubules is isosomotic with the hemolymph but has a very different ionic composition. Particularly obvious is the difference in potassium ion concentration which is several times higher in the tubule fluid than in the hemolymph. The sodium ion concentration is usually lower in the fluid than in the hemolymph, as is the case with most other ions (except phosphate). The tubule fluid, which is produced continuously, contains a number of low molecular weight organic molecules, for example amino acids and sugars; thus, it is broadly comparable with the glomerular filtrate of the vertebrate kidney, thoguh it is not produced by hydrostatic pressure.

Table 7.3. The Osmotic Pressure and Concentration (mM/litre) of Some Ions in the Hemolymph (H), Malpighian Tubule Fluid (MT) and Rectal fluid (R) in Insects from Different Habitats.

Habitat	Species (stage and conditions)	Fluid	*Osmotic* Pressure (=NaCl *solution)*	*Ions* Na⁻	K⁻	Cl⁻
Terrestrial	*Schistocerca gregaria*	H	214	108	11	115
	(adult, water-fed)	MT	226	20	139	93
		R	433	1	22	5
	Dixippus morosus	H	171	11	18	87
	(adult, feeding)	MT	171	5	145	65
		R	390	18	327	–
	Rhodnius prolixus	H	206	174	7	155
	(adult, 19-29 hr	MT	228	114	104	180
	after meal)	R	358	161	191	–
Salt water	Aedes detritus	H	97	–	–	–
	(larvae, in seawater)	MT	–	–	–	–
		R	537	–	–	–
	Aedes detritus	H	97	–	–	–
	(larvae, in distilled	MT	–	–	–	–
	water)	R	56	–	–	–
Fresh water	*Aedes oegypti*	H	138	87	3	–
	(larvae, in distilled	MT	130	24	88	–
	water)	R	12	4	25	–

The high potassium concentration in the tubule fluid and the demonstration that the rate at which tubule fluid is formed depends on the hemolymph potassium concentration led Ramsay to suggest that the active transport of potassium ions is fundamental to the

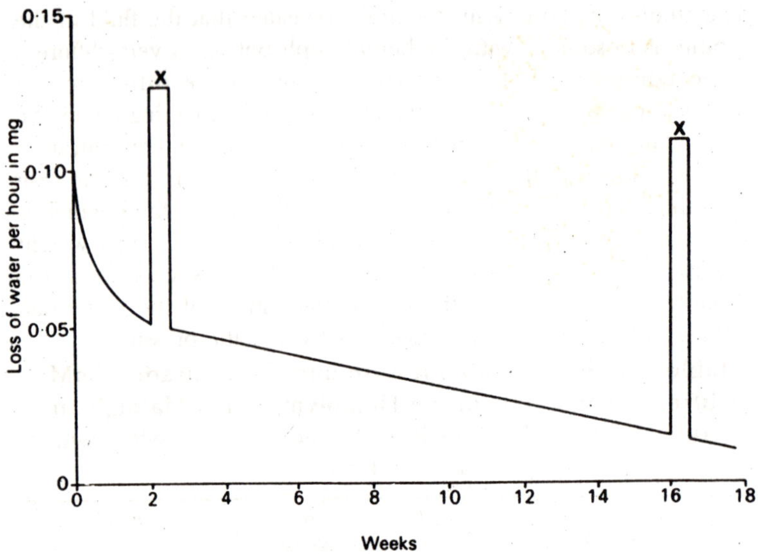

Fig. 7.12. The rate of water loss from starved Tenebrio larva. At the points indicated by X the spiracles were kept open in 5% carbon dioxide.

production and flow of the fluid. The other ions and organic molecules probably enter the tubule fluid passively. It is clear that, since the tubule fluid and hemolymph are isosmotic, the tubules are not directly concerned with regulation of hemolymph osmotic pressure. Experimental proof for this was obtained by Ramsay, who showed, using isolated tubules, that the isosmotic condition is retained over a wide range of external concentrations. Indirectly, however, the tubules are important in regulation, since the rate at which ions and water are excreted from the body is the difference between their rate of secretion into the tubule lumen and their rate of resorption by the rectum. In some insects the tubules show regional differentiation, secretion taking place in the distal part and resorption beginning in the proximal part of the tubule.

In most species, however, resorption occur mainly in the rectum, though the anterior part of the hindgut may also modify the fluid. In the rectum major changes occur in the osmotic pressure and composition of the urine. Generally the urine becomes greatly hypertonic to the hemolymph, but when much water is available a hypotonic fluid may be excreted. Wigglesworth (1931) first noted that the rectum is the main site of water resorption in many

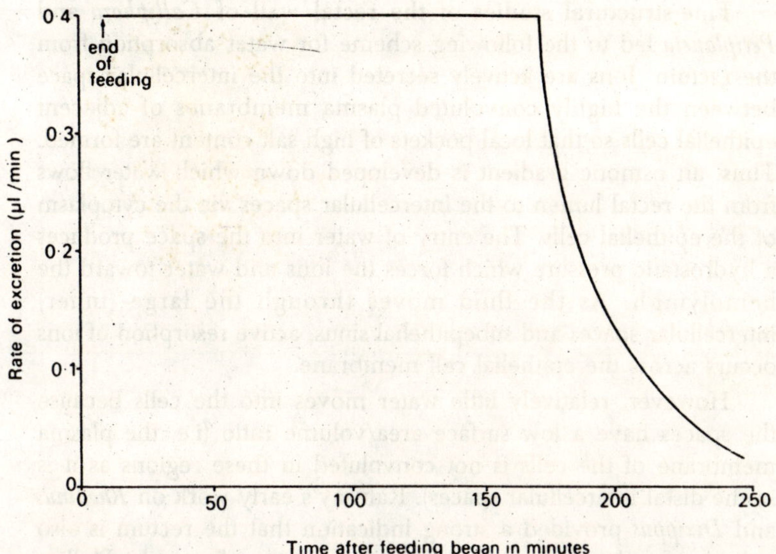

Fig. 7.13. The rate of excretion of Rhodnius immediately after feeding.

terrestrial insects. Water is resorbed against a concentration gradient; that is, it is an active process and energy is expanded.

Phillips (1964a) showed that in *Schistocerca* the rate of water movement across the rectal wall is independent of the rate of salt accumulation. The rate at which water is resorbed depends on the osmotic gradient across the wall, and, as the gradient increases during resorption, the point is reached at which the rate of active accumulation is balanced by the rate of passive diffusion back into the rectal lumen: that is, the concentration of the rectal fluid reaches a maximum value. However, this value varies according to the water requirements of the insect. Insects that have been kept in a dry environment and given strong saline to drink have a rectal fluid whose osmotic pressure is about tiwce that of insect with acess to tap water. The physiological basis of this increased ability to concentrate the urine is not known. The precise mechanism of water uptake is still unclear. Though models have been proposed in which water *per se* is actively transported across the rectal wall, there is now direct evidence that water movements occur as a result of active movements of inorganic ions, especially sodium potassium, and chloride (secondary transport of water).

Fine-structural studies of the rectal wall of *Calliphora* and *Periplaneta* led to the following scheme for water absorption from the rectum. Ions are actively secreted into the intercellular space between the highly convoluted plasma membranes of adjacent epithelial cells so that local pockets of high salt content are formed. Thus, an osmotic gradient is developed down which water flows from the rectal lumen to the intercellular spaces via the cytoplasm of the epithelial cells. The entry of water into the space produces a hydrostatic pressure which forces the ions and water toward the hemolymph. As the fluid moves through the large (inner) intercellular spaces and subepithelial sinus, active resorption of ions occurs across the epithelial cell membrane.

However, relatively little water moves into the cells because the spaces have a low surface area/volume ratio (i.e. the plasma membrane of the cells is not convoluted in these regions as it is in the distal intercellular spaces). Ramsay's early work on *Rhodinus* and *Dixippus* provided a strong indication that the rectum is also capable of reabsorbing salts, and this has been confirmed by Phillips (1964b) in *Schistocerca.* This author showed that sodium, potassium and chloride ions can be accumulated against a concentration gradient and independently of the movement of water.

Furthermore, the rate of accumulation of these ions depends on their concentrations in the rectal fluid and the hemolymph. In this way the requirements of the insect can be satisfied. In water-fed locusts ions are resorbed from the rectum as quickly as they arrive in the fluid from the tubules, and low rectal concentrations are found. At the other extreme, in saline-fed animals, the rates of resorption are low and a greatly hypersmotic fluid is produced. It appears from the limited amount of experimental work carried out that the cryptonephridial arrangement of Malpighian tubules increases the power of the rectal wall to resorb water against high concentration gradients. The system is particularly well-developed in insects that inhabit dry environments.

According to Ramsay (1964), the perinephric membrane is impermeable to water, and under dry conditions, the osmotic pressure of the perinephric cavity is raised mainly due to the presence of some unkoown electrolyte. Thus, the concentration gradient across the rectal wall is reduced, facilitating water uptake. Ramsay suggested that potassium and chloride ions are actively transported into the lumen of the perirectal tubules (which contain many mitochondria), with an accompanying movement of water. Resorption of ions and water occurs across the wall of the parts of the tubule bathed in hemolymph. An apparent lack of mitochondria in the perinephric membrane and leptophragma cells led Ramsay to conclude that ions move passively across the membrane in order to balance those removed by the tubule. However, a fine-structural and experimental study has shown this conclusion to be wrong. These authors found that the leptophragma cells have a normal complement of mitochondria. Active transport of potassium ions occurs across the cells, which are, however, impermeable to water. Chloride ions follow passively.

Brackish-Water and Saltwater Insects

The habitat occupied by brackish-water and saltwater insects can vary widely in ionic content and csmotic pressure. During periods of warm, dry weather the salinity may increase severalfold. Conversely, after heavy rains or the melting of snow in spring, the salinity may approach that of fresh water. It is not surprising, therefore, to find experimentally that such insects can regulate their hemolymph osmotic pressure over a wide range of external salt concentrations. Larvae of *Aedes detritus* and *Ephydra riparia*, inhabitants of salt marshes, can survive in media containing the equivanent of 0 to about 7-8% sodium chloride. Over this range of concentrations the hemolymph osmotic pressure changes by only 40-60%. Larvae of *Ephydra cinerea* are found in the Great Salt lake of Utah where the salinity may exceed the equivalent of 20% sodium chloride. At low salinities the hemolymph is hyperosmotic to the medium and a very dilute urine is exerted Salts are actively resorbed through the reactor wall.

At high salinities the hemolymph osmotic pressure is less than that of the environment. Water will therefore tend to move out of the insect as a result of osmosis. The water loss is counter-balanced by ingestion of the medium during feeding, but this in itself creates a problem because of hte high salt content of the ingested fluid. Annual recently, it was asumed that (1) the midgut wall was highly impermeable to salts, most of which therefore did not enter the hemolymph, and (2) as in terrestrial insects, most of the water in the tubule fluid was resorbed in the rectum, forming a greatly hypersomotic urine. However, the work of Phillps and co-workers has shown that in saline-water mosquitoes, at least, these assumptions are incorrect. The midgut wall is markedly permable both to ions and to water; indeed, almost all ions and water ingested are absorbed from the midgut into the hemolymph, a feature which Phillips *et. al.,* (1978) suggest may serve to concentrate food in the gut prior to digestion.

The excess water taken into the hemolymph is lost either by osmosis through the body wall or in the formation of Malpighian tubule fluid. As in insects from other habitats, the tubule fluid is isosmotic with the hemolymph, and the main function of the tubules in saline-water species appears to be excretion of sulfate ions which may be present in the medium in high concentration. Other ions in excess in the hemolymph are actively secreted into the lumen of the rectum by cells of the posterior rectal segment (which, interestingly, is not present in strictly freshwater mosquitoes) to form a greatly hyperosmotic urine.

Insects Living in Freshwater

The regulatory problem facing freshwater insects are the opposite of those in insects from saline conditions. Water enters the body osmotically despite the relatively impermeable cuticle and must be removed, and salts will be lost from the body and must be replaced if the hyperosmoic condition of the hemolymph is to be maintained. Freshwater insects can regulate their hemolymph osmotic pressure successfully to the point at which the external environment becomes isosmotic with the hemolymph. This is achieved by the production of urine which is hyposmotic to the hemolymph. Beyond this point regulation breaks down because freshwater insects are not able to produce a hyperosmotic urine; that is, they cannot resorb water against a concentration gradient. As in terrestrial insects, the Malpighian tubules produce a fluid which is isosmotic with the hemolymph but of different ionic composition.

Particularly obvious is the great difference in the potassium ion concentration between the two solution. When the fluid enters the rectum resorption of ions occurs. The osmotic pressure of the fluid that finally leaves the body is much lower than that of the hemolymph but not as low as would be expected from knowledge of the extent of ionic resorption in the rectum. This is because large quantities of ammonium ions appear in the rectal fluid. These ions cannot be detected in the Malpighian tubules, and it is presumed that they are secreted directly across the rectal wall, as occurs in larvae of *Sarcophaga bullata.* In freshwater insects food is the usual source of ions which are absorbed through the midgut wall.

However, in some forms ions are accumulated through other parts of the body, for example, the gills of caddis fly larvae, the rectal respiratory chamber of dragonfly and mayfly larvae, the anal gills of syrphid larvae, and the anal papillae of mosquito and midge larvae. The role of the anal papillae in ionic regulation has been particularly well-studied. In mosquito larvae a pair of papillae is located on each side of the anus. They communicate with the hemocoei and are well-supplied with tracheae. Their walls are one cell thick and covered with a thin cuticle. Koch (1938) and Wigglesworth (1938) were the first ot demostrate that mosquito larvae can accumulate chloride ions against a large concentration gradient and that the mechanism for doing so is located in the papillae. Later workers showed that sodium, potassium, and phosphate ions are also actively transported into the papillae. The ability to accumulate ions varies with the habitat in which an insect is normally found. Thus, *Culex pipiens,* which is found in contaminated water, is less efficient at collecting ions than *Aedes aegypti,* which typically lives in fresh rainwater pools. Indeed, normal larvae of the latter species can maintain a constant hemolymph sodium concentration when the sodium concentration of the external medium is only 6mM/liter. The hemolymph sodium concentration under these conditions is only about 5% below the normal level.

Hormonal Control

As we have seen above, insects possess, in the form of the Malpighian tubules and rectum, an excellent mechanism for the regulation of the salt and water content of the hemolymph. However, such a mechanism is useful only if it can be told when to start, stop, accelerate, or decelerate to suit the needs of an insect. This coordination is effected by hormones. The occurrence of diuretic hormones produced by neurosecretory cells of the brain and other ganglia is firmly established. Antidiuretic factors are also thought to occur in some insects, but the evidence for them is rather circumstantial at present. A diuretic hormone, which stimulates excretion of water, is released following feeding in many terrestrial insects. In *Rhodnius* stretching of the abdominal wall brings about hormonal release. In *Schistocerca, Dysdercus,* and other insects that feed more or less continuously, it is probably the strectching of the forget that cause release of hormone.

The hormone appears to act primarily on the malpighian tubules, stimulating them to secrete potassium ions at a greater rate, thereby creating an enhanced flow of water across the tubule wall. However, Mordue (1969) reports that in *Schistocerca* the hormone has a dual action, causing accelerated secretion through the tubules and a slowing down of water resorption through the rectal wall. Because, presumably, there is a direct relationship between the amount of food consumed, the amount of hormone released and the quantity of water removed across the tubules, it is difficult to see how such a simple arrangement will work other than in insects such as *Rhodnius* whose food has a constant water content.

The physical stimulus of "stretching" alone would not provide a precise enough mechanism for water regulation in insects whose food differs in water content. Some other control system must therefore operate. Perhaps an insect can monitor the water content of its food or, alternatively, the resorptive power of the rectal wall may be controlled directly by the hemolymph itself. Phillips (1964b) showed that the rate at which sodium and potassium ions are resorbed is dependent on the ionic concentration of the hemolymph. Almost no work has been done on the hormonal control of salt and water balance in aquatic insects. However, as a result of his experiments on *Aedes aegypti*, Stobbart (1971) has suggested that accumulation of sodium ions across the wall of the anal papillae is under endocrine control. It appears that when the sodium concentration of the hemolymph changes an abdominal monitoring center passes information to a center in the thoracic ganglia, which causes a change in the rate of hormone production.

8

INTEGRATED SYSTEM

The ability to respond the stimuli and in turn to conduct impulses to other cells is probably a general phenomenon of living cells. Specialized cells and tissues which can receive and interpret stimuli have evolved in most animals. From these specialized cells and tissues a system of control has been developed which regulates and integrates the activities of many parts of the animal. This system is the *nervous system.* An animal is formed of chemical compounds of many different kinds, but most of its components substances may be classed in two groups. Those of one group are stable compounds, the alteration of which results only in damage to the organism. These substances form the integument, the skeleton, the connective tissue, and the supporting framework of muscles, glands, and cells. Substances form the integument, the skeleton, the connective tissue, and the supporting framework of muscles, glands, and cells. Substances of the other group are labile compounds, highly unstable in their molecular structure, and some of them liable to sudden disruption on sufficient increase of stimulus. It is these substances that cause the activities of the animal. They occur principally in the secreting cells of glandular tissues, in the contractile tissue of muscles, and in the receptive and conductive parts of nerve tissue. Since the stimuli for action in the labile tissues of the animal come primarily from the environment, there must be some provision for transmitting their effect from the stable periphery of the animal to the internal tissues in which is stored the latent energy that makes action possible.

Moreover, since the normal activities of a living creature are advantageous to the organism, there must be also some provision for controlling and directing the results of the liberated energy. Anything that produces a change in the metabolic activity of the labile constitutent of living matter is called a *stimulus.* The quality of living matter that makes it responsive to stimuli is termed *sensitivity.* The effect of the stimulus on the tissue, however, does not end with the first impact; it is transmitted from molecule to molecule, and this property of progressive reaction to stimuli is known as *conductivity.*

Since it appears that reaction to a stimulus in all cases involves a destructive chemical action in the labile constituents of the tissue involved, a period of *recovery* is necessary for the restoration of the discomposed substances to their original form. *Fatigue* arises when stimuli follow in too rapid succession to allow of complete recovery. The properties of sensitivity and conductivity are presumably common to all protoplasm in some degree. They are highly developed in the protoplasm of nerve tissue; they constitute, in fact, the fundamental qualities of nerve tissue upon which the functions of the nervous system. The nervous system is derived from embryonic and is divisible into central nervous system, the peripheral nervous system and the antonomic nervous system.

The Central Nervous System

In the evolution of the annelid-arthropod nervous system it would appear that the first centralized group of nerve cells had its origin in the ectoderm at the anterior pole of the body, forming here a small ganglion associated with a sensory *apical plate.* Later there appeared several paired groups of sensory cells at the bases of tentacular or other sensory organs behind the apical plate. Then, these various primary nerve centers united in a single ganglionic mass, which is the so-called *archicerebrum* of the annelid worms. This primitive "*brain*" lies above the anterior end of the alimentary tract, where, in some of the Annelida, it is still not detached from the ecotderm. Finally, two lateroventral nerve strands, consisting of nerve cells and fibers, are developed from the ectoderm along the entire length of the body and are connected anteriorly with the archicerebrum.

Thus is established the definitive nervous system, consisting in its simplest form of a suprastomodael brain, which originates in the prostomium, though subsequenlty it may be displaced posteriorly, and of paired ventral nerve strands formed in the postoral part of the trunk. Following segmentation of the body, the *neurocytes* of the nerve strands become aggregated in the segments, producing segmentally arranged ganglia. In this stage, therefore, the central nervous system consists of an anterior median brain (*archicerebrum*) located in the head above the alimentary tract, and of two long, ganglionated nerve cords situated laterally in the ventral part of the body, in which the ganglia are united lengthwise by interganglionic *connectives* of nerve fibers, and crosswise by transverse fibrous *commissures*.

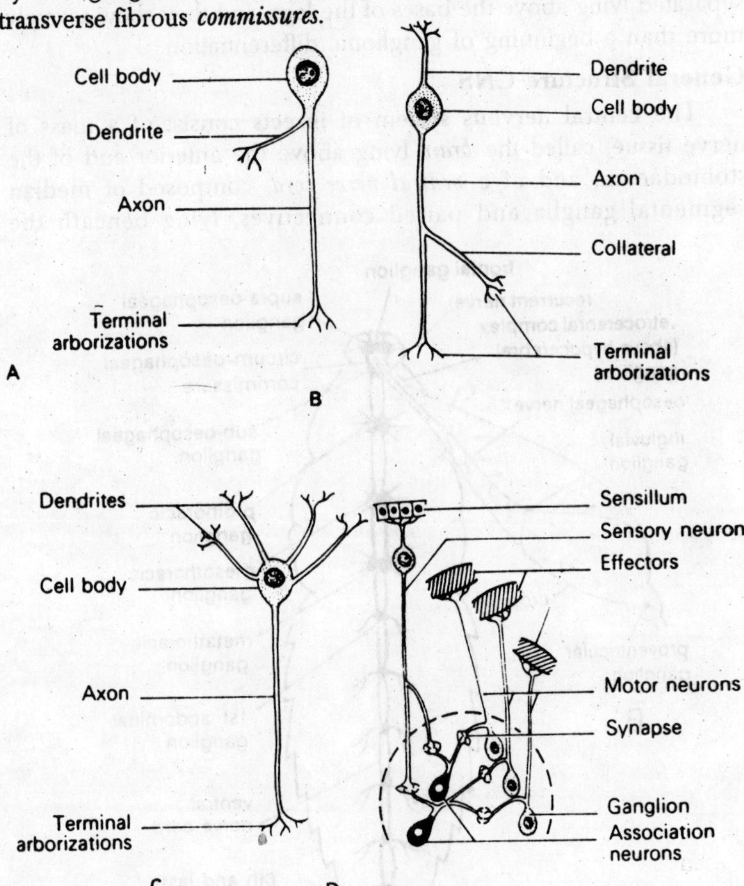

Fig. 8.1. Structural types of neurons. A–Unipolar. B–Bipolar. C–Multipolar. D–Relationship among sensory, motor, and association neurons.

Finally, in most of the Annelida and all Arthropoda, the lateral cords have approached each other, and the pair of ganglia in each segment have united to form a compound median ganglion, through the connectives in most cases remain double. Various stages in the evolution of the ventral nervous system, from a condition in which the lateral strands are widely separated to one in which they are united in a median cord, are well shown in the Annelida. In some primitive forms, moreover, the nerve tissue is not entirely detached from the ectoderm. In the Onychophora the nerve strands are widely separated lying above the bases of the legs, and they show scarcely more than a beginning of ganglionic differentiation.

General Structure CNS

The central nervous system of insects consist of a mass of nerve tissue, called the *brain* lying above the anterior end of the stomodaeum, and of a *ventral nerve cord,* composed of median segmental ganglia and paired connectives, lying beneath the

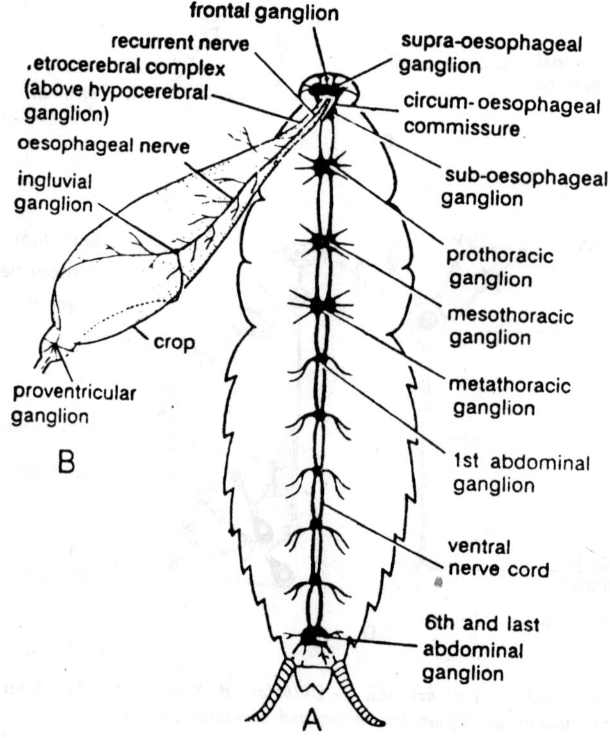

Fig. 8.2. P. americana. A–Central and peripheral nervous system. B–Stomatogastric nervous system.

alimentary canal. The two parts are joined by connectives embracing the stomodaeum. The brain is a composite structure, but there is a difference of opinion as to how many primary segmental ganglia enter into its composition. A distinct prostomial ganglion is not evident in the ontogeny of the arthropod nervous system, but the internal structure of the adult brain demonstrates, as well later be shown, that a large part of the cerebral mass, from which the optic lobes of the compound eyes take their origin, is identical with a corresponding part of the brain is known as the *protocerebrum*. If the primitive arthropods had pre antennal appendages, the nerve centers of these appendages lay immediately behind the opitc centers and are included in the protocerebrum.

The nerve centers of the first antennae constitute a distinct second section of the brain called the *deutocerebrum*. In all insects and in most crustaceans the nerve centers of the second antennae form a third part of the brain, or *tritocerebrum*. The tritocerebral

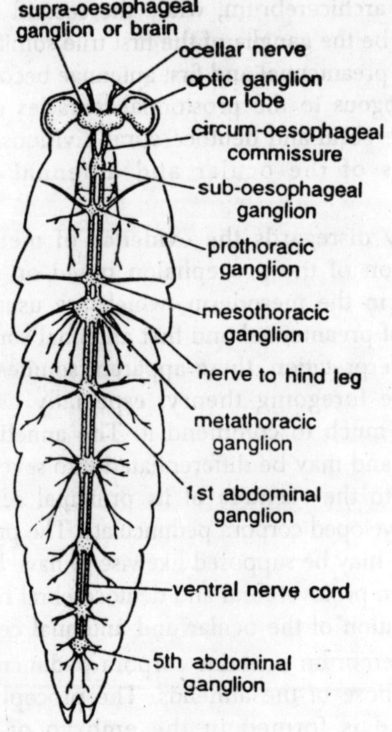

Fig. 8.3. Grasshopper. Nervous system in dorsal view.

lobes, however, are united by a commissure that always lies beneath the stomodaeum, and in some crustacea the second antennal ganglia themselves are not contained in the suprastomodael brain. The ganglia of the second antennal segment, therefore, are without question the first ganglia of the primitive ventral nerve cord, and the primitive arthropod brain included only the ganglia contained in the protocerebrum and deutocerebrum. If the preantennal and first antennal appendages represent true somites of the body, then the suprastomodael brain mass includes the highly developed archicerebrum of the prostomium, rudimentary preantennal ganglia of the first somite, and antennal ganglia of the second somite.

It seems somewhat incongruous that two pairs of segmental ganglia should be lie above the stomodaeum and have no substomodaeal connectives. A more simple concept of the brain structure, and incidentally of the procephalic segmentation, is presented by N. Holmgren (1916) and by Hanstrom, according to which the entire suprastomodael part of the brain is derived from the prostomial archicerebrum, while the second antennal centers are assumed to be the ganglia of the first true somite. As a corollary to this view the preantennal and first antennae become appendicular structures analogous to the prostomial tentacles of the Annelida, and the protocerebral and deutocerebral divisions of the brain are specializations of the ocular and antennal centers of the archicerebrum.

The theory disregards the evidence of metamerism in the postocular region of the procephalon based on the presence of paired cavities in the mesoderm, which are usually regarded as coelomic sacs of preantennal and first antennal somites. According th the other interpretation, these apparent somites are "secondary segments." The foregoing theory, especially as elaborated by Hanstrom, has much to commend it. The annelid brain is often highly evolved and may be differentiated into several distinct parts corresponding to the centers of its principal sensory nerves; it contains well-developed corpora pedunculata. The primary arthropod brain, therefore, may be supposed likewise to have been secondarily differentiated into protoverebral and deutocerebral regions as a result of the specialization of the ocular and antennal centers.

The protocerebrum contains corpora pedunculata identical in structure with those of the annelids. The procephalic part of the arthropod head is formed in the embryo of the externally

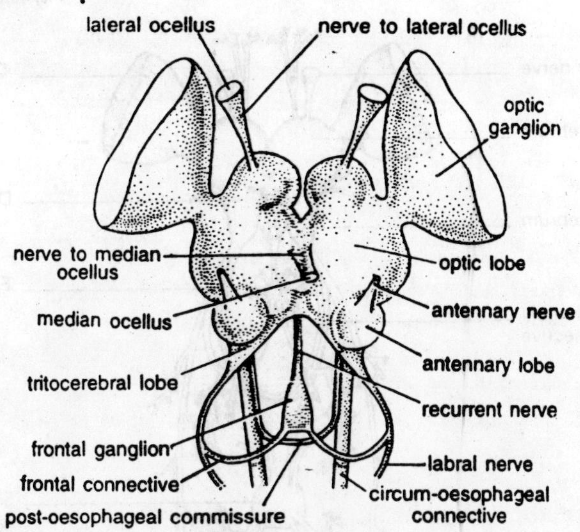

Fig. 8.4. Grasshopper. Brain in dorsal view.

unsegmented cephalic lobes, which evidently represent principally the prostomium, but it usually includes also the reduced tritocerebral segment. The tritocerebral, or second antennal, nerve centers, however, are actually the first ganglia of the ventral nerve cord, though in most arthropods they are secondarily added to the primary cerebrum.

The innervate the second antennae, the region of the mouth, and the preoral part of the prostomium and are united by connectives with both the brain and the stomodeal nervous system. With the addition of the tritocerebral ganglia to the arthropod brain, the primitive brain connectives are shortened and finally suppressed. The definitive connectives are those between the tritocerebral ganglia and the first ganglia of the gnathal region. The next three ganglia of the ventral nerve cord are those of the segments that become the gnathal region of the insect head. These ganglia are always united with one another in the mature insect to form a second composite nerve mass of the head, known as the *suboesophageal ganglion* because it lies in the ventral part of the head beneath the stomodaeum. The principal nerves of this ganglion are those of the mandibular, maxillary, and labial appendages. The thoracic region of the insect body contains three primitive median ganglia corresponding to the three thoracic segments.

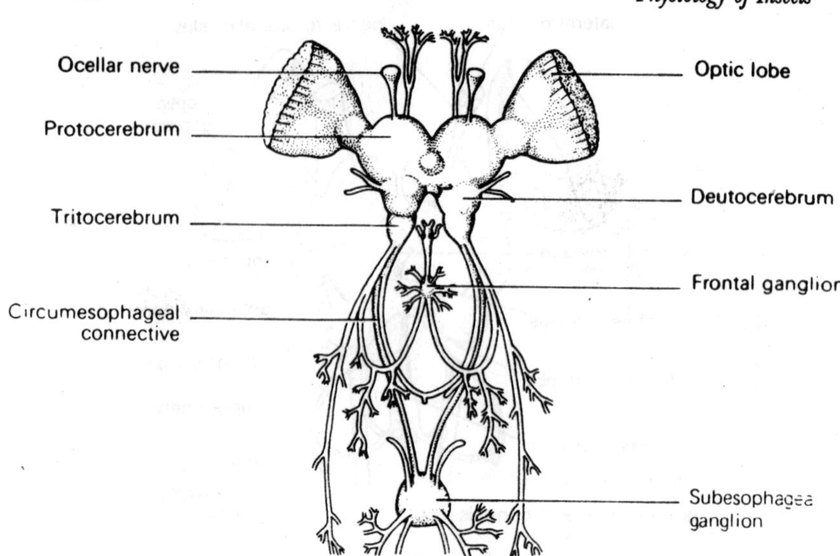

Fig. 8.5. Brain and stomatogastric nervous system of the grasshopper Dissosteira carolina. Anterior view.

Usually these ganglia remain distinct, but frequently the mesothoracic and metathoracic ganglia are united, and the definitive ganglion of the metathorax may include one or more primitive abdominal ganglia. In the insect abdomen there are at most eight definitive segmental ganglia, corresponding to the first eight abdominal somites, but the last is always a composite ganglion, since it innervates the eighth and succeeding segments. The nerve cord of the abdomen, however, is often variously shortened by the union of two or more of the posterior ganglia, and the ganglia are subject to a displacement anteriorly, in a more anterior segment. The nerves from each ganglion, however, consistently go to the segment in whcih the ganglion had its origin.

Hence, morphologically, a ganglion should be numered according to the segment in innervates, and the distribution of the nerves from a composite ganglion is usually the best index of the composition of such a ganglion. All the ganglia of the ventral nerve cord have a tendency to unite with each other in various combinations in different insects. An extreme condensation is attained in the larvae of cyclorrhaphous diptera, in which the entire ventral nerve cord, including the suboesophageal ganglion, is consolidated into an elongate mass of nerve tissue, from which the entire body is innervated.

The Brain and its Nerves

The insect brain is principally a center of association between the major sense organs located on the head and the motor neurones of the gnathal, thoracic, and abdominal regions of the body. Most of its bulk consists of a mass of neuropile tissue; but within this mass are contained, on the one hand, the roots of the nerves from the compound eyes, the ocelli, and the various sense organs of the antennae and the preoral cavity, and, on the other, the anterior terminals of nerve tracts from the suboesophageal ganglion and the ganglia of the ventral nerve cord in the thorax and abdomen. The brain, therefore, is necessary for the initiation ofall activities that are normally stimulated through the cephalic sense organs. It takes no part in the regulation of such activities, which are directly controlled from the centers of the suboesophageal and body ganglia. Hence a decapitated insect may be said to be incapable of "voluntary" action; but its vital functions continue in operation as long as its body tissues remain alive, and, if artificially stimulated through somatic receptors, many of the motor mechanisms can be set into normal activity. The insect brain contains but few motor neurones.

Internal Structure

In external form the brain varies much in different insects, but it always shows a differentiation into three successive parts, which are distinct at least in its internal organization and are usually more or less apparent in the external contour of the adult brain as three pairs of lateral swellings, or lobes. The first and largest part is the forebrain, or *protocerebrum,* the second is the midbrain, or *deutocerebrum,* the third is the hindbrain, or *tritocerebrum.* The lateral lobes of the protocerebrum and the deutocerebrum are united with each other by internal commissural tracts; the tritocerebral lobes are connected generally by a free nerve trunk; the *suboesophageal commissure,* that passes below the stomodaeum. From the tiritcerebral lobes there proceed posteriorly and ventrally the *circumoesophageal connectives* to the suboesophageal ganglion in the lower part of the head.

The lobes of the forebrain bear laterally the large optic lobes, which contain the complex visual centers of the compound eyes. The optic lobes are generally narrowed at their bases, and in some insects they are greatly elongate. The facial ocelli are connected

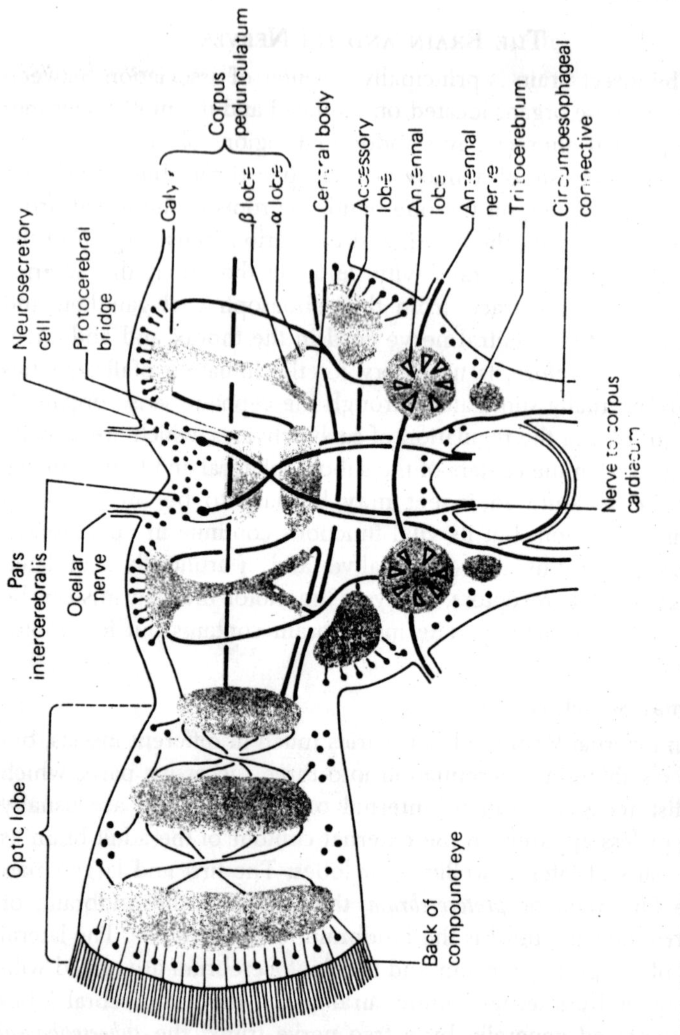

Fig. 8.6. Diagram showing major neuropilar regions (shaded) of the brain and some connections between these regions. Black dots indicate location of perikarya.

with the anterior or dorsal aspect of the protocerebrum by long slender stalks, the *ocellar pedicels* at the ends of which are conical enlargements containing the centers of the short ocellar nerves. The true nerve trunks of the brain arise principally from the deutocerebrum and tritocerebrum. The substance of the brain consists largely of a neuropile mass of intricately entangled arborizatins of association neurones, the cell bodies of which are located for the most part in the cortical region of the brain.

The only motor centers of the insect brain are situated in the deutocerebral and tritocerebral lobes, from which are innervated the antennal muscles, and probably the muscles of the labrum and some of the stomodeal muscles. In the decapod crustaceans the centers of the oculomotor muscles of the eye stalks are located in the protocerebrum. The principal features to be distinguished within the brain are special groups of cells, fiber tracts, and compact bodies formed of dense aggregations of association neurities and of glomeruli of their terminal arborizations.

The Nerves of the Brain

The principal nerves of insect brain are the nerves of the compound or simple lateral eyes, the dorsal ocelli, the antennae, the labrum, and the frontal ganglion connectives. In addition there may be present a doral tegumentary nerve, connectives with the occipital ganglia of the stomodaeal system, and sometimes other nerves.

Nervous opticus

The true nerves of the compound or simple lateral eyes are groups or retinal neurites received in the outer ends of the optic lobes, in which are located the optic centers. The optic nerves, therefore, are generally very short; but in insects having rudimentary optic centers, the optic nerves may be long trunks, as in the termites and in caterpillars. The lateral ocelli of coleopterous larvae, however, are developed in close proximity with the outer ends of the long optic lobes of the brain containing the centers of the future compound eyes.

Nervi ocellarii

The slender ocellar pedicels uniting the facial ocelli with the brain are commonly called the ocellar "nerves," but it has been shown by Cajal (1918) that the primary ocellar centers lie in the enlarged outer ends of the pedicels, since it is here that the inner ends of the retinal fibers are associated with the terminals of nerves from the brain that traverse the stalks. The true ocellar nerves, therefore, are the groups of retinal fibers that terminate in the outer ends of the ocellar stalks. The ocellar pedicels are comparable with the optic lobes of the compound eyes.

Nervous ganglio occipitals

This is short, slender nerve connective on each side proceeding from the back of the brain to the occipital ganglion of the

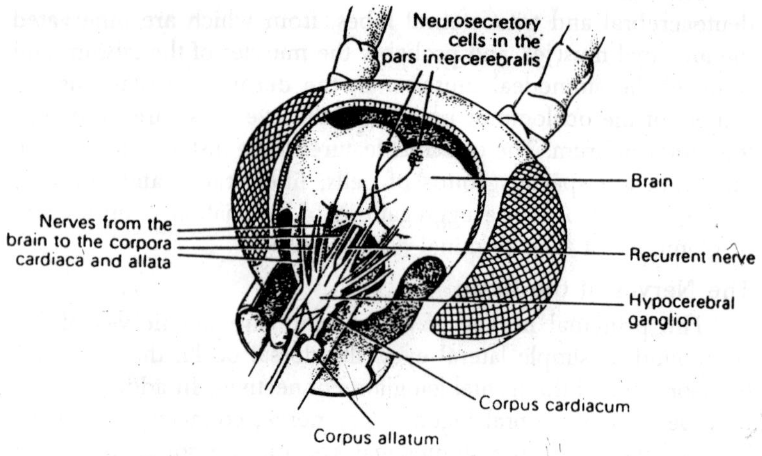

Fig. 8.7. Stomatogastric nervous system and endocrine tissues in an orthopteran species.

stomodaeal nervous system. The fibers of the occipital ganglion connectives, as shown by Holste (1923) in *Dytiscus*, originate form small groups of cells lying in the dorsal part of the protocerebrum, from which they traverse the calyx glomeruli of the corpora pedunculata to make their exist from the posterior wall of the brain.

Nervous antennalis

The antennal nerves have their roots in the deutocerebrum and are the only nerves gives off from this part of the brain in insects. Each nerve consists of both sensory and motor fibers, which are sometimes contained in a single trunk, and sometimes separated in sensory and motor branches. The sensory fibers come from the various sense organs of the antenna; the motor fibers go to the antennal muscles within the head and to those located in the scape of the appendage.

Nervous tegumentalis

A dorsal tegumentary nerve arises from the posterior or lateral surface of the brain in some insects and goes to the dorsal part of the head. The roots of this nerve in *Dytiscus*, according to Holste (1923), can be traced as far as the fibrous mass of the deutocerebrum close to the exit of the motor nerve of the antenna; but Hanstrom (1928) thinks that the dorsal tegumentary nerve must arise in the tritocerebrum, and that it belongs to the same system as the tegumentary nerve arises clearly from the base of the tritocerebrum. It goes dorsally close behind the brain and forks before the mandibular muscles into two branches distributed to the epidermis of the fastigial area between the compound eyes but apparently gives no branches to the muscles.

Nervous lateralis

This is a slender nerve present in lepidopterous larvae. It arises from the side of the brain just above the root of the circumoesophageal connective and divides into two branches. One

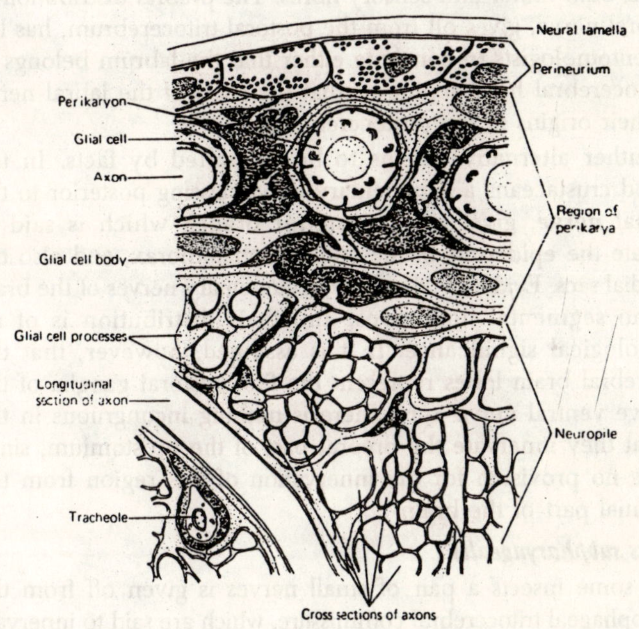

Fig. 8.8. Cross section of part of the caudal ganglion of the cockroach Periplaneta sp. (diagrammatic). Darkly shaded areas indicate extracellular spaces.

branch goes forward and ventrally to the facial region of the head lateral of the clypeal triangle, where it appears to innervate the mandibular muscles; the other branch turns posteriorly and units with the lateral occopital ganglion of the stomodaeal nervous system. Nothing is known of the central connections of this nerve or the origin of its fibers.

Nervous labrofrontalis

The labrofrontal nerve is a short trunk arising anteriorly from the tritocerebrum. It soon divided into a *frontal ganglion connective* and a *labral nerve.* The frontal connective goes anteriorly and medially to the frontal ganglion of the stomodaeal system, sometimes making a long anterventral loop, as in *Dissosteira,* from which are given off nerves to the labral muscles and the refractors of the mouth angles. The labral nerve proceeds to the labrum and probably contains both motor and sensory fibres. The preoral distribution of the labral nerve, gives off from the postoral tritocerebrum, has led some entomologists to conclude either that the labrum belongs to the tritocerebral head somite or that the roots of the labral nerve have their origins in the protocerebrum.

Neither alternative seems to be supported by facts. In the decapod crustaceans a tegumentary nerve, arising posterior to the antennal nerve, gives off a posterior branch which is said to innervate the epidermis of the entire cephalothorax, and also the nephridial sacs. Evidently, therefore, tegumentary nerves of the brain have no segmental limitations, and their distribution is of no morphological significance. If it is assumed, however, that the tritocerebral brain lobes represent the first postoral ganglia of the primitive ventral nerve cord, there is nothing incongruous in the fact that they innervate the preoral part of the prostomium, since there is no provision for the innervation of this region from the prostomial part of the brain.

Nervous subpharyngealis

In some insects a pair of small nerves is given off from the suboesophageal tritocerebral commissure, which are said to innervate the ventral dialator muscles of the stomodaeum. The tritocerebral commissure, however, is sometimes included in the circumoesophageal connectives and the suboesophageal ganglion, and in such cases the subpharyngeal nerves spring form the anterior end of the latter ganglion. In the acridid *Dissosteira* two median ventral nerves arise from the tritocerebral commissure, but they appear to innervate the neurilemma of the circumoesophageal connective and the suboesophageal ganglion.

Nervous postantennalis

Nerves of the postantennal appendages are entirely absent in insects, since these appendages are represented only by embryonic rudiments in the Hexapoda; but in Crustacea they constitute the principal nerves (second antennal nerves) of the tritocerebral ganglia.

The Protocerebrum

The forebrain, or protocerebrum, is the dorsal and largest part of the cerebral mass. It included the lateral *protocerebral lobes,* the median *pars intercerebralis,* and sometimes ventral *acessory lobes,* or *Nebenlappen.* Within the neuropile mass of the forebrain are to be distinguished groups of globuli cells, dense clusters of fibers and glomeruli forming the so-called "bodies" of the brain, and various fibrous tracts. The globuli cells of the brain are specialized association cells characterized by their small size, compact arrangement, and richly chromatic nuclei. Hanstrom (1930) distinguishes in the arthropod brain generally three primary paired groups of globuli, cells, namely, on each side, a *median* dorsal group, a posterior *lateral group,* and a ventrolateral *ventral group.*

The several globuli groups are subject to much variation in the extent of their development. In the insects the dorsal and ventral groups are reduced or usually absent, while the lateral groups become prominent elements in the brain of most Pterygota. The neurites of the globuli cells form some of the most important fibrous bodies of the brain. The fibrous and glomerulous masses of the protocerebrum include the dorsal *corpora pedunculata,* the median dorsal *pons cerebralis,* the *corpus central,* the ventrolateral *corpora ventralia,* and sometimes dorsal *corpora optica.* In addition to these bodies of the protocerebrum proper, however, there are connect with the protocerebrum the *optic centers* of the compound eyes situated in the optic lobes, and the *ocellar centers* located in the outer ends of the ocellar pedicels.

Pons cerebralis

The protocerebral bridge, lying in the dorsal and posterior part of the pars intercerebralis, is a transversely elongate body, usually horse-shoe shaped with the concavity forward or downward. The substance of the pons is mostly glomerulous; but according to Bertschneider (1921), the pons glomeruli of *Deilephila* forms tow

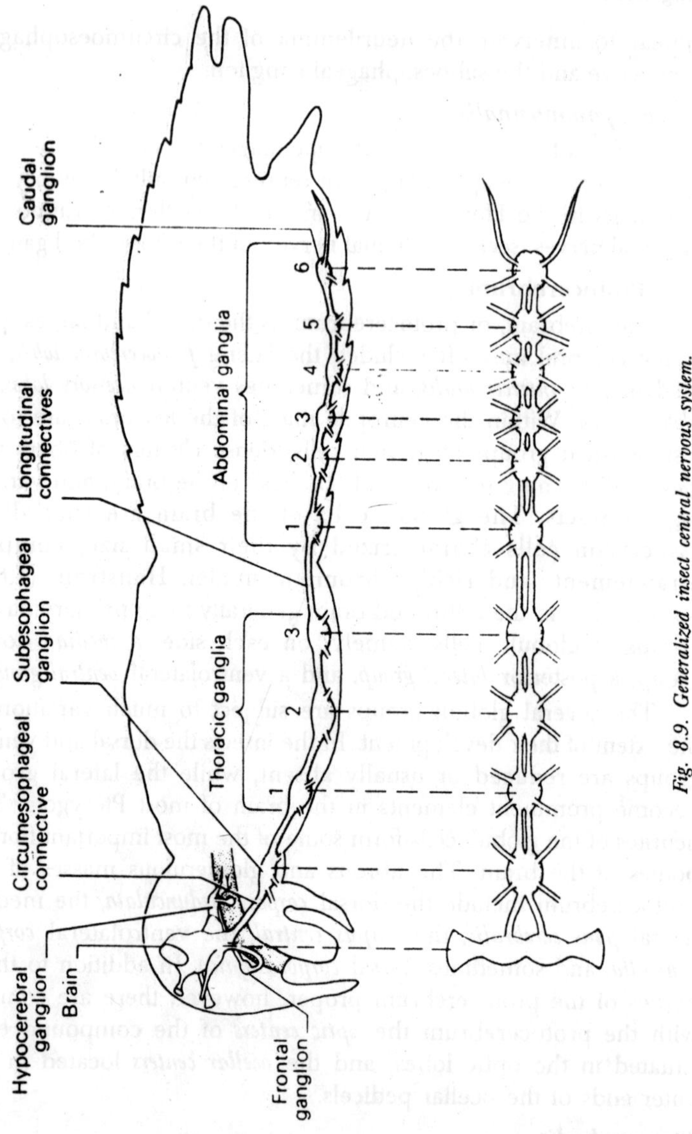

Fig. 8.9. Generalized insect central nervous system.

lateral swellings of the body with a fibrous commissure between them. The pons cerebalis is the "posterior dorsal commissure" of C.B. Thompson (1913), but it is evident from its structure that the body is an association center, since fibers enter it from many parts of the brain. In most arthropods there are associated dorsally with the pons the two cell masses of the dorsal globuli cells, and in such cases the neurites of these cells form the body of the pons. Dorsal globuli cells associated with the pons, however, are said by Hanstrom to occur among insects only in Apterygota and Ephemerida.

Corpus central

The body of the brain lies anterior or ventral to the pons. In the insects it consists of several distinct groups of glomeruli, which together form an oval or flattened mass with the long axis transverse. The subdivision of the central body constitutes the chief difference in internal structure between the brain of insects and that of Crustacea, in which the central body consists of a single mass of glomeruli. The central body has no nerve cells directly connected with it, but it is a most important center of association between the terminals of fibers form all other parts of the brain.

Corpora penduncukata

The pedunculata bodies (mushroom bodies, *pilzformigen Korper*) are situated in the dorsal part of the brain between the protocerebral lobes and the pars intercerebralis. In their typical form these bodies are mushroom shaped, as their name implies. Each consists of an expanded cap in the upper or posterior part of the brain covered with a mass of globuli cells, and of a large, thick fibrous stalk, or *pedunculus,* extending forward. The corpora pedunculata constitute the largest and most highly developed association centers in the brain of pterygote insects and are the most conspicuous features of the internal cerebral structure. The cellular caps of the pedunculata bodies are the lateral groups of protocerebral globuli cells, but in most insects and in many other arthropods each primitive cell group becomes subdivided into two or three distinct secondary groups of cells. The pedunculus of each body is formed to the anons of the globuli cells. When the globuli cells are separated into groups, therefore, each pedunculus contains as many confluent bundles of fibers as there are cell groups in the cap.

Immediately beneath the cell cap, the axons of the globuli cells give off short arborizing collaterals, which form synaptic associations with the terminals of incoming fibers from other parts of the brain. There is thus formed, corresponding to each cell group at the upper end of the peduneculus, a cup-shaped mass of fibrils and glomeruli, which is known as a *clayx.* According to the number of cell groups, each pedunculus may be surmounted by a

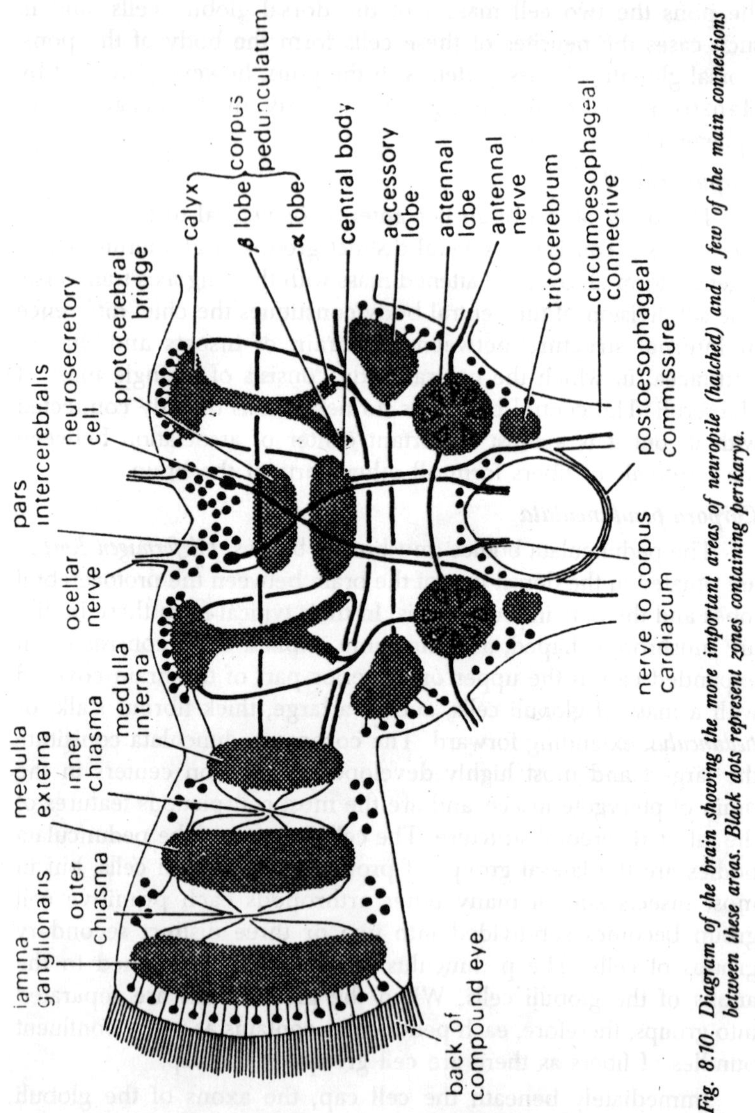

Fig. 8.10. *Diagram of the brain showing the more important areas of neuropile (hatched) and a few of the main connections between these areas. Black dots represent zones containing perikarya.*

single calyx or by two or three calyces. The pedunculi extend forward in the dorsal part of the brain and terminate in two large root branches.

One branch, the *median root* (Balken), goes inward front the main stalk of the pedunculus, and the two median roots from

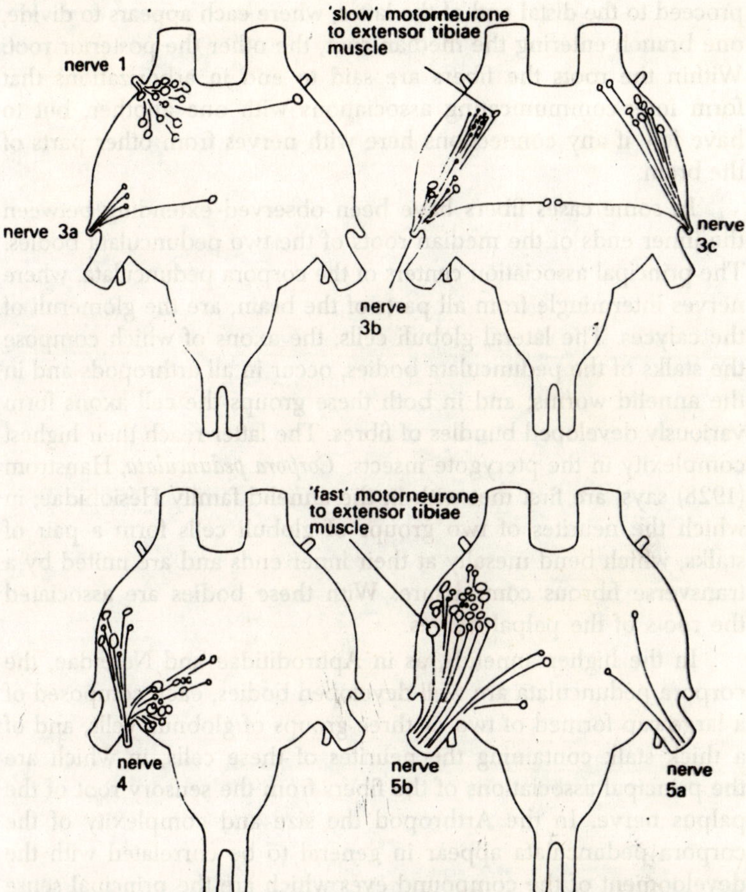

Fig. 8.11. Maps of the metathoracic ganglion of Schistocerca showing the distribution of perikarya of motorneurones associated with different peripheral nerves. Nerve 3 has three main branches. The 'axon pathways' are entirely diagrammatic and are intended only to indicate the association of the perikarya with particular nerves.

opposite sides usually end in proximity of each other, though in the Isoptera each again turns posteriorly and is extended toward the back of the brain beneath the central body and the pons. The other branch, or *posterior root* (cauliculus, *rucklaufige Wurzel*), goes posteriorly and dorsally anterior to the central body and the pons. In *Lepisma* each root of the pedunculus ends in a cluster of swelling (*Trauben*). The axons of the calyx cells entering the pedunculus proceed to the distal end of the latter, where each appears to divide, one branch entering the median root, the other the posterior root. Within the roots the fibers are said to end in arborizations that form inter-communicating associations with one another, but to have few if any connections here with nerves from other parts of the brain.

In some cases fibers have been observed extending between the inner ends of the median roots of the two pedunculata bodies. The principal association centers of the corpora pedunculata, where nerves intermingle from all parts of the brain, are the glomeruli of the calyces. The lateral globuli cells, the axons of which compose the stalks of the pedunculata bodies, occur in all arthropods and in the annelid worms; and in both these groups the cell axons form variously developed bundles of fibres. The latter reach their highest complexity in the pterygote insects. *Corpora pedunculata,* Hanstrom (1928) says, are first met with in the annelid family Hesionidae, in which the neurites of two groups of globuli cells form a pair of stalks, which bend mesally at their inner ends and are united by a transverse fibrous commissure. With these bodies are associated the roots of the palpal nerves.

In the higher annelids, as in Aphroditidae and Nereidae, the corpora pedunculata are well-developed bodies, each composed of a large cap formed of two or three groups of globnuli cells, and of a thick stalk containing the neurites of these cells, in which are the principal associations of the fibers from the sensory root of the palpus nerve. In the Arthropod the size and complexity of the corpora pedunculata appear in general to be correlated with the development of the compound eyes which are the principal sense organs directly associated with the protocerebrum; but, on the other hand, as in the Isoptera, the peduneulate bodies may be highly developed, though the eyes are small or absent.

Many comparative studies of the pedunculata bodies in insects show that the relative size of the organs gives a pretty fair index of the development of instincts and "intelligence"; and yet complex insects may be operative in larval forms, though, as Hanstrom (1925) has shown in the caterpillar, the brain centers are in a rudimentary stage of development. The presence of distinct corpora pedunculata in the brain of annelid worms, as demonstrated by Hanstrom, can lead only to the conclusion that the major part at least of the arthropod protocerebrum has been evolved directly evolved directly from a prostomial nerve mass corresponding to the archicerebrum of Annelida.

Corpora ventralia

The ventral bodies (lateral bodies, *Nebenlappen, parosmatische Massen*) lie in the ventrolateral parts of the brain just above the antennal glomeruli of the deutocerebrum. Some writers regard the ventral bodies as belonging to the deutocerebrum; but generally they are included in the protocerebrum, and they are united with each other by a transverse commissural tract that passes beneath the central body and the median roots of the corpora pedunculata. The ventral bodies, according to Hanstrom, are formed primitively of the association neurites of the ventral globuli cells, but these cells persist in only a few arthropods, and generally the ventral bodies consist only of masses of glomeruli. They are association centers having fibrous connections with the central body, the corpora pedunculata, the pons, the optic lobes, the antennal glomeruli, and other parts of the brain. The ventral bodies are usually not well-developed in the higher insects, though they appear to be of large size in both adult and larval Coleoptera, and, according to Bretschneider (1921), they are particularly large and highly elaborated in the Lepidoptera, the region containing them forming accessory lobes (*Nebenlappen*) of the protocerebrum.

On the outer surface of each of thse lobes on the back of the brain, Bretschneider says, there is a mass of cells (evidently the ventral globuli cells from which fibers stream into the ventral bodies as do those of the lateral globuli into the stalks of the corpora pedunculata. The relatively large size of the ventral bodies in the Lepidoptera Bretschneider regards as a primitive character in this order since the bodies in *Deilephila* are very similar to those of *Forficula,* and in both these insects they are the best connected parts of the brain. The size and complexity of the ventral bodies in insects generally, however, have an inverse relation to the development of the corpora pedunculata, the latter, Bretschneider believes, supplanting the ventral bodies in importance in most insects.

Corpora optica

Optic bodies are not generally present in the insect brain. In some of the Apterygota, however, according to Hanstrom (1928), there is in the dorsal part of the brain of small bodies lying above the pons cerebralis, which in *Machilis* are connected with the glomeruli of the ocellular nerves and with the medullae externae of the optic lobes. These optic bodies, therefore, are association center of both the ocelli and the compound eyes. Similar optic centers occur also in the Branchiopoda among the Crustacea.

The Optic Centers

The ganglionic centers of the lateral eyes contained within the optic lobes, are so intimately associated with the protocerebral lobes of the fully formed brain that they may be regarded as part of the protocerebrum, though they are distinct from the later their origin. As described by Wheeler (1891) in the Orthoptera, the optic lobes are formed from the outer edges of the procephalic ectoderm. Soon the scattered cells arrange themselves on each side of the primitive head region in four longitudinal rows similar to the eight median rows of neuroblasts that are to form the median part of the brain, and which are continuous with the eight rows of neuroblasts in the neural ridges of the postoral region.

The cells generated from the neuroblasts optic lobes however, Wheeler says, do not resemble those produced from the neuroblasts of the central strands and appear to multiply irregularly. The optic centers of all arthropods, regardless of the nature of the lateral eyes connected with them, have evidently had a common origin the prototype of which, or an analogous structure, is to be found in some of the annelid worms. In the polychaete family Eunicidae, Hanstrom (1926) descrbes a very simple optic center, located in an optic lobe of the brain, intervening between the eye and the cerebrum proper. The optic nerves consist of the short retinal fibers from the eye to the optic lobe. Within the latter the fibers break up into terminal arborizations that form associations with terminals from nerves of the optic tract, some of whcih arise from cells located within the optic lobe, while others have their origins in the brain itself and send their neurites into the optic center. The optic tract traverses the brain between the two optic lobes and probably has connections with other parts of the cerebrum. From this primitive optic center of the Eunicidae it is but a step to the more complicated but still very simple structure of the optic center in the branchiopod Crustacea.

Here, as shown by Hanstrom (1926); there are in each optic lobe two ganglionic bodies, a distal *lamina ganglionaris* and a proximal *medulla*, surrounded by ganglion cells. The postretinal fibers form the eye penetrate into the lamina, where their thickened terminal parts are associated with terminals from two groups of neurones. The neurocytes of one group lie distal to the lamina, and their axons extend proximally into the medulla, giving off arborizations in both optic meases; the cells of the other group are associated with the medulla and send their axons distally into the lamina, where they end in fine terminal branches. It is to be observed that there is here no crossing of the fibers between the two optic masses. Cells of another set belonging to the medulla have short fibers that end within the latter. The optic center, finally, is connected with the brain by neurones whose cell bodies lie proximal to the medulla and give off branching collaterals into the latter, while their axons form the optic tract extending proximally into the brain.

Some of the optic fibers end in the lateral part of the brain, but others go into the optic commissures situated above and behind the central body. The optic lobes of Diplopoda and Chilopoda conain likewise two optic masses; but in most Crustacea and in all Insects there are characteristically three principal association centers in each opitc lobe, namely, a distal *periopticon,* or *lamina ganglionaris,* a median *cpiopticon;* or *medulla external,* and a proximal *opticon* or medulla interna.

The connection between the eye and the periopticon remains essentially the same in all forms; but the number and variety of the opitc neurones, the structure and connection of the fibrous masses, and the associations of the fibers in the optic tract with other parts of the brain, all become increasingly complex with the progressive evolution of the compound eye and the function of "vision." The optic centers of insects are probably the most intricate nervous mechanisms developed among the arthropods. Space cannot here be devoted to a minute description of their details, and the student should consult particularly the work of Zawarzin (1914) on the optic lobes of the larva of *Aeschna,* that of Bretschneider (1921) on Lepidoptera (*Deilephila*), and that of Cajal and Sanchez (1915), in which are elaborately described the optic centers of the honey bee (*Apis mellifica*) and of the blow fly (*Calliphora vomitoria*). The grosser structure of the optic centers will be more easily understood from Bretschneider's figure of the optic lobe of *Deilephilo.*

Here it is seen that the postretinal fibers penetrate the basement membrane of the eye in bundles that enter the lamina ganglionaris. The lamina and the medulla external are connected by crossing fibers that form the *outer chiasma.* Peripheral cells in the outer part of the optic lobe send their axons into the medulla externa, which has a distinctly laminaed structure owing to the stratified arrangement of the terminals of the penetrating axons. The medulla interna is subdivided into two fibrous masses, which are connected by fibers that cross with those from the external medulla to form the *inner chiasma.* The medulla interna shows four layers of stratified fibrils within its substance, and it has elaborate fiber connections through the optic tract with various parts of the brain.

One bundle of fibers goes to the corpus ventra, another to the central body and the pons, a third forms a union with the bridge and the corpus pedunculatum, a fourth crosses beneath the central body in a commissural tract to the opposite eye, giving terminals into the central body, and a fifth rather large bundle traverses the ventral part of the brain going directly to the suboesophageal ganglion. Still other fibers end in the neuropile mass of the protocerebral lobe. The figure by Zawarzin, showing diagrammatically the relations of the nervous elements in the optic lobes of the larva of *Aeschna,* will give a clearer idea of the nature of the associations of the optic neurons within the several optic masses.

The medulla interna of the *Aeschna* larva is subdivided into four secondary parts. It will be seen here that the laminated structure of the fibrous bodies, especially of the medulla external, results from the alignment of successive groups of fibrils given off from the neurites traversing them. The fibers connecting the lamina with the medulla external form a distinct outer chiasma, and those between the medulla external form a distinct outer chiasma, and those between the medulla external and the medulla interna form a second inner chiasma. In addition to the interrupted fiber tracts extending from the eye to the brain through the three ganglionic centers, the proximal elements of which have their roots in the four parts of the medulla interna, there are also continuous fibers connecting the lamina ganglionaris and the medulla external individually with the brain. The optic centers of the blow fly, as depicted by Cajal and Sanchez, differ considerably in detail from those of the dragonfly larva and in some respects are more complex.

Most of the postretinal fibers end in the lamina ganglionaris, but some of them go through the outer chiasma and terminate in the medulla external. The medulla interna is subdivided into two parts, within which are symmetrically distributed the dichotomously branched terminals of the neurones of the lamina external. The optic tract is composed of fibers that connect with the medulla interna and with the medulla external, but not with the distal lamina. For a full description of the structure of the optic ganglia the student must have recourse to the papers above cited; the figures given here are but diagrams.

After following the wonderful maze of intricate detail in the nerve centers of the compound eyes, however, we are still at a loss to understand how the effect of light on the receptor organ is transformed by the optic apparatus into specific reflexes in the motor mechanism or into a perception of variations in light intensity, colour, form, and motion. This, fortunately, lies outside the subject matter of morphology. The lateral ocelli of endopterygote larvae are connected with the optic centers of the compound eyes, the latter being first developed in the pupa.

The lateral ocelli, therefore, appear to be temporary larval organs, as are most of the other special structures of the larva. The compound eyes are not evolved from the larval ocelli but are newly formed in the epidermis of the pupa, or in some cases in that of the larva. The retinal parts of the ocelli, at the time of metamorphosis, are withdrawn from the surface and degenerate, but in some insects remnants of them persist at the side of the optic nerve of the adult. With exopterygote insects, in which the compound eyes are formed in the embryo, the imaginal eye is often relatively larger than or of different shape from that of the nymph or larva. The enlargement of the eye involves an increase in the size of the lamina ganglionaris lying below it, and the increment in the lamina, as shown by Zawarzin (1914) in the dragonfly, is formed from a part of the latter that remains in an undeveloped embryonic conditions during the larval stage.

The Ocellar Centers

The ganglionic centers of the facial ocelli lie in the distal parts of the ocellar pedicles. It has been shown by Cajal (1918) that the inner ends of the short retinal fibers are here associated with the distal terminals of long fibers that traverse the ocellar pedicels from the brain; these fibers were mistaken by earlier writers for the ocellar nerves. The fibers from the ocellar centers, according to Cajal, go to the lower part of the brain where they are associated with the terminals of branches from the optic tracts of the compound

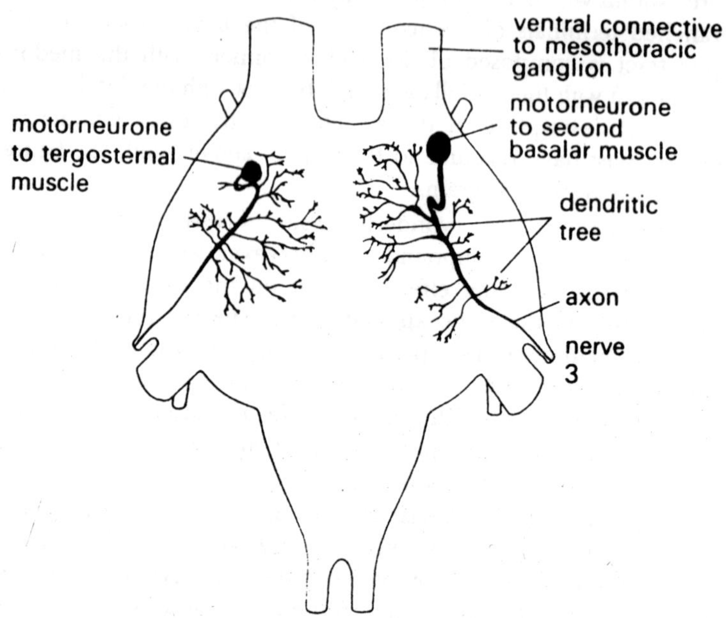

Fig. 8.12. Representation of two motorneurones with their dendritic branches in the metathoracic ganglion of a locust: left, fast motorneurone to a tergosternal muscle; right, fast motorneurone to the second basalar muscle.

eyes. Hanstrom (1928) believes that the neurocytes of the neurones of the ocellar tracts are large association cells lying mediodorsally in the pars intercerebralis. This appears to be the site of the primitive optic center of the brain, and in *Machilis* the ocellar tracts end here in lateral glomeruli. The median facial ocellus is said to have two strands of fibers in its pedicel, which in the Odonata make a chiasmatic crossing in the brain. Blackman (1912) records the occurrence in *Melanoplus femur-rubrum* of two distinct median ocelli, each complete in every respect, innervated through a bifurcate median pedicel, with a conical swelling at the end of each branch beneath the corresponding ocellus.

The Deutocerebrum

The deutocerebrum is the part of the brain containing the centers of the antennal nerves. Its lateral parts generally form a pair of distinct lobes in the adult brain, from which the antennal nerves arise. The sensory fibers of the antennal nerve trubks terminate in numerous glomeruli distributed principally in the periphery of the deutocerebral neuropile. The ganglion cells of the motor fibers going to the antennal glomeruli of opposite sides are connected by a fibrous *deutocerebral commissure* which traverses the lower part of the brain.

The Tritocerebrum

The tritocerebral part of the brain is relatively small in insects, owing to the absence of postantennal appendages and the consequent lack of nerves to those organs. The region of the tritocerebrum is usually evident as a pair of swellings or distinct lobes beneath the deutocerebral lobes, form which the circumoesophageal connectives proceed ventrally and posteriorly around the sides of the stomodaeum to the ventral head ganglion. The triticerebral lobes are connected with each other by a *substomodaeal commissure.* The commissure is often more or less united with the circumoesophageal connectives, and, when it is not evident as a free trunk, it is probably submerged in the connectives and in the anterior part of the suboesophageal ganglion.

The principal nerves of the tritocerebrum in insects are the frontal ganglion connectives and the labral nerves, but small nerves arising from the commissure must also have their roots in the tritocerebral ganglia. Morphologically, as we have seen, the tritocerebral lobes of the brain represent the first paired ganglia of the primitive ventral nerve cord and are primarily postoral in position, their union with the brain being secondary. The unique feature of the tritocerebral ganglia is their connection with both suprastomodacal parts of the nervous system, namely with the protodeutocerebral brain mass, and with the stomodaeal system.

Tł.e Fiber Tracts of the Brain

All the internal parts of the brain are intricately connected with one another by fibrous tracts formed of the axons of association neurons. Three of these tracts lying within the brain may be termed commissures, because some of their fibers at least go continuously across the brain between corresponding parts of opposite sides. A fourth tract of the same nature forms a free nerve trunk. The first commissure consists of fibers of the optic tract that connect the medullary bodies of the optic lobes with each other; the second unites the corpora ventralia beneath the roots of the pedunculate bodies; the third is the deutocerebral commissure traversing the ventral part of the brain between the antennal centers; the fourth is the tritocerebral commissure, called the suboesophageal commissure because it is usually a free nerve trunk passing beneath the stomodaeum. The other tracts run in all directions and connect the cerebral bodies with one another. The optic centers, as we have seen, are connected by fibers from the optic tract with the corpora pedunculata, the central body, the ventral bodies, and the antennal centers. The corpora pedunculata receive fibers from all parts of the brain as well as form the suboesophageal ganglion; and the central body has connections almost as extensive.

The ventral bodies are more important centers in some insects than in others; when well-developed they too have widely distributed connections. The largest tracts of the brain, however, go from the antennal centers to the calyces of the corpora pedunculata and here put the antennal sense organs in communication with fibers from all the other cerebral centers and from the ventral nerve cord. A study of the nerve associations in the brain suggests that the principal centers through which the sense organs of the head exert an influence on the motor mechanism of the rest of the body are the corpora pedunculata and the corpus centrale. As yet, however, neurologists have given much less attention to the connections between the cerebral centers and the centers of the ventral nerve cord than they have given to local associations within the brain itself.

The Ventral Nerve Cord

The ventral nervous system, as distinguished from the suprastomodael part of the brain and the stomodaeal system, consists of the postoral series of segmental ganglia and their connectives, constituting the so-called *ventral nerve cord.* Morphologically, as we have seen the ventral nerve cord begins with the tritocerebral ganglia of the brain and includes the primitive ganglia of the gnathal region of the head, as well as the ganglia of the thorax and abdomen. Since the general structure of the ventral nerve cord and the structure of the tritocerebral ganglia have already been described, we shall give attention here only to the composite suboesophageal ganglion of the head, the internal organization of a body ganglion, and the median nerves of the thoracic and abdominal ganglia.

The Suboesophageal Ganglion

The ventral nerve mass of the head is composed of the united ganglia of the primitive gnathal segments. The histology of this composite ganglion has been but little studied by precise neurological methods, which is unfortunate because of the long-standing dispute as to the number of ganglia that are contained in it. The suboesophageal ganglion innervates the mandibles, the hypopharynx, the maxillae, the labium, the salivary ducts, and at least some of the muscles of the neck. the nerve trunks contain both motor and sensory fibers.

The longitudinal nerve tracts that enter or traverse the suboesophageal ganglion are of great importance, since they contain the connective fibers between the sensory centers of the head and the motor centers of the body, but we have little detailed information concerning them. In addition to being the central organ of the gnathal and cervical nerves, the suboesophageal ganglion is also an inhibitory center of the body ganglia, since, with its removal, the somatic reflexes are found to become more readily excited by artificial stimuli.

General Structure of a Body Ganglion

For a full account of the internal structure of the ganglia of the ventral nerve cord the student must consult the detailed work Zawarzin (1924a) on the *Bauchmark* of the larva of *Aeschna*. A segmental ganglion of the thorax or abdomen is usually an oval or polygonal mass of nerve tissue, continuous anteriorly and posteriorly with the interganglionic connectives. From its sides proceed two or three principal *lateral nerves;* and in some insects a *median nerve* arises posteriorly, or also anteriorly, between the bases of the connectives. The ganglion is invested in a nucleated sheath, the neurilemma, which is continuous over the connectives and the nerves. The principal cellular elements of the ganglion are arranged peripherally, mostly in the lateral and dorsal parts. The central and ventral parts are occupied by a neuropile mass.

The lateral nerves of the ganglion that contain both motor and sensory fibers arise each from dorsal and ventral fibrous roots within the ganglion, the dorsal root containing the motor fibers, the ventral root the sensory fibers. Within the neuropile there may be distinguished five regions. Dorsally is the region of the dorsal interganglionic connective fibers. Beneath this is the motor center, or region of the dorsal nerve roots. Ventrally is situated the region of the ventral connective fibers, and immediately above it the sensory center, or region of the sensory ventral roots of the lateral nerves.

The central part of the ganglion contains the principal neurophile mass. At the ends of the ganglion only the dorsal and ventral fibers tracts are continued into the connectives. In each ganglion there are six principal groups of nerve elements: (1) the cell bodies and roots of the motor fibers of the lateral nerves; (2) the roots of the sensory fibers of the lateral nerves; (3) the cell bodies and fibers of the intraganglionic association neurones; (4) the cell bodies and collaterals of the interganglionic association neurones; (5) the cell bodies and roots of the motor fibers of the median nerve; and (6) the roots of the sensory fibers of the median nerve. The cell bodies of the motor neurones of the lateral nerves lie in the dorsolateral parts of the ganglion. Each is a large unipolar cell with a slender mediodorsal process, from which are given off finely branching collaterals into the dorsal motor center of the ganglion, while the main shaft of the fiber, or axon, turns outward to enter the dorsal part of a lateral nerve. The sensory fibers entering the ganglion from the lateral nerve trunks go the region of the sensory neuropile in the lower part of the ganglion, where some of them end in terminal arborizations, while others give off branching collaterals and then turn forward and proceed through the connectives and then turn forward and proceed through the connectives to some more anterior ganglion.

The sensory neuropile, therefore, contains fibers endings both of the lateral sensory nerves of the ganglion and of sensory fibers from the more posterior ganglia. The collaterals of some of the sensory nerves end in the side of the ganglion on which the nerve enters; others cross to the opposite side. The association neurones of the ventral nerve cord include local neurones of each ganglion and neurones whose axons from the principal fibers of the interganglionic connectives. The cell bodies of these neurones are situated in the lateral parts of the ganglion. The local, or intraganglionic, association neurones are of two types. In one type the nerve process is T-shaped, and the two branches lie in the same side of the ganglion as the cell body, one branch going dorsally, the other ventrally, to intermediate between the dorsal motor neuropile and the ventral sensory neuropile.

In the other type the neurone connects the two halves of the ganglion, a collateral being given off in one side, while the axon crosses to the other side, where it ends in terminal arborizations. The fibers of the interganglionic connectives originate from cells lying laterad of the intraganglionic neurons. The axons give off collaterals in the ganglion of their origin, some of which branch in the motor neurophile, others in the median or sensory neuropile, but the main neurites proceed either anteriorly or posteriorly through the connective tracts to other ganglia of the ventral nerve cord.

Zawarzin describes three types of connective fibers in an abdominal ganglion of the dragonfly larva: *tautomere* fibers, which leave the ganglion through the connective on the side of their origin after giving off a collateral in this side; *heteromere* fibers, which give off one collateral and then cross the ganglion to enter the connective of the opposite side; and *hekateromere* fibers which cross the ganglion but give a collateral in each side. Some of the connective fibers unit successive ganglia, others go long distances through the ventral nerve cord. The connective tracts pass superficially through the dorsal and ventral parts of the ganglia.

In the dragonfly larva Zawarzin distinguishes in each dorsal tract a median division, which contains fibers that go long distances through the nerve chain, and a lateral division containing shorter fibers; and in each ventral tract he finds an external median group of long fibers, an internal median group of short fibers, and a lateral group of short fibers. Besides these tracts of assocation connective fibers, there are two internal ventral tracts on each side which contain the sensory fibers that transverse the connections. The motor and sensory roots of the median nerve lie in the posterior parts of the ventral ganglia, but they will be described under a separate heading treating of the median nerves. The structure of a thoracic ganglion in the dragonfly larva is essentially the same as that of an abdominal ganglian, except that it is more complicated in all its details, owing to the presence of appendages on the thorax.

Since no exact study has been made on the histology of the thoracic ganglia of an adult winged insect, it is not known to what extent the nervous equipment is increased in the imago to serve the mechanism of flight. The distribution of the motor and sensory fibers from the lateral nerves in an abdominal segment of the larva of *Aeschna* is described by Rogosina (1928). The majority of the motor fibers in any one segment are derived from cells lying within the ganglion of that segment, but some of them come from the ganglion of the preceeding segment, the muscles of each segment thus having a plurisegmental innervation. In the second thoracic segment of the dragonfly larva, Zawarzin says, there ae only six pairs of motor nerve cells that supply fibers to the muscles of this segment. Since the number of muscles in the segment and its appendages greatly exceeds the number of motor cells innervating them, a single fiber must branch to several muscles.

It is interesting to observe that Rogosina finds in an abdominal segment of the same insect a corresponding number of peripheral sensory nerve cells of Type II innervating the muscles, the connective tissue, and the epidermis. The ganglia and nerves of the ventral nerve cord are abundantly supplied with trachea. In a caterpillar each ganglion receives a trachea on each side for the ventral tracheal commissure of its segment. Each ganglionic trachea divides at the root of the posterior nerve into anterior and posterior branches distributed to the ganglion, the lateral nerves and the connectives.

The Median Nerves

The ventral nerve cord of some insects, as we have been, includes a longitudinal *median nerve* lying between each pair of intergantlionic connectives. The median nerve takes its origin from the posterior part of the ganglion lying before it and gives off a pair of lateral branches that extend outward to the neighbourhood of the spiracles. In some cases the median nerve terminates at the bifureation into the lateral branches; in others it continues beyond the branches to the ganglion following. The occurrence of median nerves in different groups of insects has not been well-studied, but the nerves are commonly present in larval forms, and it is probable that where they are not present as independent trunks their fibers are buried in the interganglionic connectives and issue from the following ganglia in the anterior nerve trunk of the latter. The typical arrangement and distribution of the median nerves are well-shown in a caterpillar, in which there is a median nerve for each of the 11 ganglia of the ventral nerve cord posterior to the head.

In the thorax each median nerve appears to end at its bifurcation into the lateral branches, which are given off from a small triangular swelling; but in the abdomen a slender median filament continues to the next ganglion. The exact terminations of the fibers of the median nerves have not been determined, through the endings of the lateral nerves are usually found to be distributed to the tracheae and the spiracles. In the caterpillar each lateral branch goes outward over the inner face of the ventral muscles in the anterior end of the segment *behind* the one containing the ganglion in which the main trunk of the median nerve takes its origin. Along its course the lateral nerve gives off small branches and breaks up finally into terminal fibers distributed to the tracheal trunks in the neighbourhood of the spiracle, one of which innervates the occlusor muscle of the spiracle.

The nerve center of each segmental pair of spiracles is thus located in the ganglion of the preceding segment. The branches from the median nerve of the prothoracic ganglion go to the first pair of spiracles, which are primarily mesothoracic; those form the mesothoracic nerve go to the neighbourhood of the rudimentary metathoracic spiracles; the metathoracic nerves go to the first abdominal spiracles, and so on. If the last two ganglia of the ventral nerve cord are united, as in *Malacosoma americana* the branches of the seventh median nerve, which go to the eighth spiracles, issue from the dorsal surface of the eighth ganglion.

In *Malacosmoma* the median nerves of the abdomen branch close to the succeeding ganglion. In some caterpillars the median nerve trunk continues to the following ganglion, and its lateral branches are given off through the first pair of lateral nerve trunks of this ganglion. In the larva of *Aeschna* it has been shown by Zawarzin (1924), each median nerve contains two motor fibers and two sensory fibers. The motor axons originate from a pair of large unipolar cells lying in the posterior part of the ganglion from which the nerve proceeds. In the thorax each motor axon, after giving off numerous branching collaterals within the ganglion, turns posteriorly to enter the median nerve trunk and, at the bifurcation of the latter, divides into right and left branches, which go outward in the lateral nerves.

The sensory fibers, which are very slender and varicose, enter through the lateral branches, and those from opposite sides unite in the median nerve to form two fibers that run forward into the ganglion, where they end in fine branching terminals. In the abdomen the course of the motor fibers of the median nerve is quite different from that in the thorax. The two motor neurocytes are located superficially beneath the neurilemma, and their axons proceed posteriorly through the interganglionic connectives into the anterior part of the following ganglion. Here they turn messally and forward into the posterior end of the median nerve, which connects the successive ganglia, and finally branch in the usual manner where the lateral nerves are given off from the median trunk.

The sensory fibers, on the other hand, take the same course as in the thorax, entering the preceding ganglion through the part of the median nerve trunk lying anterior to the lateral nerves. Whether this difference in the disposition of the motor fibers of the median nerve between the thorax and the abdomen is general in insects or applies only to the dragonfly larva has not been determined. In its origin the median nerve appears to be derived from the median strand of nerve tissue formed in the embryo from the row of neuroblasts at the top of the ventral groove between the neural ridges.

According to Escherich (1902), the median nerve system in the embryo of the blow fly *Lucilia* consists at first of ganglionic cell masses located over the intersegmental lines of the body, and of intraganglionic strands traversing the segmental areas. From the posterior end of each median ganglion there is given off a pair of lateral nerves, presumably associated with the tracheal invaginations. It would thus appear that the definitive ventral nervous system of insects is derived from two distinct sources, the primarily lateral nerve strands and a primitive median nerve strand. The embryonic ganglia of the median nerve described by Escherich must eventually be included in the posterior parts of the definitive composite segmental ganglia, and in many insects the median strand is either obliterated or entirely united with the lateral strands.

The Stomodael Nervous System

The stomodael nervous system consists of sensory and motor neurones having their centers in small ganglia developed from the dorsal, or dorsal and lateral, walls of the stomodaeum. The nerve fibers are distributed to all the stomodaeal parts of the alimentary canal, and in certain insects they are continued over the entire length of the mesenteron. The latral muscles also, in some cases, appear to receive at least a part of their innervation from stomodaeal nerves, as do likewise the salivary ducts, the aorta, the corpora allata, and some of the mandibular muscles. This system centering in the stomodaeal ganglia is commonly called the "stomatogastric" or anterior "sympathetic" nervous system; but inasmuch as its ganglionic centers are derived from the stomodaeum, the term *stomodaeal nervous system* seems more fitting. Since the stomodaeal nervous system has not been fully studied from a comparative standpoint, it is impossible to give a general description applicable to all its numerous variations in different insects.

The one constant feature of the system is the presence of a median precerebral ganglion situated anteriorly on the dorsal walls of the pharynx. This is the *frontal ganglion.* The frontal ganglion is connected with the tritocerebral lobes of the brain by the *frontal ganglion connectives,* and from it there is given off a median *recurrent nerve,* which goes posteriorly on the dorsal wall of the pharynx beneath the brain and anterior end of the aorta. Sometimes one or two nerves proceed forward from the forntal ganglion to the region of the clypeus. In the back of the head there is usually a second nerve center of the stomodaeal system, which consists typically of a pair of ganglia lying just behind the brain. These ganglia may be termed the *occipital ganglia,* though they are variously called also "pharyngeal," "oesophageal," or "hypocerebral" ganglia. Each occipital ganglion is connected with the back of the brain by a short *occipital ganglion nerve* and communicates with the frontal ganglion by a branch form the recurrent nerve.

In some insects the paired occipital ganglia are united in a single median occipital ganglion situated beneath the aorta. In such cases the recurrent nerve ends in this ganglion, and the latter has two connectives with the back of the brain. Several nerves are given off from the occipital ganglia, or ganglion, but the pattern of the postcerebral stomodaeal innervation varies much in different insects. In a caterpillar, in which the occipital ganglia are widely separated, the large recurrent nerve proceeds posteriorly on the dorsal wall of the stomodaeum to the end to the crop, giving off along its course numerous lateral branches to the stomodaeal muscles. Each occipital ganglion has a connective with the back of the brain, and another with the recurrent nerve. From the second the aorta is innervated.

Laterally the ganglion gives off a short nerve to the mandibular muscles, and a nerve that goes forward and unites with the lateral nerve of the brain. From its posterior part a small nerve goes to the duct of the silk gland, and a larger nerve to the lateral wall of the crop. A more simple pattern of innervation in the postcerebral region is shown in the acridid *Dissosteira,* in which there is a single median occipital ganglion closely associated with the open, troughlike anterior end of the aorta, the latter being embraced by the short connectives between the occipital ganglion and the brain.

Three principal nerves are given off from each side of the ganglion. One goes laterally to the crorpus allatum; the second breaks up into branches distributed on the anterior part of the crop; the third and largest goes posteriorly on the lateral wall of the crop, giving off branches along its course, and ends in a lateral *ingluvial ganglion* ("gastric" ganglion) on the rear third of the crop, from which the posterior parts of the stomodaeum are innervated. According to the studies of Orlov (1924a) on the histology of the stomodaeal ganglia of the larva of *Oryctes nasicornis,* frontal ganglion and the occipital ganglion contain sensory, motor, and association nerve cells.

The distal processes of the sensory cells extend to the muscles and connective tissue of the stomodaeum. The frontal ganglion alone, however, contains association between the motor and sensory fibers. The sensory neurones of the second ganglion have no collaterals in this ganglion, but their axons extend forward through the recurrent nerve into the forntal ganglion, where they form associations with the motor neurones. The frontal ganglion not only contains the sensory-motor associations of the stomodaeal system but has connections with the brain and the ventral nerve centers by way of the frontal connectives from the tritocerebrum. It is said to be the center of the peristaltic movements of the oesophagus.

The Peripheral Nervous System

The peripheral nervous system includes the nerve trunks radiating from the ganglia, and the distal branches and terminal organs of the motor and sensory fibers contained in the nerve trunks. A full description of the peripheral nervous system, therefore, should contain an account of the distribution of all the nerves in the body; but since a subject of such magnitude could not be treated in a general text, we shall consider here only the terminal of the motor nerves and the distal endings of the sensory nerves.

The Sensory Neurones

When the sensory fibers of a nerve trunk are traced outward from the central ganglia, they are found to end in cells lying either within the epidermis of the body wall, or immediately beneath it, or on the somatic muscles or the wall of the alimentary canal. These cells appear to be the true neurocytes of the sensory neurones, for in insects there are no other nerve cells in the course of the sensory fibers, such as those of the spinal ganglia of vertebrates. The peripheral cells of the sensory nerves are either bipolar or multipolar. The distal processes of most of them go direct to specific ectodermal sense organs. Cells of this kind, which are always bipolar, are those designated *sensory cells of Type I.*

The others, which may be either bipolar or multipolar, but which are typically multipolar, are provided with one or more distal processes that branch elaborately and end with five varicose fibrils, which terminate on the inner surface of the body wall, on the somatic muscles and connective tissue, and on the muscles and wall of the alimentary canal. Cells of this kind are those distinguished as *sensory cells of Type II.* According to the nature of their neurocytes, therefore, the sensory neurones themselves may be classed as of Type I or Type II. The two groups of sensory neurones appear to be morphologically distinct. Those of the second type are undoubtedly the older, since they form the principal sensory innervation in the annelid worms, which consists of a diffuse branching of the terminal processes of the neurocytes on the inner surface of the epidermis. Sensory neurones of Type I are most numerous in arthropods having sclerotic plates in the body wall and are evidently developed as a means of circumventing the loss of sensitivity to external conditions, which would otherwise result from the hardening of the cuticula.

Sensory Neurones of Type I

The neurocytes of the first type of sensory neurones are the so-called sense cells of the specific ectodermal sense organs. They are always bipolar, and their distal processes are immediately connected with the cuticular parts of the receptors. The proximal processes are the centripetal axons which end in terminal arborizations within the central ganglia. In most cases the sense cell (or cells) of a sense organ lies within the epidermis in close association with the constructive cells of the receptor; but in various larvae the sense cells of the tactile setae distributed over the body lie beneath the epidermis, and their distal processes penetrate the basement membrane to enter the receptor. The development history of the intraepidermal sense cells has been carefully studied, and all investigators agree that these sense cells take their origin from undifferentiated ectodermal cells.

It seems certain, also, that the sense cells must in all cases be the neurocytes of the sensory nerves proceeding from them, and yet the growth of centripetal axons from the epidermal sense cells has not been demonstrated, and some investigators claim that the connection between the sense cells and the sensory nerves is established secondarily, which would imply, therefore, that the sense cells of insects are *secondary sense cells,* as are most of the sensory cells in the epidermis of vertebrates. The subject is fully discussed by Hanstrom (1928), who concludes that the arthropod sense cells are *primary sense cells,* but the exact origin of the nerves of the ectodermal sense organs of the Arthropoda appears yet to need further elucidation from the standpoint of development. Furthermore, the relation between the intraepidermal and subepidermal sense cells of Type I has not been determined.

Sensory Neurones of Type II

Sensory neurones of this type are particularly abundant in the annelida and in soft-skinned arthropods, such as the larvae of holometabolous insects, but they occur also in arthropods, including crustacea and insects, that have a sclerotized integument. Neurones of Type II are nerve connected with specific sense organs. Their neurocytes lie on the inner face of the body wall, on the muscles,or on the wall of alimentary canal, and their finely branching distal processes end in free terminals which innervate the epidermis, the somatic muscles, connective tissue, and the muscles and epithelium of he alimentary canal. The ontogenetic origin of these neurones from the ectoderm has not been determined. The integumentary innervation of the crayfish in well-known. That of *Astacus fluviatilis* has recently been studied in detail by Tonner (1933), who finds that practically the entire inner surface of the body wall is covered by a network of branching and uniting fibers from numerous multipolar nerve cells lying in the connective tissue beneath the epidermis.

In addition to this multipolar cell net, however, Tonner finds in *Astacus* also an inner integumental plexus of fibers branching from nerves given off from the ganglia of the ventral nerve cord. The two systems, moreover, are united by connecting fibers and together constitute an elaborate *integumentary nervous system.* Among insects, sensory cells of Type II are particularly abundant in soft-skinned holometabolous larvae, where, in some forms, they give rise to an elaborate subepidermal nerve net. In the larva of *Aeschna,* Rogosina (1928) finds in each abdominal segment just 12 cells of this type, there being on each side of each segment one cell located on the sternal region, three on the lateral region, and two on the tergal region. The axons of these cells enter the ventral ganglion of the segment through the first and second lateral nerve trunks.

The subepidermal innervation of the larva of *Melolontha vulgaris* is minutely described by Zawarzin (1912a). It consist of a network of large and small nerve branches distributed over the entire inner surface of the body wall, including the appendages, but is particularly developed on the middle of the back. The principal nerves of the net are the distal processes of irregular bipolar and multipolar sensory cells of Type II. The processes branch dichotomouly into the fibers of the larger meshes, and these ramify to form the threads of the finer meshes. The fibers of the larger meshes in *Melolontha* are relatively smooth, but the finer branches are characteristically varicose, presenting numerous small swellings along their courses, a feature noted by most writers who have studied the subepidermal innervation of other insects.

The actual endings of the fibrils have perhaps not been observed, but the terminal branches appear to end free on the inner surface of the basement membrane. Little is known concerning the sensory innervations of the muscles of the body wall and appendages of insects. Orlov (1924) suggested that the skeletal muscles may be innervated from the subepidermal nerves, and Rogosina (1928) has found in the lateral region of the abdomen of an *Aeschna* larva the distal process from a sensory cell of Type II branching both to the epidermis and to a muscle fiber. The terminals of the sensory muscle nerves within the muscle fiber differ by their tufted structure both from the sensory terminals of the epidermis and from the motor nerve endings in the somatic and visceral muscles.

In addition to the muscle innervation, Rogosina describes also an innervation of connective tissue in the *Aeschna* larva proceeding form the two sensory cells of Type II found in the tergal region of each abdominal segment. A sensory innervation of the alimentary canal has been described by Zawarzin (1916) in *Periplaneta americana* and by Orlov (1924) in larvae of scarabaeid beetles. Rogosina (1928) says there are no sensory nerves on the alimentary tract in the larva of *Aeschna*. Zawarzin finds numerous sensory cells of Type II, mostly multipolar, distributed over the walls of the crop of the cockroach. The distal processes break up into fine varicose fibers that innervate the epithelial cells of the crop and the neurilemma of the stomodaeal nerves and ganglia.

The axons of these cells go to ganglia of the stomodaeal nervous system, where they terminate in neuropile arborizations. The alimentary canal of scarabaeid beetle larvae, as descried by Orolov, is innervated both from the stomodaeal system and from the abdominal ganglia of the ventral nerve cord, there being in each group of nerves both motor and sensory fibers. The sensory neurocytes of the stomodaeal system are multipolar cells, the distal processes of which branch into varicose fibrils forming a network on the stomodaeum and ventriculus innervating the muscles and connective tissue. The sensory nerves of the proctodaeum, which have their roots in the abdominal ganglia, are distributed principally to the connective tissue of the anterior and posterior narrow part of the rectum.

The Motor Nerve Endings

From the few studies that have been made on the terminations of the motor nerves in insects, it appears that there are two types of motor innervation of the somatic muscles. In one type, probably characteristic of the more generalized insects, the ends of the nerve fibers branch diffusely and the branches run lengthwise upon the muscle fibers or wind around them. In the cockroach, according to Marcu (1929), the nerves do not enter the muscle fibers, the terminal branches ending free between them; but in a dragonfly larva, Rogosina (1928) says the nerve terminals, as shown in cross section, penetrate the sarcolemma and end among the fibrillae in the peripheral part of the muscle fiber. In the second type of muscle innervation the motor fiber ends against the muscle in a small flattened or conical body, sometimes called the *"end plate,"* or "Doyeres cone." These structures, however, as shown by Marcu in the somatic muscles of Coleoptera and Diptera, are merely the places on the muscle where the nerve suddenly breaks up into a brushlike group of fine branches that enter the muscle fiber directly and penetrate between the myofibrillae.

A single nerve fiber in this type of muscle innervation goes to but one muscle fiber; in the simpler type a nerve fiber may branch to several neighbouring muscle fibers. The motor nerve endings on the muscles. Thus, in the larva of *Aeschna,* according to Rogosina (1928), the nerves branch diffusely on the muscles. In the larvae of scarabaeid Coleopoptera, on the other hand, the motor nerves of the alimentary canal, as described by Orlov (1924), end in swelling within which each breaks up into a group of small varicose fibrils.

Synapses

Synapses occur where glial cells are absent so that the axons lie very close together, their membranes being separated only by a narrow synaptic gap. Synapses are believed to be characterised by foci of electron dense material in area 150-500μm long close to the cell membrane where the two axons are adjancent. There foci are found particularly in what is presumed to be the presynaptic fibre. Also indivative of a synapse, although occurring elsewhere, are small synaptic vesicles, again usually found in the presynaptic fibre. These vesicles are of two types, some about 200 to 500 A in diameter, which tend to aggregate in clusters and may contain the transmitter substance and others up to 1000A across which resemble neurosecretory droplets.

Mitochondria are most abundant in axons where synaptic vesicles occur. Synapses of another type, perhaps involving rather than chemical transmission, occur between the perikarya in the corpora pedunculata of *Formica* (Hymenoptera). There are gaps in the glial cells surrounding the perikarya so that the membranes of adjacent cells approach each other very closely, perhaps forming a compound membrane, but no accumulations of vesicles are present in the adjacent cytoplasm.

Nerve muscle junctions

The motor nerve supplying an insect muscle makes contact with each muscle fibre at a number of points. The nature of the ending varies considerably. In the simplest form, such as occurs, for instance, in the flight muscles of Diptera, the fine nerve branches pass longitudinally over the surface of the muscle or sometimes, as in *Tenebrio* (Coleoptera), the terminal axon is completely invaginated into the muscle fibre so that it makes contact with the muscle all round its circumference. In Orthoptera the axon divides at the surface of the muscle and the branches, with their sheaths, form a clawlike structure. Each junction may contain only one axon, as in *Tenebrio,* or more than one, as in *Blatta* (Dictyoptera). There is no known difference between the junction of fast and slow axons and it is probable that multiaconal junctions include both. The fine structure appears to be similar in all these forms. Glial cells lacking along the nerve muscle interface so that the axon and muscle fibre plasma membrane are close together, separated only by a synaptic gap of about 100 A. the Terminal axoplasm contains synaptic vesicles some 250-450 A across.

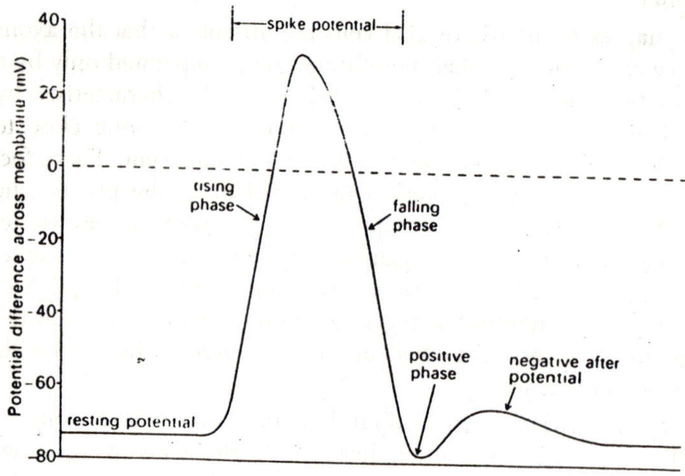

Fig. 8.13. Diagram of the changes in potential difference across the plasma membrane of an axon occurring during the passage of an impulse.

Physiology of the Nervous System

Stimuli may be perceived in a number of ways depending on the nature of the stimulus and the characteristics of the sense organs. The energy received by the sense cells as a result of stimulation is then transformed transduced to electrical energy and this leads to the production of a nerve impulse which travels along the nerve axon to the central nervous system. Here the impulse crosses a synapse and, directly or via one or more interneurons, continues along a motor neuron, crossing a final synapse before producing some response from an effector organ, usually a muscle.

Reception and Transduction of the Stimulus

Different types of stimuli, mechanical, chemical or visual, are perceived in different ways and involve different sense organs. Mechanical stimuli appear to cause some mechanical distortion of the receptor dendrite; chemical stimuli may act in a number of unknown ways, but it is suggested that sugars form complexes with specific receptor molecules at the receptor site; while visual perception probably involves the breakdown of some light sensitive pigment. Whatever the method of perception, the energy received by the sense cell is transformed to electrical energy. In some way not understood the stimulus affects the permeability of the plasma

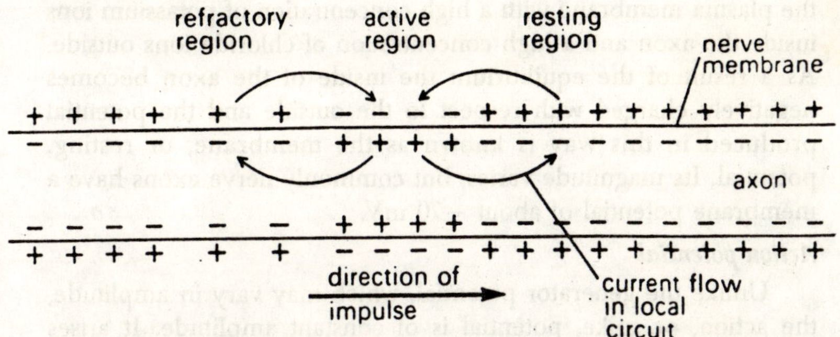

Fig. 8.14. Diagram of impulse conduction along a nerve axon. Conduction does not occur towards the left because the membrane is in a refractory state due to recent passage of the nerve impulse.

membrane of the dendrite so that it becomes depolarised. The potential produced in the dendrite by this depolarisation is called a receptor potential.

It varies in size, or is graded, according to the strength of the stimulus, a weak stimulus producing only a weak receptor potential. The receptor potential leads to the development of a generator potential, although it is probable that in insects these are often not differentiated from each other. The generator potential that is believed to arise in the region of the perikaryon. Like the receptor potential it is graded and if it exceeds a ceertain theshold value it triggers off the production of the all-or-none nerve impulse in the initial segment of the axon.

Nerve Impulse

The production and conduction of a nerve impulse in insects is not fully understood, it appears to be similar to the process in other animals. The account of the theory of conduction is therefore based to a large extent on the general concepts of nervous conduction.

Membrane potential

The ionic concentrations within an axon differ from those in the adjacent extracellular fluid despite the fact that the plasma membrane is freely permeable. Sodium is actively pumped out of the axon so that it concentration inside is much lower than outside and this movement is linked with an inward movement of potassium ions. Ionic movement is also influenced by large indivisible organic anions within the axon and a Donnan equilibrium is set up across the plasma membrane with a high concentration of potassium ions inside the axon and a high concentration of chloride ions outside. As a result of the equilibrium the inside of the axon becomes negatively charged with respect to the outside and the potential produced in this way is known as the membrane, or resting, potential. Its magnitude varies, but commonly nerve axons have a membrane potential of about – 70 mV.

Action potential

Unlike the generator potential, which may vary in amplitude, the action, or spike, potential is of constant amplitude. It arises from a depolarisation of the axon membrane associated with a change in permeability. When a nerve impulse is initiated the change in permeability is produced by the generator potential, but as the impulse passes along the axon the change is self-regenerative. The first change which occurs in a brief, but very marked, increase in permeability to sodium as a result of which sodium ions flow into the axon down the concentration gradient. This produces a rapid positive swing in the charge on the inside of the membrane, amounting to 80-100 mV. in the cockroach, representing the rising phase of the action potential.

Adjacent areas of axon are negatively charged so that a current flows in local circuit away from the point of depolarisation inside the axon and towards it on the outside. Where this current reaches an area of resting membrane it produces a slight depolarisation, of the order of 20 mV., so that the permeability to sodium rises and the charge on the inside of the fibre swings towards positive, becoming increasingly permeable as it does so. In this way a wave of increased permeability, and hence a nerve impulse, is propagated along the fibre without decrement. The period of permeability to sodium is short-lived and is follwed by a period of increased permeability to potassium as a result of which potassium flows out of the fibre which again becomes negatively charged on the inside. This is the falling phase of the action potential. Thus the total duration of hte spike or action potential is very brief, only one or two milliseconds. After the potential has returned to its resting level it overshoots slightly because of the high permeability to potassium. This is known as the positive phase and after it the potential swings back again to a level slightly higher than normal. This phase, the negative after-potential, results from the potassium which is released in the falling phase of the action potential

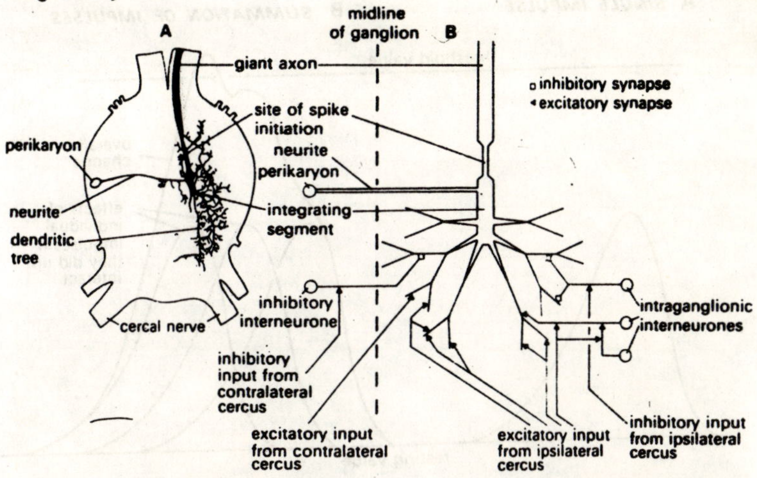

Fig. 8.15. A–Terminal abdominal ganglion of Periplaneta showing the origin of one of the giant fibres. B–Diagram showing the diverse sensory input to a giant fibre from both ipsilateral and contralateral sensilla and interneurones.

accumulating just outside the axon membrane so that the tendency for potassium to move out is reduced.

The negative after-potential persists for several milliseconds, but finally the membrane potential returns to normal. In insects the negative after potential is short-lived compared with vertebrates, probably because the lacunae between the glial folds provide relatively large spaces into which the potassium can diffuse quickly. Following the development of an action potential the ionic composition within the axon is altered, the sodium concentration has increased and the potassium concentration decreased. The ionic quantities involved are very small and it has been estimated that an axon 500m in diameter loses about a millionth of its potassium ions in the passage of one impulse. Nevertheless, if the axon is to continue functioning over long periods a recovery mechanism is required to bring the ionic concentrations back to their original values. This involves the sodium pump which actively extrudes sodium ions, probably in exchange for potassium. In this way the concentration of sodium is slowly lowered, while that of potassium is raised to its original level. Normally the stimulation of a nerve does not lead to the production of one nerve impulse, but of many.

Since these are all of the same amplitude, information concerning the stimulus can only be conveyed in the numer and

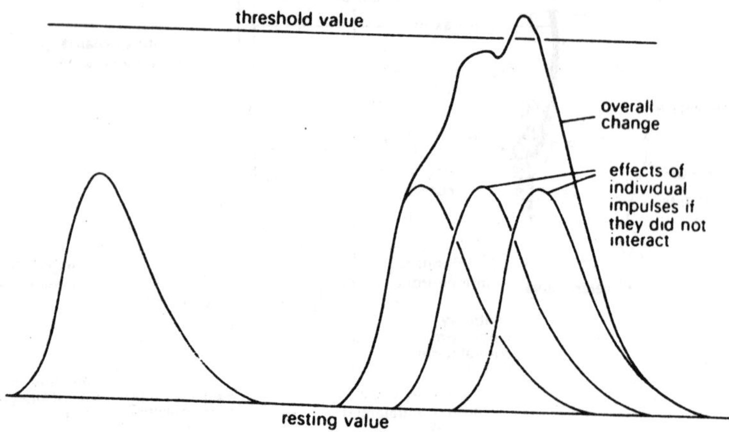

Fig. 8.16. Diagram illustrating the changes in potential of a postsynaptic membrane. A–A single impulse in the presynaptic fibre produces a rise in the postsynaptic potential, but this does not reach the threshold value and so no action potential is produced. B–The increases in potential produced by three impulses arriving in rapid succession are summed so that a large overall increase is produced, exceeding the threshold and leading to the development of an action potential.

frequency of the impulses, the latter being proportional to the size of the generator potential. There is, however, a limit to the frequency at which impulses can follow each other. In the presence of an action potential no further stimulation can initiate another impulse, nor can an impulse from elsewhere pass through the area. The period for which this is true is known as the absolute refractory period; it lasts for two or three milliseconds. Immediately after this an impulse can be produced, but only by a very strong stimulus and as the nerve recovers progressively weaker stimuli serve to produce impulses until the level of excitation returns to normal. This relative refractory period lasts for 10-15 msec. The velocity of conduction in a nerve fibre is proportional to its diameter. Giant fibres, with diameters of 8-50 m, have conduction speeds of 3-7m./sec.; normal fibres, of about 5m diameter, conduct at speeds of 1.5-2.3 m/sec. The velocity of conduction varies with temperature.

Effects of haemolymph composition on conduction

Since the passage of impulses in the nerve involves the movement of ions into and out from axon it follows that this movement, and hence the impulse itself, is influenced by the concentration of ions in the medium bathing the axon. If the level of potassium in the external fluid is raised there is less tendency for potassium to flow out of the axon and hence the membrane potential is reduced. Similarly a low level of sodium is the external medium reduces the height of the action potential. In many insects the concentration of sodium in the haemolymph is high, while the concentration of potassium is low, but in some particularly in herbivorous insect such as *Carausius,* the converse is true and there is much more potassium than sodium.

It might be supposed that this would influence nervous conduction in some way, but this is not so because the concentrations of ions in the extracellular fluid within the nerve sheath are quite different from those in the haemolymph, the sodium concentration greatly exceeding the potassium concentration. Even in *Periplaneta,* where there is more sodium than potassium in the haemolymph, the ionic concentrations within the nerve sheath are different. Thus, the nervous tissues lies in its own micro-environment which resembles, in ionic composition, the body fluids of other animals. The ionic concentrations within the nerve sheath of *Periplaneta* can be accounted for on the basis of a Donnan equilibrium between the fluid within the sheath and the haemolymph outside, but, apart from the calcium and magnesium concentrations, this is not the case in *Carausius.* Here sodium is actively pumped in from the haemolymph by the perineurium, but the mechanism by which potassium is maintained at a high level is not understood

Table 8.1. Ionic Concentration (in mM/litre in the Nerve Cord and Haemolymph of Carausius and Periplaneta

Ion	*Haemolymph conc*	*Nerve cord concentrations (estimated)*	
		Extracellular fluid	*Cell water*
		Carausius	
Na	20	212	86
K	34	124	556
Ca	6	12	62
		Periplaneta	
Na	157	284	67
K	12	17	225
Ca	4	18	14
Cl	184	107	–

Calcium is essential for efficient nerve conduction, being bound to molecules on the nerve membrane. When the calcium is displace, as it is temporarily on stimulation, the permeability of the membane increases. the calcium in the membrane is in equilibrium with that in the surrounding medium and lowering the level of external calcium results in progressive depolarisation of the membrane and complete nerve block. Thus calcium is an essential constituent of the haemolymph. In *Carausius* magnesium is also an essential constituent of the haemolymph, possibly being able to replace sodium to some extent in the development of an action potential. In other species high concentrations of magnesium block conduction.

Transmission at the Synapse

When an impulse has passed along an axon it must cross a synapse in order to stimulate another neuron or an effector. Transmission across synapse involves a chemical and it is suggested that substance concerned is stored in the synaptic vesicles. On the arrival of an impulse the membranes bounding the vesicles fuse with the cell membrane of that the transmitter substance is released into the synaptic gap. The transmitter substance becomes attached at receptor sites on the postsynaptic membrane and, through a change in permeability, cuases it to become depolarised. The magnitude of the depolarisation is presumed to be proportional to the number of vesicles released. Normally they are released in small numbers in a random fashion, producing miniature potentials in the postsynaptic dendrite. These miniature potentials, which in insect muscle average only 0.25 mV., are too small to initiate a nerve impulse or muscle contraction, but when a nerve impule arrives at a synapse it is believed that large numbers of synaptic vesicles are released simultaneously so that the postsynaptic membrane is strongly depolarised.

The resulting potential is normally big enough to initiate a nerve impulse or the activity of an effector. The depolarisation of the postsynaptic fibre is only short-lived because the chemical transmitter is rapidly hydrolysed by an appropriate enzyme. Within the nervous system the chemical transmitter is probably acetylcholine, but some other substance is involved at the nerve muscle junctions. Acetylcholine is destroyed postsynaptically by the enzyme acetylcholine, esterase; it is synthesised in the pre-synaptic fibre in a two-stage reaction, the second involving the acetylation of choline from acetyl coenzyme A in the presence of choline acetylase. The postsynaptic receptor sites of insects appear to be less sensitive than those of other animals, requiring much higher concentration of transmitter substances to effect transmission. This may reflect the high amino acid content of the extracellular fluid.

Amino acids exert both inhibitory and excitatory effects on synapses and it is essential that transmission should only be effected by postsynaptic changes greater than those produced by the amino acids normally present. Relative to the rate of conduction in the axons, the impulse takes a long time to cross a synapse. Synaptic delay frequently varies from about one to five milliseconds and, for instance, in the cockroach the passage of the single synapse in the system from the tip of the cercus to the front end of the giant fibres takes about 25% (1.5 msec. in 5.8 msec.) of the total conduction time. Delays may be much longer, as at the synapse between cockroach giant fibers and the motor fibers to the legs, where the average delay at the synapse is over twice as long as the total time taken for all the other events in the reflex.

Table 8.2. The Sequence of events and their Durations in the Startle Response of the Cockroach

Event		*Time in msec.*
1. Probable excitation time of cercal receptor		0.5
2. Conduction time in cercal nerve		1.5
3. Synaptic delay in afferent-giant junction		1.5
4. Conduction time in giant fibre		2.8
5. Estimated synaptic delay in giant-efferent junction		
	average	38.2
	minimum	13.2
6. Conduction time in fast motor nerve		1.5
7. Neuromuscular transmission		4.0
8. Development of contraction		4.0

Spontaneous Discharge

The production of an impulse in a nerve fibre normally results from stimulation of a sense cell, but many axons also discharge spontaneously in the absence of any sensory input. This discharge may be increased or decreased by sensory input, or it may remain unaffected, continuing in intermittent bursts over long periods. In

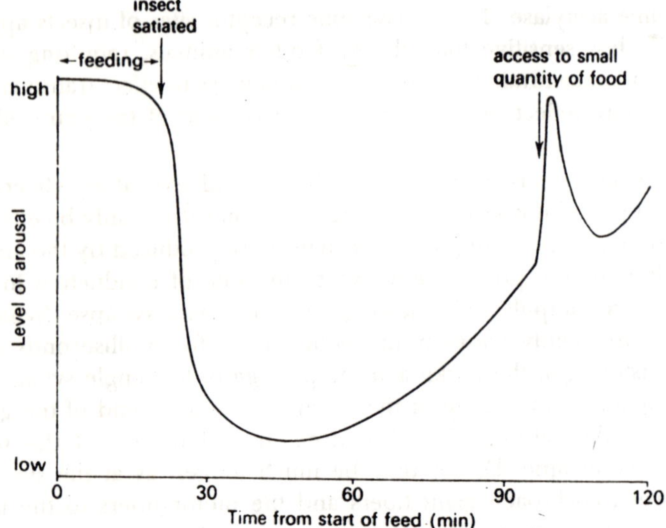

Fig. 8.17. Diagrammatic representation of the changes in the level of arousal of a locust in relation to feeding.

some cases this spontaneous discharge may be concerned with the maintenance of muscle tonus, but in general its effect is probably to maintain the system in a highly active stage so that it reacts readily to input and is more easily excited.

Stimuli which would be subthreshold in an unexcited fibre may produce a response in a fibre already discharging spontaneously. The rate of spontaneous discharge varies with temperature. For instance the output from an isolated ganglion of the cockroach increases with temperature up to a maximum and then falls off. The temperature at which the maximum occurs depends on the temperature to which the insects have been exposed previously, and the ganglia of insects kept at 22°C, have a maximum output at 22°C., while in insects kept at 31°C. the maximum is at 31°C. Sudden changes in temperature produce different effects in different fibers.

9

RECEPTORS

A continuous flow of information of the *change* in the environment must reach the central nervous system to maintain the normal life and to adopt according to the need of the moment. Instruments for registering and measuring such changes are the *receptor* and the system dealing with these receptors is called the *sensory system.* The sensory structures use the energy to do work, namely, to generate a message that can be conducted to a decoding area, the central nervous system, so that an appropriate response can be initiated. The message is, of course, in the form of a nerve impulse. Sensory structures are generally specialized so as to respond to only one energy form and are usually surrounded by accessory structures that modify the incident energy. As Dethier (1963) points out, the small size and exoskeleton of insects have had marked influence on their sensory and nervous system. Smallness and, presumably, short neural pathways would provide for a very rapid response to stimuli.

However, it would also mean relatively few axons and, therefore, a limitation in the number of responses possible to a given stimulus. This has led to a situation in insects where stimulation of a single sense cell may trigger a series of responses. Further, almost all insect sense cells are *Primary* sense cells, that is, they not only receive the stimulus but initiate and transmit information to the central nervous system; in other words, they are true neurons. In contrast, in vertebrates, almost all sensory systems include both a specialized (*secondary*) sense cell and a sensory neuron that transmits information to the central nervous system. The cuticle provides protection and support by virtue of its rigid, inert nature, yet sense cells must be able to respond to very subtle (minute) energy changes in the environment. Thus, only where the cuticle is sufficiently "weakened" (thinner and more flexible) will the energy change be sufficient to stimulate the cell. An insect, therefore, must strike a balance between safety and sensitivity.

In contrast to mammalian skin, which has millions of generally distributed sensory strcutures, the surface of an insect has only a few thousand such structures, and most of these are restricted to particular regions of the body. Two broad morphological types of sense cells are recognizable, those associated with cuticle (and therefore including invaginations of the body wall) (Type I neurons)

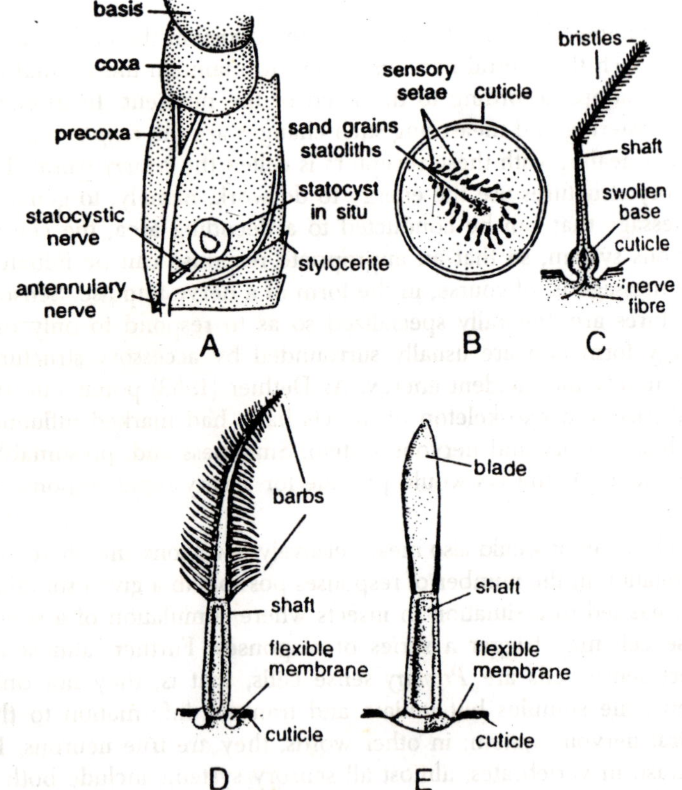

Fig. 9.1. Palaemon. A–Precoxa of antennule showing statocyst in situ. B–T.S. of statocyst. C–A single receptor seta (highly magnified). D–A tactile seta. (tangoreceptor). E–An olfactory seta (chemoreceptor).

and those which are nerve associated with cuticle and lie on the inner side of the integument, on the wall of the gut, or alongside muscles or connective tissue where they function as proprioceptor (Type II neurons). A type I neuron and its associated cells are derived embryonically from the same epidermal cell. They and the associated cuticle form the sensillum (sense organ). All types of sensilla, with the possible exception of the ommatidia of the compound eye, are homologous and derived from cuticular hairs. It is convenient to consider insect sense organs under the following six main headings but a few unusual forms of sensory perception are discussed separately.

1. Mechanoreceptors.
2. Auditory Organs
3. Chemoreceptor
4. Temperature Receptors
5. Humidity Receptors
6. Visual Organs

It may be noted that auditory organs are actually only specialized types of mechanoreceptor while humidity perception may ultimately prove reducible to one of the other mechanisms.

Mechanoreceptor

These are the receptors for the sensation of touch, pressure and pain etc. The members of this diverse group have the common property that they are excited by processes involving the mechanical deformation of some part of the receptor. As such they may mediate the sense of touch, including contact with solid objects or curents of air or water; they may also respond to mechanical stresses set up in the cuticle and so function as proprioceptors, including the specialized organs of equilibrium; and, as mentioned above, in certain cases they respond to changes in air pressure or the displacement of air-particles and so act as sound receptors. Four main structural types of mechanoreceptor may be distinguished:

Articulated Sensory Hairs

Each sensory hair or *trichoid sensillum* comprises the usual trichogen and tormogen cells which secrete respectively the hair and its socket together with a bipolar neuron which is produced distally into a specialized structure containing arrays of microtubules and in close contact with the base of the hair. Such hairs are widely distributed over the insect body, being stimulated to produce a

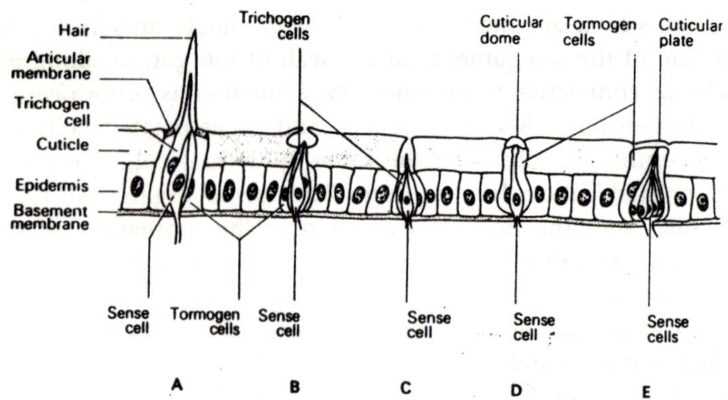

Fig. 9.2. Types of sensilla related to the trichoid type (diagrammatic). A–Trichoid. B–Coeloconic. C–Ampullaceous. D–Campaniform. E–Placoid.

nervous impulse on movement of the hair in its socket. *Trichoid sensilla* are often grouped together into 'hair plates' or 'bristle fields' where they act as proprioceptors and thus subserve many different biological functions.

On the vertex of *Locusta* and *Schistocerca* they form a cephalic air-flow receptor that helps to regulate flight, and in *Grylius* males a pad of sensory hairs on each fore wing controls their stridulatory movements. In ants and other Hymenoptera, bristle fields occur on the neck, antennae, petiole and legs, enabling the insects to orient themselves with respect to gravity; in *Apis* workers cervical hair plates and sensory hairs on the mendibles and antennae are needed for normal comb construction. Reflexes involving the trichoid sensilla of the tarsi and the coxal joints are important in the nervous control of posture and wallking.

Companiform Sensilla

These are comparable to the innervated hairs in that two specialized epidermal cells and a bipolar neuron are associated with each sensillum; structures intermediate between the two forms occur in *Drosophila* mutants. A companiform sensilum consists essentially of a dome-shaped region of relatively thin cuticle, oval in surface view and set a little above or below the general level of the integument, with a denser rim of cuticle surrounding it. The distal process of the neuron is packed with microtubules, partly arranged into a cilium-like structure; its tip is covered with a dense secretion bound tightly to the cuticle of the dome. Stresses in the cuticle are thought to be resolved into displacement of the dome which stimulate the distal process, though the detailed mechanism of transduction (as in other mechanoreceptors) is not fully explained. Campaniform sensilla occur on various parts of the body, including the cerci, wings and legs.

In *Periplaneta* there is one at the base of each large tactile spine on the legs and in *Schistocerca* the companiform sensilla near the base of the subcostal vein play a part in regulating the twisting of the wing during flight. Large numbers of companiform sensilla occur arranged in parallel rows over the base of he haltere of Diptera and are of major importance in flight control. In the ant *Atta cephalotes* three groups of campaniform sensilla at the junction of the trochanter and femur can perceive vibrations of the substrate occurring with frequencies between 0.05 and 4 KHz.

Chordotonal Organs (Scoloparia)

These are usually compound structures composed of a number of specialized sensilla (*Scolopohores* or *scolopidia*) which each contains a relatively conspicuous sensory body, the *scolop* or *scolopale*. They

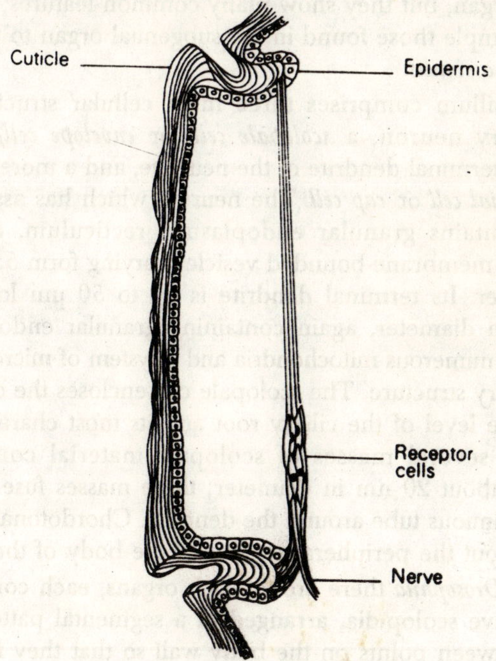

Fig. 9.3. Simple chordotonal organ in a beetle larva.

are sub-cuticular, often with no external sign of their presence. Each scolopidium consists essentially of three cells arranged in a linear manner; the neuron, and enveloping, or scolopale cells, and an attachment, or cap cell. The scolopidia in the locust tympanal organ also possess a fibrous sheath-cell round the base of the dendrite. In this case, at least, the dendrite ends in a cilium-like process containing a peripheral ring of nine double filaments and with roots extending proximally within the dendrite. The tip of the cilium lies in a hollow of the extracellular scolopale cap. Within the scolopale cell is is the tubular scolopale.

In *Locusta* this is made up of five to seven rods of fibrous material arranged in a ring and enclosing an extracellular space round the cilium. The contents of the extracellular space are not known. The attachment cell connects the sensillum to the epidermis. The two types may be distinguished respectively as *integumental* (*amphinematic*) and *subintegumental* (*mononematic*) chrodotonal organs though the distinction is probably not very important. The detailed structure of the scolopophore varies somewhat, even within a single chordotonal organ, but they show many common features; we can take as an example those found in the subgenual organ to the fore leg of *Gryllus assimilis*.

Each sensillum comprises three main cellular structures: a bipolar sensory neuron, a *scolopale cell* (or *envelope cell*) which surrounds the terminal dendrite of the neurone, and a more distally placed *interstitial cell* or *cap cell*. The neuron, which has associated glial cells, contains granular endoplasmic recticulum, a Golgi apparatus and membrane-bounded vesicles varying form 55 to 105 nm in diameter. Its terminal dendrite is up to 50 µm long and about 2 µm in diameter, again containing granular endoplasmic reticulum with numerous mitochondria and a system of microtubules forming a ciliary structure. The scolopale cell encloses the dendrite from about the level of the ciliary root and its most characteristic inclusion are several masses of scolopale material containing microtubules about 20 nm in diameter; these masses fuse distally to form a continuous tube around the dendrite. Chordotonal organs occur throughout the peripheral regions of the body of the insect.

In larval *Drosophila* there are 90 such organs, each containing from one to five scolopidia, arranged in a segmental pattern and suspended between points on the body wall so that they function as proprioceptors. In *Melanoplus* (Orthoptera) there are 76 pairs. In the thorax of many insects there are large chordotonal organs, containing about 20 scolopidia; these record the movements of the head on the thorax. Others in the wing bases of some insects, but not Orthoptera, record some of the forces which the wings exert on the body. In *Apis* there are three such organs, each with 15-30 scolopidia, at the base of the radial and subcostal veins and in the lumen of the radial vein. Typically four chordotonal organs occur in each leg.

The first is attached proximally within the femur and distally is inserted into the knee joint. In *Machilis* this organ contains seven scolopidia; in grasshoppers there are about 300. Proximally in the tibia is the subgenual organ and distally another chordotonal organ, which in *Apis* contain about 60 scolopidia, arises in the connective tissue of the tibia and is inserted into the tibio-tarsal articulation. Finally, a small organ with only about three scolopidia extends from the tarsus to the pretarsus. These organs function as proprioceptors monitoring the positions of the leg joints.

Subgenual Organs

The subgenual organ is a chordotonal organ usually containing between 10 and 40 scolopidia in the proximal part of the tibia. It is not associated with a joint. Processes from the accessory cells at the distal ends of the scolopidia are packed together as an attachment body which is fixed to the cuticle at one point, while the proximal ends ae supported by a trachea. Often the organ is in two parts, one more proximal, called the true subgenual organ by Debaisieux (1938), and the other slightly more distal. Both are present in Odonata, Dictyoptera and Orthoptera where in Tettigoniodea and Grylloidea the distal organ probably gives rise to the intermediate organ and the crista acoustica. In Homoptera, Heteroptera, Neuroptera and Lepidoptera only the distal organ is present, while in *Machilis,* Coleoptera and Diptera no subgenual organ is present at all. These organs are sensitive to vibrations of the substratum and to airborne sounds if these are of sufficiently high intensity to cause the leg or substratum to vibrate. They are extremely sensitive and in *Periplaneta,* for instance, respond to a displacement of only $10^{-9} - 10^{-7}$ cm.

The subgenual organs of *Periplaneta* are sensitive to vibrations at frequencies up to 8 kc./sec., with an optimum at 1500 c./sec. *Callophora,* which has no subgenual organs, on the other hand, only responds to displacements greater than 10^{-5} cm. within the frequency range 50-1000 c./sec. In this case the sensilla which respond to vibration are the tibio-tarsal chordotonal organ, the tarsal hairs and hair beds between the joints of the legs. The response of the subgenual organs of *Periplaneta* is synchronous with the stimulus frequency up to 50c./sec, but at higher frequencies is asynchronous. In nature it is unlikely tht pure tones occur, rather the insect is subject to pulse-like vibrations transmitted through the substratum and including the high frequencies as transients. Hence, it is important that the organs should be able to perceive the high frequencies. The mechanism by which the subgenual organs are stimulaed is not understood.

Possibly vibrations of the leg create vortices in the haemolymph within the leg and these move the chordontonal organ so that the sense organs is stimulated. Alternatively, the vibrations of the leg may cause the organ to vibrate with its own natural frequency, but, because the attachment cells are bound together and attached to the cuticle, they have a different natural frequency from the scolopale cells which are relatively free. Hence the proximal and distal part of the organ will vibrate at different frequencies so that rapid and complex changes occur in the forces acting at the junction between the two parts and these rapid changes serve to stimulate the sense cells.

Johnston's Organ

Johnston's organ is a chordotonal organ lying in the second segment of the atnenna with its distal insertion in the articulation between the second and third segments. It occurs in all adult insects except Collembola and Diplura, and, in a simplified form, is present in many larvae. It consists of a single mass of several groups of scolopidia and is most highly developed in the males of Culicidae and Chironomidae where the pedicel is enlarged to house the organ. In Culicidae the base of the antennal flagellum forms a plate from which processes extend for the insertion of the scolopidia. The latter are arranged in two rings all round the axis of the antenna and in addition there are three single scolopidia which extend from the scape to the flagellum. Johnston's organ perceives movements of the antennal flagellum.

In *Calliphora* most of the sensilla comprising the organ give phasic responses, potentials only developing during and immediately after the movement so that a single to and fro movement of the flagellum produces an 'on' and an 'off' response. The amplitude of the 'on' response increases with stimulus intensity due to different units having different thresholds and, if the stimulus is of very short duration, the 'off' response may be completely suppressed. Hence at high frequencies of stimulation small changes in stimulus pattern may produce major changes in excitation. Some of the sensilla respond to movement in any direction, others only if they are moved in a particular direction, stretchig being the effective stimulus. Since movement of the flagellum may result from a number of causes Johnston's organ may serve a variety of functions in any one insect. In *Calliphora* Johnston's organ acts as a flight speed indicator and is concerned with the maintenance and control of flight speed. Wind blowing on the face cuases the arista to act as a lever, rotating the third antennal segment on the second. This stimulates Johnston's organ which, because of the phasic nature of most of the sensilla, responds primarily to changes in the degree of rotation of the third segment.

Even in a steady airflow the antenna trembles, perhaps because of the setting up of eddies, so that Johnston's organ is stimulated. Since with an increased angle of rotation more scolopidia are stimulated it is possible that, as a result of the trembling of the antenna causing continuous restimulation, Johnston's organ might five a measure of the degree of static deflection of the third antennal segment as well as changes in its position. Johnston's organ also perceives sounds carried through the air to *Calliphora*, but its use as an organ of hearing a best known in Chironomidae and Culicidae where it enables the males to locate the females by their flight tone. The males of these insects have plumose antennae with may fine, long hairs arising from each annular joint. These hairs are caused to vibrate by sound waves and their combined action produces a movement of the flagellum.

The amplitude of flagellar movement is greatest near its own natural frequency which approximately corresponds to the flight tone of the mature female and stimulation at this frequency leads to the seizing and clasping response in mating. *Aedes aegypti* males are most readily induced to mate by frequencies between about 400 and 650 c. sec., but the limits to which they respond become wider as the insect gets older and are wider in unmated than in mated males. Sounds at other frequencies and high intensites produce a variety of reactions: cleaning movements; jerking, flight or freezing. In order to find the female, the stimulated male must be able to determine the direction from which the sound is coming. It is suggested that if the sound waves are parallel with the flagellum the base plate is pushed in and out.

The sensilla of the inner ring of Johnston's organ respond when they are stretched, that is when the plate is pushed in, so that their output frequency will be the same as the stimulus frequency. The output of the outer ring of sensilla, however, will be double this because some cells will respond to the 'in' contain the basic stimulus frequency an its first harmonic. If the sound waves are at right angles to the flagellum, the latter rocks on its axis so that the output from both rings of sensilla will be vary, the basic frequency. At intermediate positions the output will vary, the basic frequency progressively disappearing as the parallel component of the sound waves becomes less. By this means the insect might be able to orientate to the sound. The dominant forces acting on the antenna of ants are perceived by Johnston's organ, as presumably they are in any insect, and this may play a part in their orientation to gravity.

Johnston's organ is concerned with the orientation of *Notonecta* in the water. An air bubble extends between the head and the antenna so that when the insect is correctly orientated on its back the antenna is deflicted away from the head. If, however, the insect is the wrong way up the antenna is drawn towards the head and Johnston's organ registers the position. Gyrinid beetles are able to perceive ripples on the water surface, apparently due to displacements of the antennal flagellum, so that they are able to avoid collisions with other insects and, by echolocating by means of their own ripples, also avoid the sides of their container. In order to do this the insect must also be able to detect the direction from which the ripples are coming.

Auditory Organs

Though noises may be transmitted through and solid media, little is known of their reception by insects in such cases and it is convenient here to regard sound as consisting of air-disturbances of low intensity, irrespective of whether they fall within the range of human hearing. Such disturbances produce both a local pressure increase and a displacement of air-particles away form the source of the sound, and it is probable that some insect auditory organs act as displacement receptors while others may be pressure receptors. They are, in fact, mechanoreceptors which differ from tactile sense organs only in that they respond to disturbances of much lower intensity and further study may well reveal receptors of an intermediate character. An important elements in hearing is the time constant of the sense organs concerned: that is, the time taken for return to the original state after the stimulus ceases.

For the human ear this is about 20 milliseconds, so that sounds that follow one another at intervals of less than about one-tenth second will not be distinguished. The time constant in insects may be as low as 2-milliseconds or less, and they can distinguish sounds separated by intervals as small as one-hundredth second. The insect ear is therefore well-adapted to recognize the intervals between very rapid pulses of sound. In spite of their underlying similarity, four types of sound receptors may be recognized.

Tympanal Organs

These are paired structures always composed of a thin cuticular membrane, the *tympanum,* associated with tracheal airsacs and chordontonal sensilla. They occur in the Orthoptera, the Cicadidae and some Lepidoptera. In the Acridoidea there is a tympanal organ on each side of the first abdominal tergum, with an externally visible tympanum surrounded by a cuticular rings. Internally, a group of numerous scolopophores form a swelling known as Muller's organ, applied to the inner surface of each tympanum and connected by the auditory nerve to the metathoracid ganglion. Two sclerotized processes and a pyriform vesicle filled with a clear liquid are intimately associated with Muller's organ; they probably transmit the tympanal vibrations to the sensilla. The first abdominal spiracle, near the anterior margin of the tympanum, gives off an air-sac which is applied to the inner surface of that membrane. Two additional air sacs arise form the ventral tracheal trunk on each side of the 2nd abdominal segment and lie internal to and in close contact with the other sac.

The ultrastructure of the scolopophore is essentially similar to that of the mechanoreceptors discussed above. The cell body and axon of the bipolar neuron are enclosed in a Schwann cell and a fibrous sheath cell is wrapped around the basal part of the dendrite; a total of 60-80 chordontonal units is arranged in four groups. The tympanal organs of the Acridoidea show directional sensitivity to sound and are able to recognize individual sound pulses when these occur at rates of up to 90-300 per second, depending on the species. They respond over a wide frequency range, from below

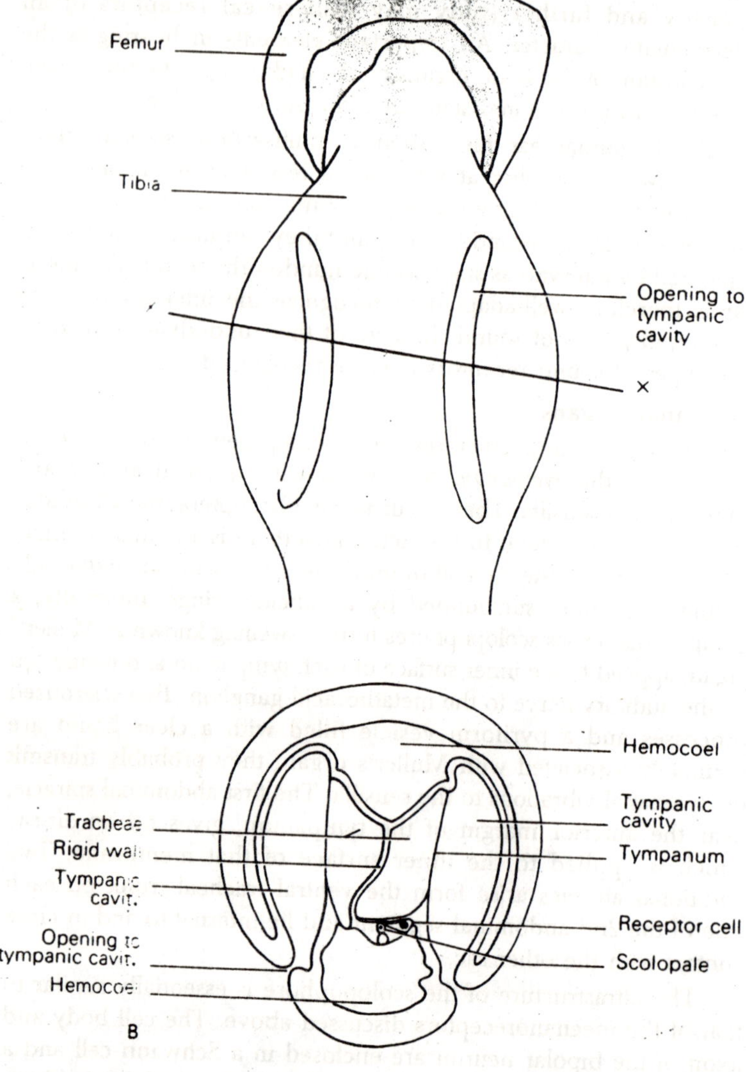

Fig. 9.4. Tympanic organ on the foretibia of a long-horned grasshopper. A–Anterior view of the tibia. B–Transverse section at the level indicated approximately by the line X–AX in A, epidermal cells omitted.

1.5 KHz to over 80 KHz. The electrophysiological discharge is related to the intensity of the sound but the response to pure tones is asynchronous, i.e., it does not depend on the frequency of the sound. This has led to the theory that insects with such tympnal organs do not discriminate between sounds of different pitch, though when the sound varies regularly in intensity (i.e., when it undergoes *amplitude modulation*) the responses synchronize with the modulation frequency; the frequency of the 'carrier wave' is immaterial provided it falls within the auditory range of the species. There is, however, more recent evidence that the Acridoid tympanum can, to some extent at any rate, distinguish different frequencies. This seems to be partly due to differences in the sensitivity of the four groups of sensilla and partly to the existence in the tympanum of spatially distinct centers of resonance.

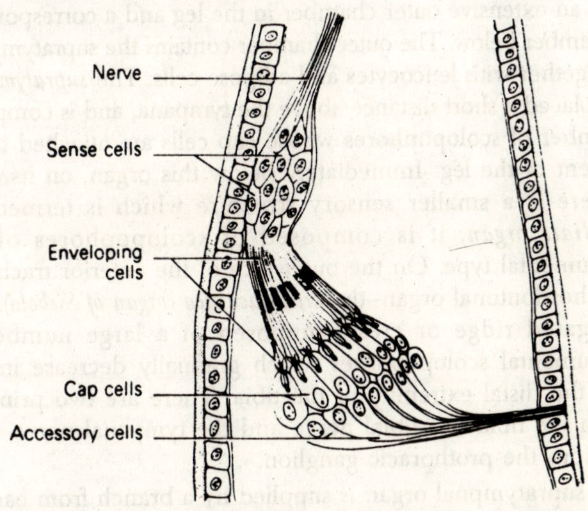

Fig. 9.5. Subgenual organ of an ant exposed by section of the tibia.

Further, in *Chorthippus* Howse *et. al.* (1971) have shown that the normal response in the auditory nerve does not follow the amplitude modulation envelope but rather depends on the existence in the stridulatory song of ultrasonic *transients* (i.e., very rapid changes in sound intensity). In the Tettigoniodea and Grylloidea there is often a pair of tympanal organs at the base of each fore tibia. In many genera they are easily seen, but in others each organ is concealed by citucular fold and comes to lie in a cavity which communicates with the exterior by a slit-like opening. These organs attain great complexity of structure of structure and most of what is known concerning them is due to the researches of

Schwabe (1906) and others. In *Decticus verrucivorus* the tympanal organs are of the concealed type. The trachea supplying the leg is greatly modified and on entering the tibia it becomes inflated and divides into an anterior and a posterior branch, which reunite below the auditory organ. Each trachea is closely applied to the tympanum of its side, which thus has air of the trachea on its inner surface.

It is noteworthy that these tracheae communicate with the exterior by a special orfice on either side, in close proximity to the prothoracic spiralce, and these orifices are only present in species with tympanal organs. In a transverse section of the tibia it will be observed that the two tracheae occupy the area between the tympana. There is an extensive outer chamber in the leg and a corresponding inner chamber below. The outer chamber contains the supratympanal organ together with leucocytes and adipose cells. The *supratympanal organ* is placed a short distance above the tympana, and is compsoed of a number of scolopophores whose cap cells are attached to the integument of the leg. Immediately below this organ, on its outer side, there is a smaller sensory structure which is termed the *intermediate organ;* it is composed of scolopophores of the subintegumental type. On the outer face of the anterior trachea is a third chordontonal organ–the *crista acustica* (*organ of Siebold*). It is an elongated ridge or crest composed of a large number of subintegumental scolopophores which gradually decrease in size towards the distal extremity of the tibia. There are two principal nerves in the tibia–the tibial nerve and the tympanal nerve–both arising from the prothoracic ganglion.

The supratympnal organ is supplied by a branch from each of those nerves, while the two remaining organs are innervated by the tympnal nerve. Rather surprisingly, Friedmann (1972b) has found that in *Gryllus assimulis* the ultrastructure of the cells in the tympanal organ is quite unlike those of the subgenual and intermediate organs. The tympnal cells here are apparently specialized epidermal cells with arborescent processes and they lack all association with scolopophores. They do, however, contain scolopale-like structures and longitudinally arranged microtubules but their innervation is not understood. Further comparisons of Tettigonioid and Grylloid tympanal organs are needed since Schwabe's account has been supported by Michel (1974). In general the known physiology of the Tettigonioid tympnal organs is similar to that of the Acridoidea.

The organs show directional sensitivity, their range of response exends to ultrasonic frequencies, they show asynchronous discharge and they respond to transients in the stimulus. In *Homorocoryphus* a trachea from the prothoracic spiracle conducts sound to the rear of the tympanum and the insects are more sensitive to sound directed at the open spiracles than at the tympani openings on the tibia. In many families of Lepidoptera conspicuous tympanal organs occur at each side of the metathorax or at the base of the abdomen, their structure differing in the different families that possess them. Each organ is lodged in a cavity formed by invagination of the cuticle and its external opening may be guraded by an integumentary fold. A tympanum, tracheal air-sacs and two chordontonal sensilla are present.

The latter have the typical ultrastructure and are connected by the tympanal nerve to the metathoracic ganglion. The physiology of these auditory organs is best known in some Noctuidae, where they respond to frequencies form 3 to 240 kHz, the two sensilla differing in sensitivity. Their function is to detected the cries of predatory bats; sonic and ultrasonic bat cries cause electrical activity in the auditory nerve and the organs also enable the moths to detect the direction of a sound in three-dimensional space. Interneurons in the brain are concerned with central processing of the auditory information. Tympanal organs are also found in other insects but have been less extensively studied. They occur in both sexes of the Cicadidae at the base of the abdomen, where a group of about 1500 chordotonal sensilla is stretched across an auditory cavity bounded by a transparent tympanum, the 'mirror'.

In *Chrysopa* small swelling near the base of the radial vein of each force wing are bounded ventrally by thin cuticle and contain chordotonal organs. Although the cavity of the origin is filled with blood rather than by a tracheal air-sacs, there is experimental evidence that they perceive ultrasonic vibrations between 13 and 120 kHz and can detect sound pulses delivered at up to 150 per second. Chordotonal organs in the mesothorax of *Notonecta, Nepa* and *Corixa* also react to air-borne sound in the range from about 100 to 10000 Hz.

Auditory Hairs

Minnich (1925, 1936) has shown in many Lepidopteran larvae that responses to sound are mediated by hair sensilla which react to low frequencies. Pumphrey and Rawdon-Smith (1936) found that hairs on the cerci of some Orthoptera act as displacement receptors of sound vibrations having a frequency of less than about 3000 cycles per second and that, at least in the lower parts of the range, the impulses in the cercal nerves synchronize with the frequency of the stimulus.

Johnston's Organ

Roth (1948) has shown that in males of *Aedes aegypti,* sound waves can set in motion the hairs on the antennal flagellum and this in turn causes movements of the whole flagellum which stimulate the sensilla of Johnston's organ. In this and other species the males can hear in the frequency range 150 to 550 Hz with the maximum sensitivity around 380 Hz, a frequency that corresponds to the flight tone of the females. Males do not perceive their own more highly pitched flight sounds nor do females respond to sound inthe above range, their less plumose antennae being mainly chemoreceptive.

Pilifer of Choerocampine Hawkmoths

A unique auditory organ, sensitive to untrasonic frequencies, has been found in the head of several species of Sphinfidae belonging to the subfamily Choerocampinae. In these insects sound perception depends on contact between the median wall of the second segment of the labial palp and the distal lobe of the pilifer, which contains the sensory transducer. When stimulated experimentally at the optimum frequency of 30 to 70 kHz the pilifer lobe responds to displacements as minute as 0.02-0.1 nm.

Campaniform Sensilla

In the companiform sensilla no outgrowth from the cuticle remains. The terminal filament from the sense cell ends in a sense rod or scolopale, which is inserted into a dome-shaped area of relatively thin cuticle, which are usually oval in shape, having a long diameter of 20-30m. The dome is thickened along its long diameter and inserted into it is the dendrite of a single neuron, often enclosed in a scolopale. These sensilla often occur in groups, all those within a group having the same orientation and possibly connecting to the same nerve fibre so that they function as a unit.They occur in all parts of the body subject to stress and are concentrated near the joints as at the base of the wing or the haltere in Diptera. On the leg of the cockroach there are four groups on the trochanter, each containing 15-20 sensilla, one at the base of the femur, another at the base of the tibia and one on each of the tarsal segments. In the insect skeleton all stresses can be expressed as shearing stresses in the plane of the surface. Such stresses produce changes in the shape of the campaniform sensilla and, because of the thickened bar across the dome, compression forces along the length of the sensillum raise the dome while extension in the same direction lowers the dome.

The nerve is stimulated when the dome is raised, that is by compression forces along the length of the sensillum and, for instance, most or all of the companiform sensilla on the leg are so orientated that they are stimulated when the foot is on the ground and the leg bears the weight of the insect. The longer the axis of the dome the greater the sensitivity of the sensillum and in groups of campaniform sensilla there is usually a range of sizes, perhaps giving a range of sensitivity. The response of campaniform sensilla is determined by the properties of the cuticle its elasticity, thickness and curvature; by the forces exerted on the cuticle by muscles; and by the pattern of forces due to gravity or inertia. They thus function as proprioceptors, but, unlike vertebrate proprioceptors, they do not respond to the movements of individual muscles, but to the resultants of a variety of contending strains on the cuticle. Thus some companiform sensilla monitor the movements of the wings and halteres, others are concerned in the control of leg movement as, for instance, in *Periplaneta* where the leg depressor reflex is initiated by stimulation of the companiform sensilla of the trochanter. Like all proprioceptors they are slow adapting.

Stretch Receptors

Stretch receptors differ from other insect sensillo in consisting of a multipolar neuron with free nerve endings, while all the others contain a bipolar neuron with a dendrite associated with the cuticle. These types of neuron are differentiated as Type II and Type I neurons respectively. Stretch receptors occur in connective tissue or associated with muscles. In *Periplaneta* there is a pair in the dorsal region of each of abdominal segments two to seven above the bands of longitudinal muscle. The neuron is embedded in fibrous connective tissue which is connected to an intersegmental membrane at one end and to the dorsal body wall and the dorsal muscles at the other.

In *Blaberus* (Dictyoptera) the ends of the dendrites inside the connective tissue are not sheathed in Schwann cells, as is the rest

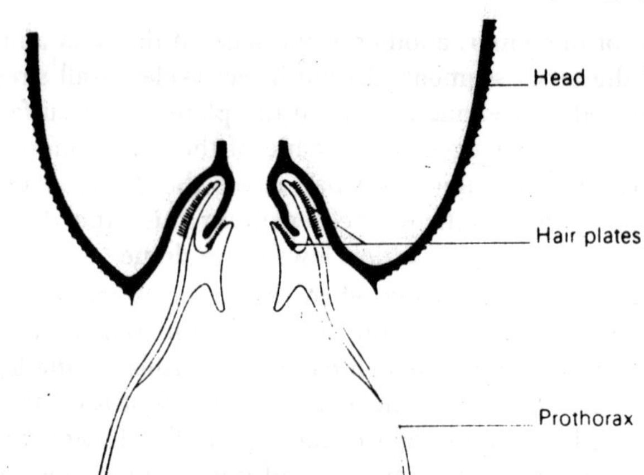

Fig. 9.6. Transverse section of the head of the ant Formica polyctena, showing hair plates on the prothorax (diagrammatic).

of the neuron, and are only 0.1-0.2m in diameter. The connective tissue consists of a matrix with fibrils embedded in it, but without any limiting membrane so that it is in direct contact with the haemolymph, and plasmatocytes are closely applied to the surface. The inelastic fibrils serve to support the neuron and, perhaps, prevent it from being stretched too far. Acrididae also have stretch receptors dorsally in the abdomen, but the dendritic endings are associated with a slender muscle fibre which runs from the anterior end of one segment to the anterior end of the next, forming a part of the dorsal longitudinal muscles.

Stretch receptors associated with muscles also occur in Lepidoptera larvae pupae and adults, and in Trichoptera and Neuroptera, but here the muscle strand is separated from the other muscles and has a separate innervation. It contains typical muscle fibrils, but these are spares in the centre of the fibre which contains a large amount of sarcoplasm and a giant nucleus. Attached to the muscle and partly embedded in it is a tube of connective tissue known as the fibre tract, which contains reinforcing fibres set in a matrix. These fibres are secreted by the tract cell which lies within the connective tissue and also has a giant nucleus. The neuron gives rise to two to four main dendrites which run along the length of the fibre tract, remaining free at the centre, but bound to the tract by connective tissue at the ends. From these main dendrites side branches, which at the tips are not clothed in Schwann cells, pass to the outside, and insinuate their way inside the fibre tract.

Larval *Antheraea* (Lepidoptera) have a pair of stretch receptors in the first nine abdominal segments just above the dorsal longitudinal muscles.

At metamorphosis the dorsal longitudinal muscles of larval Lepidoptera disappear, but, except for those in the eighth and ninth abdominal segments, the stretch receptors persist. They do undergo a metamorphosis which involves dedifferentiation and subsequent dedifferentiation of the associated muscles. These sensilla are proprioceptors and are stimulated by stretching. In the complete absence of tension there is no output, but under normal tension an output of 5-10 impulses per second is maintained for hours. With further stretching the impulse frequency rises and the adapted frequency is proportional to the length of the receptor. The muscle thus has a tonic response, but superimposed on this is a phasic response which appears while the length of the receptor is changing.

The impulse frequency increases with the velocity of stretch and, in addition, a burst of spikes is produced by any acceleration of stretching, either at the start or in the course of stretch. When stretching is complete the impulse frequency falls to its new tonic level. The stretch receptor can thus, through its tonic output, provide information on the position of one part of the body with respect to another, while the phasic response signals changes in these positions. Since at low frequencies, up to about five per second, the output is synchronised with stretching these receptors may also monitor relatively slow rhythmic movements such as ventilatory movements.

The frequency with which the receptor responds to cyclical stimulation may be limited by the viscosity of the connective tissue matrix in which the dendrites end, but the muscle associated with the lepidopteran stretch receptor may offset this effect to some extent by taking up the slack when the receptor is suddenly released. Relaxation causes a drop in the output of the receptor and this leads to a brief increase in the frequency of impulses to the muscle so that it contracts. Conversely, when the stretch receptor is stimulated the tonic stimulation of the muscle is inhibited so that it relaxes. This perhaps has the effect of preventing overstimulation of the receptor during rapid stretching.

Statocysts

Well-developed statocysts are unusual in insects, but in *Dorymyrmex* (Hymenoptera) there is a statocyst on the matathorax just above the coxa. It consists of an invagination of the cuticle lined with tactile hairs. Within the cavity are one or two sand grains which become supported by two cuticular projections in such a way that, although they cannot make large movements, they are free to move if the insect changes its orientaiton. As a result of a change in orientation they press on some of the hairs, different hairs being stimulated with different orientations so that the organs function as gravity receptors. Comparable organs occur in the head of *Anoplotermes* (Isoptera), and another but with a cuticular statolith instead of a sand grain, on the prosternum of *Dorymyrmex.* An organ, known as Plamen's organ, which may function as a statocyst occur in the head of larval and adult Ephemeroptera. This consists of a circular nodule at the junction of four tracheae mid-dorsally behind the eyes. No special innervation of this organ is known, but behaviour is disturbed if it is destroyed.

Pressure Receptors

Most aquatic insects are buoyant because of the air which they carry down under the water when they submerge, but *Aphelocheiru* (Heteroptera), living on the bottoms of streams and respiring by a plastron is not buoyant. Although the plastron can withstand considerable pressure it only functions efficiently in water with a high oxygen content such as normally occurs in relatively shallow water. Hence some mechanism of depth perception is an advantage to *Aphelocheirus,* although it is unnecessary in the majority of other, buoyant insects. On the ventral surface of the second abdominal segment of *Aphelocheirus* is shallow depression containing hydrofuge hairs which are much larger than the hairs of the plastron. They are inclined at an angle of about 30 to the surface of the cuticle and dispersed amongst them are thin-walled sensory hairs.

The volume of air trapped by these hairs depends on the balance between the pressure of air inside and the pressure of water outside. If the insect moves into deep water the increase in water pressure reduces the volume of air so that the hairs are bent over, carrying with them the sensilla which are thus stimulated. The insect responds to such an increase in pressure by swimming up, but there is no response to a decrease in pressure. Changes in the gas tension in the water, by influencing the exchange of gases, may also influence the tracheal system and hence could affect the volume of air trapped by the hydrofuge hairs.

However, it is possible that such small changes may be damped out by compensating expansion or contraction of an air-sac on the trachea near the spiracle leading into the depression housing the receptor. Pressure receptors are also present in *Nepa* (Heteroptera) which has a pair on the strerna of each of abdominal segments three to five. Each consists of a number of mush-room-shaped plates enclosing an airspace. Never endings occur in the plates, and sensory papillae on the inner wall of the airspaces are sitmulated by the plates pressing on them. A spiracle opens into the space so that the spaces in the three organs of one side are connected through the tracheal system.

The receptors give no general response to an increase in pressure, but differential stimulation of the organs of one side produces a response. If the head is tilted upwards the air in the system of pressure receptors tends to rise towards the head end and as a result the anterior mushroom-shaped plates will tend to be pushed out while those on the posterior organ will collapse against the papillae so that the organ is stimulated. The converse is true if the head is tilted downwards, and in this way the insect obtains information on its orientation.

Chemoreceptors

Nerve endings of many kinds are doubtless sensitive to irritant chemical substances. From this has been evolved the common chemical sense of primitive invertebrates. In insects and invertebrates this sense has become further differentiated into taste and smell. Far in insects both are sub-served by primary sense cells; their distribution varies from species to another. Despite considerable variation in detail, the chemoreceptor sensilla of insects show many common structural feature. They are apparently dervied from unspecialized clothing hairs and their cuticular parts are secreted by equivalents of the trichogen and tormogen cells, though other epidermal cells may also contribute. Forming part of each sensillum is a group of bipolar neurons (occasionally a single neuron, sometimes up to 60 or more).

The perikaryon of the neuron is normal, with Golgi bodies, endoplasmic reticulum and numerous mitochondria. Where the dendritic process of the neuron leaves the cell body it has a ciliary ultrastructure and is often surrounded by a 'cuticular' sheath of uncertain composition. Distal to the ciliary region the dendrite contains only microtubules and small vesicles and it may undergo extensive branching in the fluid-filled vacuole left after retraction

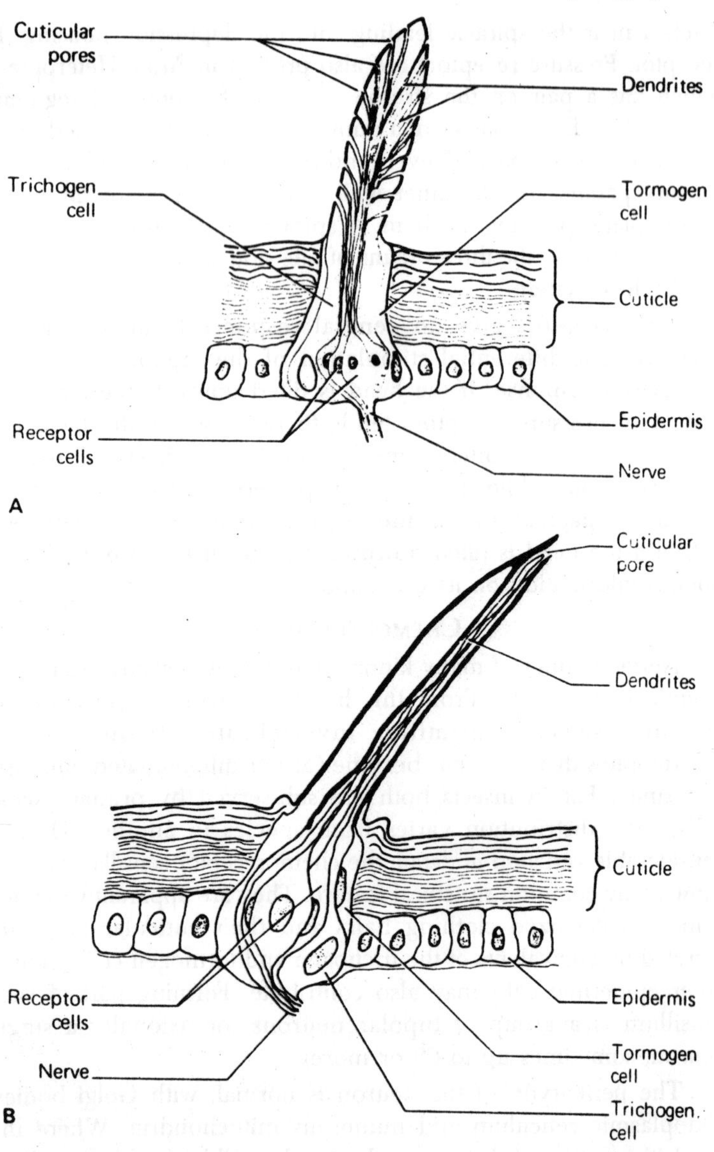

Fig. 9.7. Chemoreceptors. A–Olfactory sensillum. B–Contact chemoreceptor.

of the trichogen cell. The cuticle surmounting olfactory sensilla is pierced by many complex pores through which the chemical stimulus reaches the nerve endings; contact chemoreceptors, perceiving taste stimuli, have one or two larger pores near their tip. The exact nature of the pores, the tubules by which they extend inwards, and their relations with the dendritic branches are still under active investigation.

There is evidence that in some sensilla the individual sensory neurons respond to different chemical stimuli and in several cases a single sensory hair is known to act both as a mechanoreceptor and a chemoreceptor. According to the gross structure of the cuticular parts it is possible to distinguish several main types of chemoreceptor sensilla, as well as a few less common forms. The recent classification of Slifer (1970) lays primary emphasis on the thickness of the sensillar cuticle, but the more familiar scheme based on the shape of the sensillum is retained here for the major divisions.

Sensilla Trichoida

These are hair-like structures which may be thin-walled or thick-walled (when they are sometimes known as *sensilla chaetica* and may be difficult to distinguish superficially from mechanoreceptors or non-innervated clothing hairs). They are the most abundant and widespread type of chemoreceptor, occurring on various parts of the body. Many examples are found on the labella and tarsi of Diptera, where they mediate taste. Others occur on the maxillay palps of Acridids where they play a part in food-plant selection, and on the antennae of may species, where they have an olfactory function. They are the most numerous olfactory sensilla on the antennae of *Bombyx mori* and are also found on the ovipositor of *Musca,* where they are presumably gustatory.

Sensilla Basiconica and Styloconica

These are peg-like or cone-like organs, distinguished form the trichoid sensilla by the more thick-set appearance of the projecting portion. Like the trichoid sensilla they can have a thin or a thick cuticular wall (which may be grooved or variously sculptured) and they respond to various chemical stimuli. Examples are known form the maxillary palps of Lepodopteran larvae, where they respond to solutions of sugars, alcohols, amino acids and other plant constituents. Others, on the ovipositor of *Phormia,* reach to inorganic ions. On the maxillary palps of female Culcidae they are sensitive to carbon dioxide and they are widely distributed as antennal olfactory receptors.

Sensilla Coeloconica

These are less common and differ from thin-walled basiconic sensilla in that the cuticular peg in sunk below the general surface of the cuticle, whence their alternative name of pit-pag organs. In some the pit is very deeply sunk and connected with the surface by a more or less elongate tube–the so called *sensilla ampullacea.* Coeloconic sensilla were described from the antennae of Coleoptera, Lepidoptera and Diptera by older authorities. More recently they have been stuided on the antennae of *Apis mellifera,* where some of them are carbon dioxide receptors and on the antennae of *Hippelates, Anopheles, Bomyx, Leucophaea* and *Lasius.*

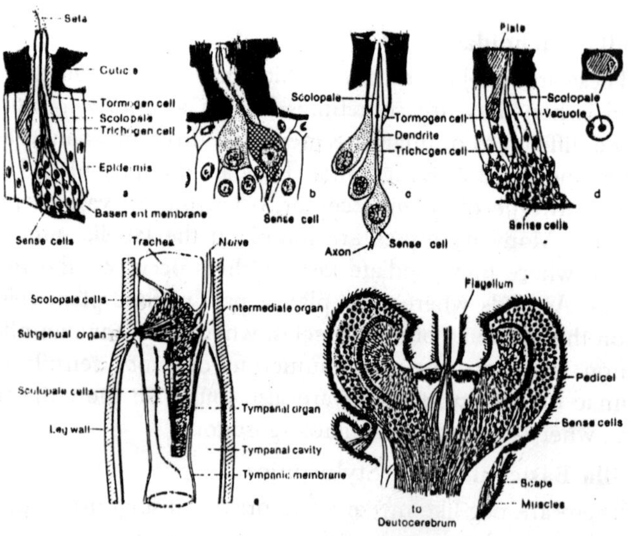

Fig. 9.8. Sensilla; a–trichoid chemoreceptor; b–campaniform mechanoreceptor; c–coeloconic chemoreceptor; d–plate organ; e–chordotonal acustical organ in tibia of right leg of tettigoniid; f–Johnston's organ in antenna of the fly Chaoborus.

Sensilla Placodea

This rather distinctive form of sensillum appears as a thin, elongate, oval or circular plate of cuticle which is flat or raised into a slight dome. Pores are present over the surface of the plate or around its periphery. Sensilla placodea are best known from the antennae of *Apis* where they are very numerous and act as olfactory receptors. On the antennae of Aphidoidea there are a small number of specialized sensilla placodea in which an outer, perforated cuticular plate is separated from an inner plate by a space containing many fine dendritic branches.

Atypical Sensilla

There is undoubtedly far greater diversity of structure among chemoreceptor sensilla than is indicated above. Among others which have attracted detailed attention may be mentioned the following; (i) the *sensilla coelosphaerica* found on the last antennal segment of *Nicrophorus;* they are olfactory receptor in which the usual pore tubules are replaced by a complicated network of filaments; (ii) the elaborately looped *circumfila* found on Cecidomyid antennae, which are essentially thin-walled chemoreceptors with the pores lying among a labyrinth of fine surface ridges; (iii) the much folded sensory plaque organs of the Fulgorid *Pyrops candelaria,* which have numerous bipolar neurons arranged in groups and may have evolved from a cluster of basiconic sensilla; (iv) the three pairs of tufted sensilla found on the thorax of *Calliphora* larvae and which are probably chemoreceptors, as are also the papilla-like sensilla occurring among the pseudotracheae on the labellar lobes of the adult. Because of the high specificity of chemoreceptor sensilla, and even of their constituent neurons, it might be most logical to regard them as mediating a great variety of individual senses.

It is, however, convenient to distinguish two main groups, the olfactory senses (small) and the gustatory senses (taste). The differences between these seem relatively clear in terrestrial insects but need further elucidation in aquatic species and those inhabiting such moist environments as the soil or plant and animal tissues.

Olfactory Senses (Smell)

The sense of smell in insects is stimulated by low concentrations of the vapour phase of a great variety of substances which are relatively volatile at ordinary temperatures. The olfactory receptors are trichoid, basiconic, coleoconic and placoid sensilla, usually if not always provided with many cuticular pores. Olfactory stimuli have been intensively studied through the search for attractant and repellent substances in economic entomology, but while chemical and behavioural aspects of the problem has received great attention much less is known of the detailed mechanism of perception and very little of the processes of transduction.

Measurements of insects olfactory acuity vary considerably in accuracy, but the olfactory thresholds of many simple chemical substances are of the same order for the honeybee as for man. A few examples of threshold concentrations, measured in molecules per cc of air for various insect species and substances are: *Nicrophorus,* skatol, 4.17 x 10; *Pieris rapae,* benzaldehyde, 3.28 x 10; *Apis mellifera,* nerol, 3.2 x 10. Much greater sensitivity is shown towards certain highly specific biological attractants, such as bombykol, the female sex-attractant of *Bombyx mori,* a single molecule of which is enough to elicit a nervous response in an antennal sense-cell. Many factors such as age, sex, nutritional status and previous conditioning, as well as temperature, humidity and the rate of air-flow, ae known to affect olfactory thresholds and in some cases a reversal of response may occur naturally during the life of the insect or can be induced experimentally.

For example, the newly-emerged females of *Pimpla ruficollis*–a parasite of the Pine-shoot Moth, *Evetria buoliana*–are repelled by the essential oils of pine, but when sexually mature they become attracted to them and return to the pines where their hosts are available for oviposition. Again, Thorpe (1939), by larval conditioning, has induced *Drosophila* adults to react positively to the normally repellent odour of peppermint oils. There is electrophysiological evidence that excitatory odours induced dendritic polarization of the sense-cell (the receptor potential) followed by nervous impulses in the axon. Inhibitory odours hyperpolarize the cell and depress the impulses. Among the sense-cells of olfactory receptors there seem to be two major kinds.

Odour specialist cells respond identically to highly specific, biologically important odours like the pheromones discussed below. *Odour generalist cells,* on the other hand, differ in the spectra of their responses to a variety of odours and perhaps provide a peripheral basis for central nervous discrimination of odours. Attempts to related the olfactory properties of substances to their physiochemical characteristics have not yet been very successful, though there are several interesting theories, such as that of Wright (1966) who postulates a limited number of 'primary odours' whose attractiveness to insects can be inferred from the infra-red spectra of the various compounds tested. One of the most remarkable features of olfactory responses is their considerable specificity and the role played by specific attractant or communicating stimuli in the life of the insect. For many such substnaces the term 'pheromone' was introduced by Karlson and Butenandt (1959) and they have become the subject of active study. A pheromone may be defined as a substance secreted externally by an individual and received by other nonspecific individuals, in which it elicits some specific behaviour pattern or developmental process. Such substances are generally olfactory stimulants and may be classified by their biological functions into the following groups:

Trail-marking pheromones

Odour trails have been described in a few species of termites and in many ants. They are laid on the ground by successful foragers returning to the nest and consists usually of short-lived, undirected streaks of material secreted from the gut or various epidermal glands. Comparable pheromones are secreted by the mandibular glands of Meliponinae workers and male *Bombus* spp. to mark stationary objects on their flight-paths.

Sex attractants

These are now among the most fully investigated pheromones and though best known from the substances released by the abdominal glands of some female moths to attract males that lie down-wind, female sex attractants have also been found in at least nine other orders of insects. Many have been identified chemically, some can now be synthesized, and those from economically important insects have been used in their control. Examples of female sex attractants include bombykol (*trans*-10, *cis*-12-hexadecadien-1- ol) from *Bombyx mari,* disparlure (*cis*-7, 8-epoxy, 2-methyloctadecane) from *Porthetria despair,* propylure (10- proplyl-trans-5, 9-tridecadien-1-ol) from *Pectinophora gossypiella,* and musculature (tricos-9-ene) from *Musca domestics.*

Male insects may also porduce sex pheromones that attract females and induce them to mate. These include the substances produced by the tergal glands of some male Blattaria, those from the hair-pencils, coremata and androconia of various Lepidoptera and from comparable structures in various species of Hemiptera, Coleoptera, Hymenoptera, Mecoptera and Diptera. Female cotton boll weevils, *Anthonomus grandis,* are attracted by a mixture of terpenoids secreted by the males.

Olfactory markers and surface pheromones

These attract other individuals to the secreting insect or to sties which they frequent, but they are not epigamic in function.

They include the caste-recognition scents and colony odours of social Hymenoptera and the secretions of the Nassanoff glands in the abdomen of *Apis* spp. The latter include geraniol, and geranic acid and attract other workers to newly-located foraging areas.

Assembly and aggregation pheromones

These induce the formation of large, temporary or persistent aggregations of a species and include the pheromone that causes hibernation aggregations of the Coccinellid beetle *Semadalia undecimnotata*. The cohesion of a swarm of honeybees is due to the production of 9-hydroxydec-trans-2-enoic acid from the mandibular glands of the queen. In several species of Scolytidae attractant pheromones are secreted by the male or female according to the species and attract both sexes, though sometimes unequally. The situation is complicated as the species are attracted to a tree initially by olfactory stimuli, several pheromones are involved, and it is debatable whether they should be regarded as sex attractants or aggregation pheromones.

Alarm pheromones

These include a variety of 20 or more relatively simple compounds, of which 4-methyl-3-heptanone and 6-methyl-5-hepten-2-one are the most widespread, though citral, limonene and formic acid also function in this way. They have been found in may Formicidae but also in *Apis, Trigona* and a few Isoptera, and are produced by the mandibular glands and various abdominal glands according to the species concentrated. Alarm pheromones cause the insects perceiving them to run or fly more actively and to show aggressive behaviour, but their effects vary considerably and may depend on the concentration of the pheromone.

Morphogenetic pheromones

Used in a broad sense, this term denotes phermones which affect development, including the post-embryonic and postmetamorphic growth of the reproductive system needed to attain sexual maturity. Matue males of *Schistocerca gregaria,* for example, secrete material that promotes the sexual maturation of other males. In *Apis mellifera* 9-oxodec-*trans*-2-enoic acid, together with an unidentified scent, make up the 'queen substance' that is secreted by the mandibular glands of the queen bee and inhibits the development of the ovary rudiments of the workers. It also has behavioural effects on the workers, preventing them from constructing queen cells an rearing new queens.

In the Isoptera, the extensive morphogenetic changes that make up caste differentiation appear to be controlled by several pheromones of unknown composition. Allied to the pheromones are other biologically important odours produced by one species, but of benefit mainly to the other species that perceives them. These are the 'Kairomones' of Brown *et. al.* (1970) and they are included in the oviposition, food and host odour which attract insects to the plants, animals and other sites on which they feed or lay their eggs.

Oviposition attractants

The deposition of eggs in places suitable for their further development is often the result of the female insect being attracted by scent from the oviposition site. Many parasitic insects, for example are attracted to their hosts in this way. Insects with saprophagous larvae are attracted by odours such as ammonia and skatol arising from the bacterial decomposition of organic materials. Among phytophagous species, females of *Hylemyia antiqua* are attracted to and oviposit in sand treated with *n*-propyl mercaptan, an odours constituent of onion plants on which oviposition normally occurs and on which the larvae develop. The attraction of adult females of the moth *Chilo suppressalis* to rice plants or plant extracts and their oviposition on the plant is related to the concentration of *p*-methyl acetophenone present, though other attractants may also be involved. In neither of the last two examples do the adults feed on the plants concerned.

Food-plant attractants

Gustatory responses play a major role in food-plant selection, as indicated below, but olfactory attraction also occurs. Phytophagous species are attracted more or less specifically by the small of essential iol and other non-nutritive substances, locusts are attracted by the smell of grass, and the larvae of *Hylemyia* and *Chilo* in the experiments cited above were attracted bythe same plant odours as were the adult insects. The aggregation of *Dendroctonus ponderosae* and other Scolytid beetles depends not only on the attractant pheromones mentioned above but also on the release of volatile substances in the resinous exudate from attacked pine trees. Experiments on the caterpillars of *Manduca sexta* have shown that each receptor cell of the antennal basioconic sensilla has only a limited specificity and that the specificity patterns of different cells overlap so that the larva can recognize a greater variety of volatile plant constituents than if each receptor were sensitive to only one odour.

In many insects it must be remembered that olfactory responses to plant stimuli operate only over distances of a few centimeters in still air; they may then be more important in discrimination than in attraction from a distance, and in some insects olfactory stimuli may play little part in host-plant selection. In the aphids, for example, plants seen to be discovered by random dispersal, aided by the effects of physical features on air-movement. After the insect has alighted it discriminates between hosts and unsuitable plants through responses that are more likely to be visual or gustatory than olfactory. For further information on the whole subject of olfactory responses in host-plant selection.

Attraction to animal host

There is much behavioural evidence that olfactory stimuli play some part in enabling blood-sucking insects such as Culicidae and *Glossina* to locate their hosts from a distance, but there is relatively little precise information on the chemical and physiological basis of the responses. Some species of mosquitoes can respond to the respond to the presence of calves or an equivalent source of carbon dioxide at distances of 15 metres or more, though other species do not respond at half this distance. A combination of L-lactic acid and carbon dioxide attracts female *Aedes aegypti* and a rather miscellaneous variety of substances are attractive to other mosquitoes, though none equals the odours from the host animal itself. In *Glossina morsitans* carbon dioxide and host odours are attractive under laboratory conditions and in the forest-inhabiting species. *G. medicorum* an olfactory response involving up-wind orientation is probably an important factor in hostlocation.

Gustatory Senses (Taste)

These are defined by responses to relatively high concentrations of non-volatile stimulatory substances which usually come into contact with the receptors in aqueous soluiton. For this reason the gustatory sensilla are sometimes known as contact chemoreceptors and they mediate important responses in feeding, the rejection of unpalatable substances and in oviposition by adults with endoparasitic larvae. The receptors include trichoid, basiconic and styloconic sensilla as well as special forms like the interpseudotracheal papilla of Diptera.

Contact chemoreceptors are commonly distributed on the tarsi, the maxillary and labial palps (including the labella of Diptera) and less often on the antennae. Concentrations of such sensilla are found on the mouthparts and the walls of the pre-oral food cavity, e.g. Thomas (1966), Galic (1971), and Moulins (1971). In blood-sucking Diptera they form functionally important groups in the wall of the cibarium and elsewhere and in Hemiptera they are localized in a special gustatory organ. There are wide differences in insect taste-thresholds according to the substance, species and receptor investigated. In general sugars are acceptable; some–such as fructose, glucose, sucrose and maltose–are percieved at low concentrations while others like galactose, mannose and arabinose stimulate only at high concentations.

A wide variety of substances, including acids, salts, alcohols, esters, amino acids and oils, are all rejected if the concentration is sufficiently high, but dilute solutions of acids and salts are sometimes preferred to water or dilute sugars solutions. In the well-studied gustatory hairs on the labella of *Phormia* there are five sense-cells in each sensillum. One of these is a mechano-receptor, but the other four comprise two 'salt cells' sensitive respectively to chlorides and the sodium salts of fatty acids, a 'water cell' and a 'sugar cell'. Evidence is now accumulating for a similar specialization among the individual sense-cells of other gustatory receptors and the biophysical basis of transduction is under active study. The complexity of gustatory perception is illustrated by studies on the responses of the styloconic sensilla on the maxilla of Lepidopteran larvae. Each of these two sensilla has 4 chemoreceptor cells and in most species some of the cells are individually sensitive to salts, amino acids, inositol and sucrose. Others are more specifically sensitive to substances characteristic of particular food-plants, such as mustard oil glucosides, populin or anthocyanins, while still further cells respond to various unrelated deterrent substances.

There are also additional soruces of diversity between species and receptor cells that need to be considered. Those cells that are generally sensitive to a certain substance may differ quantitatively in their thresholds and sensitivity spectra; inhibitory and synergistic effect on the receptor occur through interactions between the stimuli; and no doubt these various peripheral processes are further modifed to determine feeding behaviour through the central integrative action of the nervous system. Food-plant selection and discrimination between different foods seems to be complex processes in which varying roles are played by nutritionally important constituents of the plants and by non-nutritional 'secondary plant substances' with specific attractant or deterrent properties.

In a simplified experimental analysis of feeding in *Schistocerca* found that contact receptors on the palps detected sucrose, which stimulates feeding, while a deterent (azadirachtin) was perceived by receptors near the apex of the labrum. Continued feeding on acceptable material depends on the stimulation of receptors on the labrum, clypeus and hypopharynx. Among the specific gustatory attractants and feeding stimulants of various insects are the mustard-oil glucosides or cabbage, hypericin from *Hypericum hirsutum*, and various essential oils, alkaloids, catalposides and other compounds. For full accounts of the role of gustatory physiology in food plant selection and feeding, and for the special case of plant-sucking Hemiptera se Miles (1958) and Mittler (1967). Less is known of the part played by taste receptors in controlling the feeding behaviour of blood-sucking Diptera, though there is little doubt of their importance. In mosquitoes, for example labellar receptors are sensitive to sugars, water and deterrent substances while labral receptors respond to blood.

The ingestion of liquids is, however, controlled by cibarial receptors which are stimulated by blood, sugar and deterrents. Gustatory senses are also involved in the oviposition behaviour of parasitic insects and in the discrimination they show in finding hosts. The Braconid *Orgilus lepidus,* for example, can distinguish between healthy and previously parasitized hosts through contact chemoreceptors on the medial and lateral stylets of its ovipositor. Even more remarkable is the Cynipid hyperparasite of aphids, *Charips victrix,* which apparently uses sensilla at the tip of its ovipositor to discriminate among the parasite larvae lying in the aphid haemocoele.

Thermo Receptors

All insects are sensitive to high temperatures and avoid places which are unduly hot. Heat may be transferred to or from an insect by radiation, convection or conduction and also by evaporation and condensation. Only the first three methods are important as stimuli to behavioural changes and experimental assessments of their relative importance have been made in some cases. In grassphopper this sensitivity is distributed over the entire body, but is most marked in the proximal half of the antennae and on the pulvilli and tarsi of the hind and fore legs. Responses to radiant heat alone have been shown in few insects. *Schistocerca gregaria* displays a postural response, seems principally to depend on radiant energy. Specialized areas on the body of Acrididae have been identified as thermoreceptors, but this now seems unlikely.

Blood-sucking insects are often attracted by the heat of their mammalian hosts or by artificial bodies of similar temperature, probably responding through antennal receptors to convective transfer of heat in the air. Temperature receptors on the antennae, maxillary palps and tarsi of several insect are discussed by Gebhart (1953) and there is now electrophysiological evidence of cold receptors in a few species. In *Periplaneta americana* they occur on the arolia and tarsal pulvilli (and become more active when temperature is lowered below 13°C), and also on the antennae, where a few small trichoid sensilla on the ventral side of the flagellar segments respond both to the temperature at a given instant and to the rate of temperature change. The third antennal segment of Lepidopteran larvae also contain receptors, perhaps styloconic sensilla, that are very sensitive to falls in temperature. Many experimenters have exposed insects in an apparatus providing a gradient of temperature and found that they tend to congregate in a preferred zone (e.g., mainly from 24-32°C. for *Stomoxy calcitrans;* Nieschulz, 1934).

Krumbiegel (1932) found that different races of *Carabus nemoralis* have preferenda correlated with temperatures characteristic of the districts which they inhabit and in some species the zone depends on the previous history of the specimens. Bees can be trained to come to the warm places; they can remember a temperature difference as small as 2°C. The lice *Pediculus* and *Haematopinus* respond to warmth and follow a warm glass rod in all directions; the sense is distributed over the body but its chief site is the antennae. Not all experiments with a temperature gradient are satisfactory since the method fails to distinguish between convective and conductive transfer and humidity variations within the apparatus are not always eliminated.

Humidity Receptors

Some insects certainly choose their resting places according to the humidity and react to differences in humidity but little is known to the sensory mechanism concerved. Some insect orientate by the vapor from a distant soruce of water while others, by avoiding high or low humidities, tend to congregate in a preferred zone. This may be similar to the humidity of their natural habitat but behaviour is also affected by the insects water-balance. In *Stomoxys calcitrans* the probing responses induced by a rapid increase in relative humidity may be more important in normal feeding behaviour than any olfactory response to host odours. Inferences from the behavioural of experimentally treated insects have led to humidity receptors being identified on the antennae, sometimes as basiconic, trichoid or placoid sensilla or, in *Pediculus,* as peculiar tuft-like sensilla. Electrophysiological recording has shown that basiconic sensilla on the antennae and maxillary palps of *Aedes aegypti* respond to water vapor, as do three receptor cells near the median basiconic sensillum on the third antennal segment of *Manduca sexta* caterpillars. The honey bee and the blood sucking Muscids *Stomoxys* and *Lyperosia* are able to perceive water from a distance, perhaps by means of special receptors in their antennae. The mode of action of humidity receptors is not clear.

Perception of Miscellaneous Stimuli

Responses to a few unusual stimuli have been reported occasionally. Species of the Buprestid beetle *Melanophila* are attracted to forest fires, apparently through the action of infra-red receptors consisting of some 70 sensilla in a pit near the mid-coxal cavities. They are most sensitive to radiation with wavelengths from 2.4 to 4 mm and are perhaps modified temperature receptors. A possible role in the detection of infra-red radiation has also been ascribed on structural grounds to unusual placoid sensilla on the antenna of the Braconid *Coloides burnneri.* In *Componotus* (Formicidae), Martinsen and Kimeldorf (1972) have found behavioural responses to X-rays and in *Calliphora* and *musca* the direction of landing after flight is apparently related to the axis of the horizontal component of the earth's magnetic field. The earth's magnetic field has also been found to affect the direction of the waggle dance' performed by worker honeybees on the vertical comb.

PHOTORECEPTOR

Responses to light (though not necessarily to the same wavelengths as are perceived by the human eye) are mediaed by (a) *Dermal receptors,* (b) *Dorsal ocelli,* (c) *Lateral ocelli* or *stemata* and (d) *Compound eyes.* Typically, the imago possess compound eyes and dorsal ocelli though the latter may be absent; lateral ocelli occur only in Endopterygote larvae. Reduction or loss of photoreceptors is common among species which live habitually in dark situations (e.g. endoparasitic and cavernnicolous insects and those inhabiting the nests of termites or ants or burrowing in soil or plant tissues). It is also characteristic of many ectoparasites (Mallophaga, Siphunculata, Siphonaptera, most Pupiparan Diptera).

Dermal Light Sense

Several insects (Lepidopteran larvae, *Periplaneta, Tenebrio* larvae) react to light even after the eyes and ocelli have been removed covered with opaque material. The general body surface appears to be sensitive to light and localized receptors have not been identified. In *Schistocerca* and *Locusta,* rhythmic deposition of cuticle can be uncoupled from the 'circadian clock' that controls it through a direct, long-term effect of light–especially the wavelengths between 435 and 520 nm in the epidermal cells.

Dorsal Ocelli

The structure and functions of the dorsal ocelli have been reviewed by Goodman (1970). When typically developed there are three drosal ocelli on the fronts and vertex of the head, though the median one may be missing. In the Blattaria they are usually absent or represented by a pair or reduced, though light-sensitive, fenestrae. In the Siphonaptera Watchman (1972a) has shown that the only visual organs have a structure quite unlike the ommatidia of a compound eye; they may perhaps be the paired dorsal ocelli, as Hanstrom (1927) first suggested. As a general rule the ocelli are absent in apterous insects and present in winged forms though a few strong ocelli.

The vary considerably in their detailed structure in different groups of inects but the following major features can be recognized; (i) *The cornea.* This is usually a thickened, more transparent region of cuticle that surmounts the ocellus externally to form a lens. In the Ephemeroptera the cornea is convex but not thickened and the lens is formed from a mass of polyhedral cells lying beneath the corneagen layer. (ii) *The corneagen layer* consists of modified epidermal cells; they are colourless and transparent and secrete the lens, sometimes forming a distinctive layer of vitreous cells, apparently dioptirc in function. (iii) *The retina.* This commonly consists of some 500-1000 primary sense-cells which form a shallow cup and are sometimes arranged into groups of 2-5 cells, the

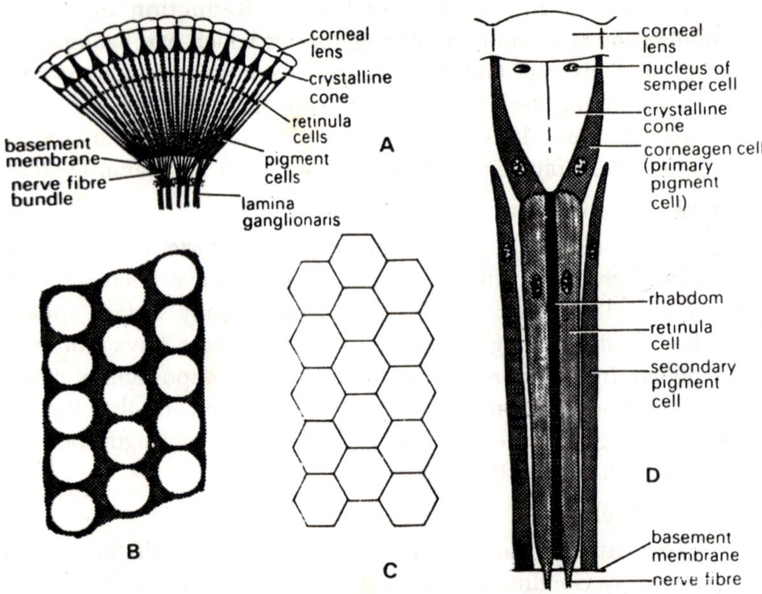

Fig. 9.9. Diagrams illustrating the structure of the compound eye. A–Section through part of an eye showing the arrangement of ommatidia. B–Surface view of part of the eye of an aphid, which consists of a small number of ommatidia, showing the facets well separated by unmodified cuticle. C–Surface view of part of the eye of a syrphid, which consists of a large number of ommatidia with the facets crowded together. D–Detail of a single ommatidium.

retinulae. Proximally the retinal cells form short axons synapsing with second-order neurons whose cell-bodies lie in the pars intercerebralis of the brain. Some part of the surface of each retinal cells is specialized as a light-sensitive rhabdomere, composed of closely-packed microvilli which measre, in *Schistocerca,* about 0.5 mm in length. Within the retinal cells there may be conspicuous rough endoplasmic reticulum as well as multivesicular bodies about 50 nm in diameter and much larger 'onion bodies' composed of numerous concentric membranes. (iv) *Pigment cells.* These are absent from *Periplaneta* and vary considerably in other species. They may invest the whole ocellus or form an iris-like ring of cells; pigment migration has been reported in a few species. Their man function is to prevent light enterting the ocellus except through the lens. (v) *Central nervous connections.* The ocellar nerve consist of the short axons of the retinal cells and longer ones from protocerebral neurons.

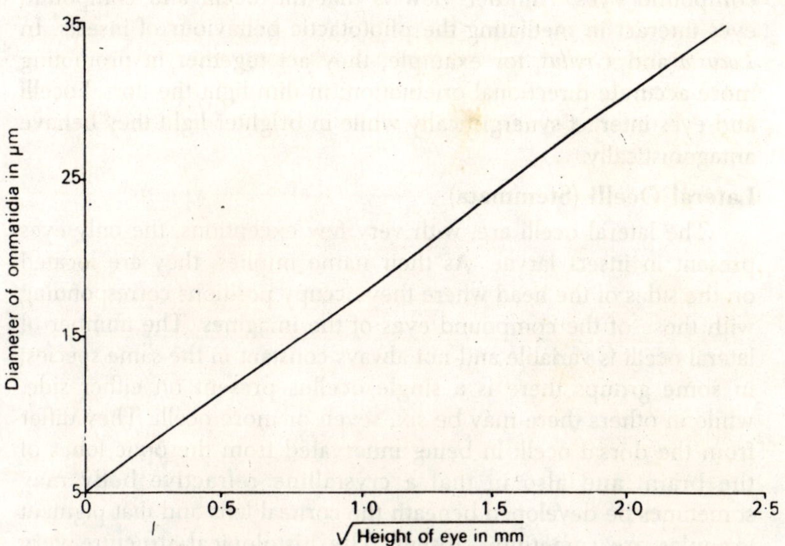

Fig. 9.10. The relationship between ommatidial diameter and the size of the eye in Hymenoptera.

Partial decussation occurs in the protocerebrum and tracts appear to run to the calyces of the corpora pedunculata with further descending connections to the thoracic ganglia in the ventral nerve cord. The functions of the dorsal ocelli are not entirely clear. Their visual fields are overlapped by those of the compound eyes and they are unlikely to mediate the perception of form. This is because the principal focal plane of their lens falls below the level of the retinal layer and because of the convergence of the very many retinal cells on to a much smaller number of ocellar nerve fibers. Despite claims to the contrary, the dorsal ocelli are probably unable to recognize polarized light and they do not seem to be implicated in colour vision or in the entertainment of circadian rhythms.

On the other hand, there is electrophysiological evidence that they signal the level of light intensity and changes in that level; the electrical response of a stimulated ocellus is a complicated process resulting from the presence of up to four different components. Occluding the ocelli usually leads the insect to move less rapidly, especially in bright light. This has favoured the view–not now generally accepted–that the dorsal ocelli may be 'stimulatory organs' serving to raise the excitatory level of the insect with respect to other visual stimuli perceived through the compound eyes. Another view is that the ocelli and compound eyes interact in mediating the phototactic behaviour of insects. In *Locusta* and *Gryllus,* for example, they act together in promoting more accurate directional orientation; in dim light the dorsal ocelli and eyes interact synergistically while in brighter light they behave antagonistically.

Lateral Ocelli (Stemmata)

The lateral ocelli are, with very few exceptions, the only eyes present in insect larvae. As their name implies, they are located on the sides of the head where they occupy positions corresponding with those of the compound eyes of the imagines. The number of lateral ocelli is variable and not always constant in the same species; in some groups there is a single ocellus present on either side, while in others there may be six, seven or more ocelli. They differ from the dorsal ocelli in being innervated from the optic lobes of the brain, and also in that a crystalline refractive body may sometimes be developed beneath the corneal lens and that pigment granules are sometimes absent. The histological structure very varied, but four main types may be considered.

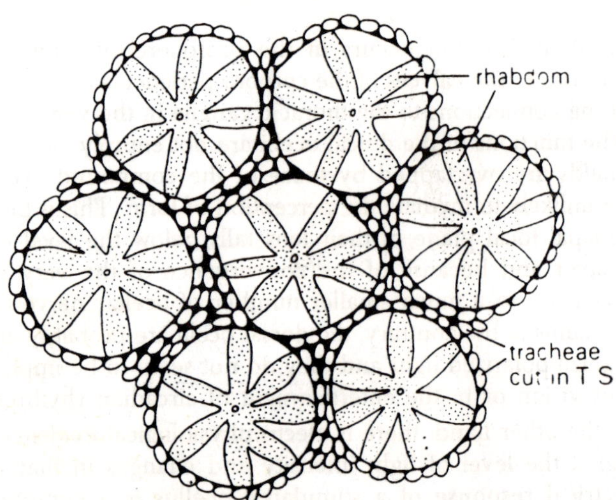

Fig. 9.11. Cross-section of a group of ommatidia of a moth showing the interommatidial spaces packed with tracheae.

(i) In the Tenthredinidae the single lateral ocellus is not unlike the typical dorsal ocellus in structure. The cuticle forms a lens-like cornea secreted by a thick underlying layer of corneagen cells (modifiied epidermis). Beneath the latter is a retina composed of many retinulae, each made up of four cells whose apposed rhabdomeres form a rhabdom. A broadly similar type of structure is found in *Tipula, Cicindela* and *Acilius.*

(ii) In *Dytiscus* (Gunther, 1912), *Euroleon* and *Sialis,* a lens-like crystalline body is secreted beneath the cornea but the structure otherwise resembles the first type.

(iii) In the Lepidoptera and Trichoptera, each larval ocellus has a cornea and crystalline body but only seven retinal cells are present, forming a single retinula with their apposed rhabdomeres constituting a single axial rhabdom. The resulting organ is strikingly similar to each ommatidium of a compound eye.

(iv) In the larvae of several Nematoceran Diptera the lateral ocelli may be simplifed. There are few retinulae and a corneal lens is reduced or absent but the corneagen cells of some species have a vitreous appearance recalling that of a more or less degenerate crystalline body. In the unpigmented ocellus of *Chironomus* even these vitreous cells are absent and a single retinula lies directly beneath unmodified cuticle. In *Aedes aegypti* larvae the retinular cells have the kind of ultrastructure expected in insect eyes, with rough endoplasmic reticulum, rhabdomeric microvilli and a zonula adherence junction between continuous cells. Pinocytosis occurs at the base of the microvilli and protein taken up in this way is sequestered in multivescicular bodies, themselves later transformed into lamellar structures. The larvae of Cyclorrhaphan Diptera have a pair of very simple photoreceptors which are probably degenerate lateral ocelli. In *Musca* there is a small group of light-sensitive cells on each side of the pharyngeal sclerities and invisible externally. They are most sensitive to green light and apparently unable to perceive red.

The physiology of the lateral ocelli of Lepidoptera has been studied by Dethier who showed that both cornea and crystalline body can form a more or less distinct, inverted image which, irrespective of the distance of the object, falls somewhere on the rhabdom. Acting together, the six pairs of ocelli form twelve visual fields with little or no overlap and so provide a very coarse mosaic of intensities. By moving the head from side ot side as it advances, the larva can examine a larger field and is at least capable of orientating itself towards the boundary between light and dark parts

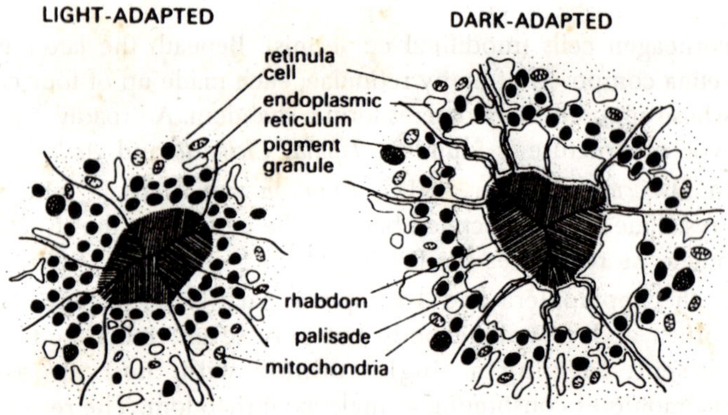

Fig. 9.12. Diagrams of cross-sections of the rhabdom and parts of the retinula cells from an ommatidium of Periplaneta in the light-adapted and dark-adapted states.

of its environment. The behaviour of caterpillars seeking food and pupation sites suggests that they have some colour vision and in *Bombyx mori* larvae the spectral sensitivity curves of the lateral ocelli, obtained by electrophysiological recording, show peaks the near ultraviolet and green regions that suggest a two-receptor system of colour vision.

Compound Eyes

The principal feature distinguishing compond eyes from ocelli is that in the former the cornea is divided into a number of separate facets, whereas there is only a single facet to each ocellus. Compound eyes are aggreations of separate visual elements known as ommatidia, each corresponding with a single facet of the cornea. Like lateral ocelli, they are innervated from the optic lobes of the brain. The number and size of the facets of the compound eye vary greatly. In workers of the ant *Ponera punctatissima,* each eye is a single facet; there are 6-9 facets in the same caste of *Solenopsis fugax,* while among other ants the number varies between about 100 and 600 in the worker, 200 and 830 in the females, and 400 to 1200 in the males. In *Musca* the eye consists of about 4000 facets. In honey-bee there are 4,900 ommatidia in queen, 6,300 in worker and 13,000 in drones. In some Lepidoptera from 12000 to 17000 and in Odonata between 10,000 and 28,000 or more.

In most insects the facets are very closely packed together and hexagonal but where they are fewer and less closely compacted, they are circular. The facets are not always of equal dimensions over the whole area of the eye. Thus, in males of *Tabanus* they are often larger over the anterior and upper parts of the eye, but the two fields are not sharply demarcated. In males of other Diptera such as *Bibio* and *simulium,* the two areas of different sized facets are very distinctly separated, each eye appearing double. An extreme condition is attained among certain Coleoptera and Ephemeroptera (*Cloeon*), where the two parts of the eye are separate so that the insect appears to possess two pairs of compound eyes. In *Cloeon* the anterior division of each eye is elevated on a pillar-like outgrowth of the head, while the posterior division is normal.

Occurrence and Structure of Compound Eyes

Most adult insects have a *pair of compound eyes,* one on either side of the head, which bulge out to a greater orlesser extent so that they give a wide field of vision in all directions. In Notonecta, for instance, the visual field extends through 246 in the horizontal plane and 360 in a vertical plane, the fields of the two eyes

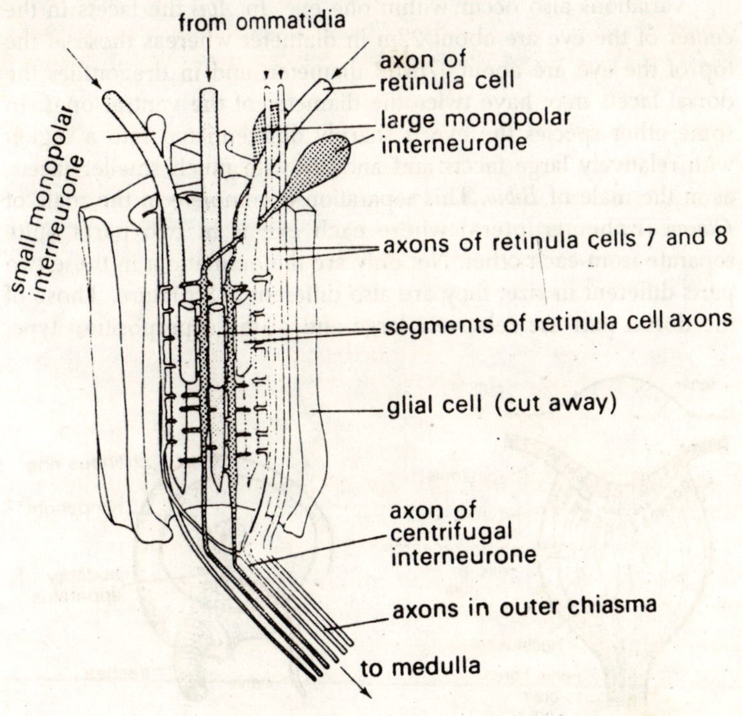

Fig. 9.13. Diagram of an optic-cartridge in the lamina ganglionaris.

overlapping so as to give binocular vision in front, above and below the head. In some insects, such as Anisoptera and male Tabanidae and Syrphidae, the eyes extend dorsaly and are contiguous along the midline, this being known as the holoptic condition.

The compound eyes are strongly reduced or absent in parasitic groups such as the Siphunculata, Siphonaptera and *female coccids* and this is also true of cave-dwelling forms. Each compound eye is an aggregation of similar units known as *ommatidia,* the number of which varies from one in the worker of the ant *Ponera punctatissima* to over 10,000 in the eyes of dragonflies. When only a few ommatidia are present the facets which they present to the outside are separated from each other by narrow areas of cuticle are around; more usually, with larger numbers, the facets are packed close together and assume a hexagonal forms. Ommatidia vary in size from insect to insect and within the Hymenoptera facet size is proportional to the square root of the height of the eye.

Variations also occur within one eye. In *Apis* the facets in the center of the eye are about 22m in diameter whereas those at the top of the eye are about 17m in diameter, and in dragonflies the dorsal facets may have twice the diameter of the ventral ones. In some other species the eye is sharply differentiated into a region with relatively large facets and another with much smaller facets, as in the male of *Bibio.* This separation is complete in the male of *Cloeon* (Ephemeroptera) where each eye is in two parts quite separate from each other. Not only are the ommatidia in these two parts different in size, they are also different in structure. Those of the dorsal part are relatively large and of the superposition type,

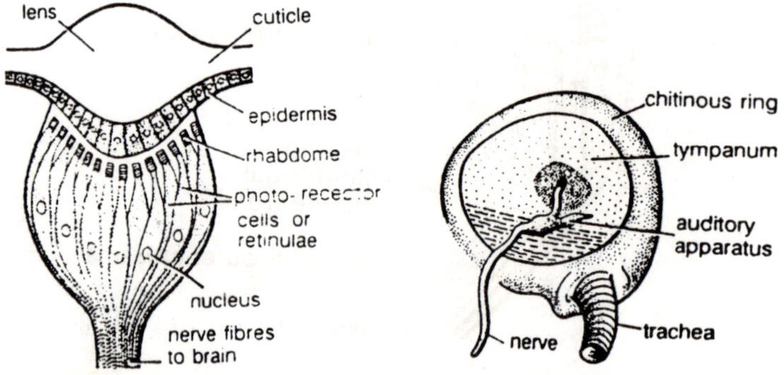

Fig. 9.14. Grasshopper. Ocellus in V.S.

Fig. 9.15. Grasshopper. Interior view of tympanum and adjacent trachea.

while in the lateral part the ommatidia are smaller and of the apposition type. The eyes are also divided into two in the aquatic beetle Gyrinus (Coleoptera) wher the dorsal eye is above the surface film when the insect is swimming and the ventral eye is below the surface.

Structure of an Ommatidium

Each *ommatidium* consists essentially of an optical, light-gathering part and a *sensory part,* perceiving the radiation and *transforming* it into electrical energy. The optical part of the system usually consists of two elements, a *cuticular lens* and a *crystalline cone.* The cuticle covering the eye is transparent and colourless and usually forms a biconvex corneal lens at the outer end of each ommatidium. It is *these lenses which* in surface view form the facets of the compound eye. In *Aleyrodes* all the lenses are not colourless, but each colourless one is surrounded by six yellow lenses. Some insects have the outer surface of the corneal lens produced into mintute conical nipples about 0.2 µ high arranged in a hexagonal patter with a m spacing. It is supposed that these projections decrease reflection from the surface of the lens and so increase the proportion of light transmitted through the facet.

The cornea, like the rest of the cuticle, is secreted by epidermal cells, each lens being produced by two cells, the corneagen cells, which later become withdrawn to the sides of the ommatidium and form the primary pigment cells. *Beneath the cornea are four cells,* the *Semper cells,* which, in most insects, produce the crystalline cone. This is a hard, clear intracellular structure bordered laterally by the primary pigment cells. *Eyes in which the crystalline cone is present are called eyes.* Immediately behind the crystalline cone in eucone eyes are the sensory elements. These are *elongate nerve cells known as retinula cells* in each of the which the margin nearest the ommatidial axis is differentiated to form a rhabdomere which extends the whole length of the cell.

Primitively each ommatidium probably contained eight retinula cells arising from three successive divisions of a single cell. This number is found in some insects, such as *Apis,* but in many there is a tendency to reduction of this number of six or seven, the other one or two persisting as short basal cells in the proximal region of each ommatidium. the cytoplasm of the retinula cells contains pigement granules which are especially concentrated at the edge of the rhabdomere, but these granules do not contain the

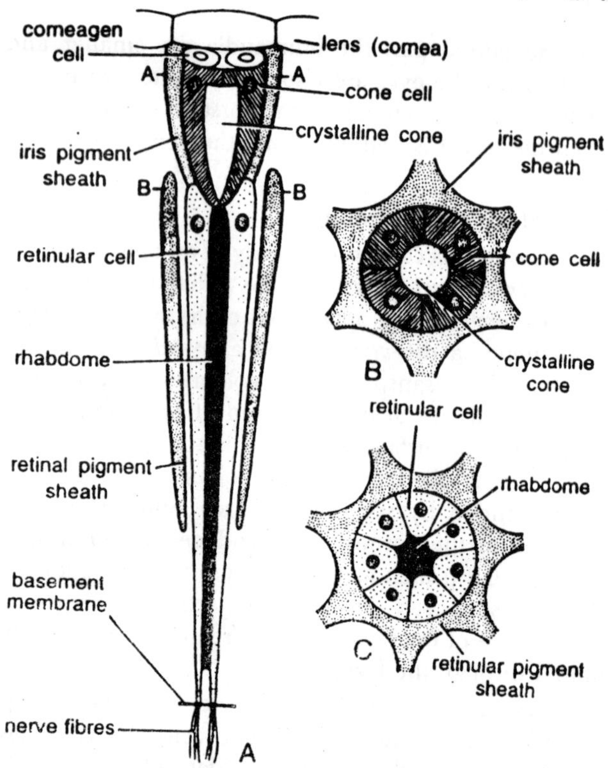

Fig. 9.16. Periplaneta. A–L.S. of an ommatidium; B–T.S. through cone at A–A; C–T.S. through rhabdome at B–B.

visual pigment. Arising from each cells is a nerve axon which passes out through the basement membrane at the back of the eye into the optic lobe. The *rhabdomere* consists of *close-packed microtubules* or microvilli about 500 A across, hexagonal in cross-section and extending towards the central axis of the ommatidium at right angles to the long axis of the retinula cell.

The microtubules of each retinula cell are all parallel with each other and roughly aligned with those of the retinula cell opposite but they are set at an angle to those of adjacent retinula cells. In *Drosophila* each rhabdomere is 60 μ long and 1.2 μ in diameter, while in Sarcophaga their diameter is about 0.5 μ. Collectively the rhabdomeres of each ommatidium form the rhabdom. In Diptera the individual rhabdomeres remain separate, grouped around a central matrix which may have a fluid consistency, but more often they are contiguous an may fuse together. In *Apis* they are fused in pairs while some Orthoptera all the rhabdomeres are fused into a single unit. The retinula cells are shorted by 12 to 18 secondary pigment cells which isolate each ommatidium from its neighbours. Tracheae may pass between the ommatidia, but in Apis no tracheae penetrate the basement membrane of the eye.

Modification of the Ommatidial Structure

In some Hemiptera, Coleoptera and Diptera the *Semper cells* do not form a cystralline cone, but become transparent and undergo only a little modification. Ommatidia of this type are described as acone ommatidia. The acone condition may be primitive and eyes of may Apterygota are of this type. In other Apterygota, such as *Lepisma* (Thysanura) and *Orchesella* (Collembola), there is a very simple crystalline cone, while the Semper cell of most Diptera and some Odonata produce cones which are liquid-filled or gelatinous rather then crystalline. Ommatidia of this type are described as pseudocone. Finally, in some beetles, such as Lampyris, the lens is formed from an inward extension of the cornea, not from the Semper cells which form a refractive structure between the cuticle and the retinula cells. This is an exocone ommatidium. In the eyes of many

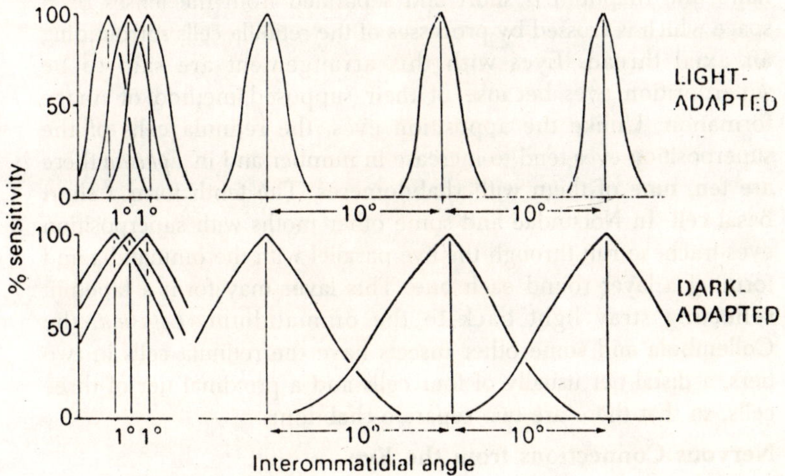

Fig. 9.17. Visual fields of ommatidia in light- and dark-adapted eyes, with interommatidial angles of I° (left) and Io° (right).

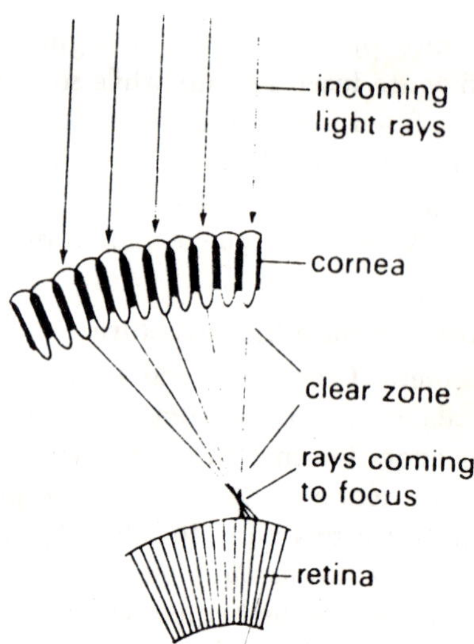

Fig. 9.18. The manner of formation of a superposition image in a clear-zone, well-focused eye.

insects the rhabdom is very long and extends from the back of the lens almost to the basement membrane. Because of the supposed method of image formation these are known as apposition eyes.

In some Coleoptera and nocturnal Lepidoptera, on the other hand, the rhabdom is short and separated from the lenses by a space which is crossed by processes of the retinula cells surrounding an axial thread. Eyes with this arrangement are said to be superposition eyes because of their supposed method of image formation. Unlike the apposition eyes, the retinula cells of the superposition eyes tend to increase in number and in *Ephestia* there are ten, nine of them with rhabdomeres. The tenth form a short basal cell. In Noctuidae and some other moths with superposition eyes tracheae run through the eye parallel with the ommatidia and forming a layer round each one. This layer may form a tapetum reflecting stray light back to the ommatidium. *Lepisma*, the Collembola and some other insects have the retinula cells in two tiers, a distal tier usually of four cells and a proximal tier of three cells, so that there are two separate rhabdoms.

Nervous Connections from the Eye

There is no optic nerve in insect, the eye connect directly with the optic lobe of the brain. This consists of three ganglionic layers, the lamina ganglionaris, medulla externa and medulla interna, connected to each other by chiasmata. Each ganglionic layer consists of an inner region of neuropile with the cell bodies at the periphery. Behind the eye the fibres from the retinula cells may decussate, as in *Calliphora,* or they may run in parallel strands, as in *Schistocerca.* Two kinds of axons run from the retinula cells, numerous short fibres to the lamina ganglionaris and occasional long fibres, perhaps from the basal cells, to the medulla externa. The cells of the lamina ganglionaris are situated distally. They are monopolar and their axons pass inwards to the medulla externa, first making dendritic synapses with the retinula axons. There appears to be good deal of convergence of the nerve fibres, fibres, fibres from several ommatidia connecting with each ganglion cell, but the degree of convergence differs in different parts of the eye. Other fibres, arising from cells in the medulla externa, run, but do not necessarily conduct, centrifugally and make connections in the lamina ganglionaris.

Functioning of the Eye

If the insect is do more than differentiate between light and dark, that is if it has any degree of form vision, it must possess an optical system capable of forming a suitable image and a series of receptors capable of perceiving the image.

Image Formation by the Optical System

Apposition eyes

The classical mosaic theory of insect vision supposes that each ommatidium forms an image of a limited part of the visual field. Each image has a given overall intensity which varies from one ommatidium to the next depending on the amount of light reflected from the object so that collectively the ommatidia produce a series of sports of light of different intensites which together form a picture of the object. In general, experimental evidence lends support to this theory, although with some modifications of detail. Each ommatidium perceives light coming from a wide angle, which, in *Locusta* and Calliphora, is about 20, and not simply from the field delimited by the ommatidial angle, the angle which the ommatidium subtends at the basement membrane, often 1 or 2. Hence the visual fields of adjacent ommatidia overlap, but the amount of light transmitted through the lens system falls off sharply as the angle of incidence increases, so that the greater part of the light entering each ommatidium does come from a limited area.

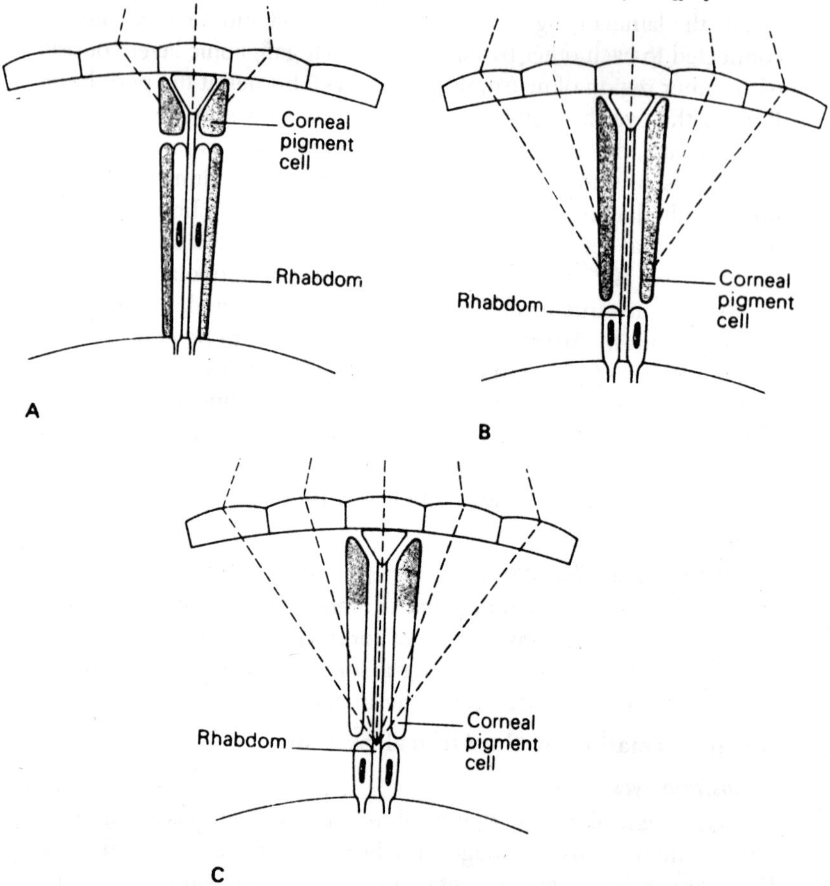

Fig. 9.19. Image formation in compound eyes (diagrammatic). A–Apposition type. B–Superposition type, light adapted. C–Superposition type, dark adapted. Dashed lines represent rays of light originating from a common point.

In the apposition eye, each lens system of corneal lens and crystalline cone capable of forming an inverted image on the rhabdom. This image may have some significance, but Kuiper suggests that the function of the lens system is simply to concentrate the light into a narrow beam entering the rhabdom. The pigment cells of apposition eyes optically isolate the ommatidia so that little light passes from one to another, but the significance of this is not clear.

It was believed that such isolation was necessary in order to obtain good resolution, but mutants of *Calliphora* which lack the screening pigment have equally good resolution of a moving pattern and are more sensitive than normal flies, requiring less incident light energy to produce a retinal response. This arises from the fact that light moving obliquely through the eye is not absorbed by the screening pigment as it is in normal eyes, so that it is able to stimulate adjacent ommatidia. This, however, leads to the rapid decay of the visual pigment and it may be that the function of the screening pigment normally present is to reduce the amount of light reaching the inner part of the eye so that the rate of decay of the visual pigment is reduced.

Superposition eye

In the superposition eye the screening pigment moves. In the light it extends inwards and separates the ommatidia, but in the dark it condenses distally round the lenses. Not only does the pigment move whithin the secondary pigment cells, but the cells themselves also migrate proximally and distally to some extent. The primary pigment cells do not move, nor does the pigment within them. As a result of these pigment movements, the light-adapted eyes, with the ommatidia separated from each other by pigment, functionally resembles an apposition eye. In poor light, however, when the pigment is contracted, light might pass obliquely through the eye without being absorbed.

As a result each rhabdom might receive light form a number of lens systems, not only that of its own ommatidium, and the image formed as a result is called a superposition image. In this way more use is made of the available light since very little of it is absorbed by screening pigments. This is an advantage in crepuscular and nocturnal insects which are active when little is available. In *Lampyris* and some other beetles a single upright image is formed by the eye which, morphologically, is of the superposition type, but the image is formed below the basement membrane and so can have no functional value.

In other insects there is so little convergence of light form adjacent ommatidia that no image is formed. This, together with other evidence, sugessts that superposition images, if they are formed at all, may not be important to insects and it is possible that superposition eyes from images by apposition. This would be facilitated by a refractive substance between the cone and the rhabdom acting as a wave guide and funneling light to each rhabdom from its lens system. Such a guide might be provide by processes of the retinula cells, or the Semper cells in *Lampyris*, which extend from the lens to the rhabdom, but they will not act in this way if their refractive index is low as seems likely. If superposition images are not formed by these eyes, the observed pigment movements must have some significance other than that suggested above. Possible expanded condition in daylight affords protection to the visual pigments, while contraction in the dark increases the sensitivity of the eye because less light is absorbed by the pigment. Not all workers are agreed that superposition images are not formed since it has been shown that a series of images may be formed at progressively greater depths in the eye as a result of diffraction from the lens systems. These diffraction images are superposition images formed by groups of ommatidia acting together so that the effecitive diameter of the lens system is increased.

It is suggested that this might from the basis of the insect's ability to discern stripes with a very low angular separation and, at the same time, would account for the great depth of the sensory system of the eye. The image so formed may be extremely complex and distorted compared with the object, but it is possible that these patterns could be used as a means of recognition where this was not very precise. This theory is criticised on various grounds. McCann and MacGinitie (1965), for instance, suggest that small errors in the experimental pattern presented to the insect might cause small changes in the intensity of light falling on the eye, implying that the exceptional acuity associated with diffraction images is not a reality. Other criticisms are that diffraction images are only seen in bright light, otherwise the light deeper in the eye is absorbed by the pigments, and it is also doubtful if the insect can distinguish and utilise these images at the sensory level.

Reception of Light

The rhabdom is presumed to be the site of photoreception and it probably acts as a wave guide, trapping most of the light which enters it. The refractive index of the rhabdom is higher than that of the surrounding cells (1:5:1:33) so that, unless the light enters it at a very oblique angle, the light is totally reflected at the interfaces. In vertebrates the conversion of light energy into a nerve impulse involves a photo receptor pigment. This is a chromoprotein known as rhodopsin which consists of retinene, the adlehyde of vitamin A, conjugated with a protein. The production of rhodopsin is a *continuous process,* but in the light it is continually bleached and it dissociates, liberating retinene with an altered molecular configuration. This process leads to the production of a nerve impulse. In darkness no bleaching occurs so that the rhodopsin accumulates. Evidence is accumulating which suggests that the process is basically similar in insects.

Retinene has been isolated from the heads of Hymenoptera, Diptera and Orthoptera as well as other groups and it is reversibly converted to vitamin A by the action of a dehydrgenase. It has also been found that in the dark adapted head of Apis the ratio of retinene; vitamin A is 4:1 whereas in the light adapted head it is 1:2 suggesting that retinene is converted to vitamin A in light. The absorption spectrum of the visual pigment may vary according to the protein with which the retinene is conjugated and it is probable that in some insects there are three pigments with different spectral sensitives.

Visual Responses

Taxis

A taxis is a movement of orientation to a source of stimulation, which may be light or any other stimulus. Often the orientation is coupled with locomotion so that the animal may move towards or away from the source. Many insects, such as locusts, orientate towards a source of light, a positive phototaxis, so they if they walk they move towards the light; others, such as blowfly larvae, orientate and move away from a source of light, a negative phototaxis. This basic orientation depends on a tendency to maintain symmertrical stimulation of the two eyes, although some insects are still able to orientate with one eye blackened. The normal taxis may be modified by other factors. At temperature below 16°C. *Apis* is negatively phototactic; at higher temperatures it exhibits a positive response. High light intensities also tend to produce negative reactions. Often these changes are lined to the biology of the insect and may depend on the physiological condition. *Ips* (Coleoptera), for instance, reacts positively to light when about to fly and negatively when about to feed.

Under natural conditions the complex interaction of different factors probably means that simple phototactic reactions do not occur, but orientations to dark objects against a light background, sometimes differentiated as skototaxis, frequently do occur. This is involved in movements towards solid objects such as food plants.

A taxis which is normally of some importance is the dorsal light reaction, the tendency to orientate the head so that the dorsal ommatidia of the two eyes are equally strongly illuminated. This reaction plays an important part in the maintenance of stability in the rolling plane during flight. *Notonecta*, which normally swims on its back, has a ventral light reaction.

Some insects may orientate so that they move at a constant angle to the light source with the result that the sense organs of the two sides are unequally stimualted. If the light source is moved the insect alters its path so that its angle to the light source is moved the insect alters its path so that its angle to the light source remains the same. This is called a menotaxis and is exhibited by ants and the caterpillar of *Aglais uriticae* (Lepidoptera). Menotaxis forms the basis of astrotaxis in which the orientation to the light source, in this case usually the sun, is constantly altered so as to compensate for the apparent movement of the sun. As a result the insect maintains a constant compass direction. Orientations of this type ae learned and bees, for instance, need to make several collecting trips before they learn to orientate accurately. An astrotaxis is imporant to social Hymenoptera as a means of finding the eay back to the nest. In insects, such as *Melolontha* (Coleoptea), which have no nest, an astrotaxis probably serves merely to keep the individual insect on a steady course.

Kineses

Locomotory reactions which are not oriented, but in which the speed of movement or rate of turning are related to the intensity of stimulation are known as kineses. Locusts, for instance, are more active in the light than in the dark, while cockroaches are more active in dark. Under natural conditions changes in light intensity are nearly always associated with changes in temperature and the latter are probably of greater general importance, but photokinetic movements do occur. For instance, locusts being to move about on the grass soon after it gets light, but before sunrise so that there is no corresponding increase in temperature. This movement is almost certainly a photokinesis to start with, although the influence of temperature soon becomes over-riding.

Optomotor Reaction

The optomotor reaction is a behavioural response to pattern of stimulation moving over the eye. In experimental work the pattern usually consists of vertical stripes and the response of a turning movement tending to keep the images in the eye as stationary as possible. In the field the reaction is brought about by the apparent movement of environmental features as the insect moves. The passage of images across the eye from behind forwards indicates to the insects that it is moving backwards, while image movement from front to back indicates forward movement. A flying insect appears to perfer images to pass over the eye from front to back at a certain moderate speed, it has a preferred retinal velocity.

Usually if an insect flies into the wind. An upwind orientation is maintained so long as it is able to make headway against the wind. If, however, the wind is too strong and the insect is carried backwards, as indicate by the forward movement of images in the eye, it lands. Hence orientation with respect to the wind involves an optomotor reaction although stimuli other than visual ones may also be important. Some stream-dwelling insects hold their position in the current by an optomotor reaction. *Notonecta,* for instance, orientates upstream and swims strongly as the current tends to drift it downstream so that it tends to keep its visual field constant and as a result maintains its position. In a tank without any landmarks, or if the eyes are blackened, *Notonecta* is unable to keep station and is swept downstream.

Form Perception

Behavioural responses indicate that insects are able to separate objects with an angular separation of 1 or 2 and it follows that larger objects should be clearly visible to most insects. *Locusta* responds to a pattern of black stripes on a white ground, being attracted to the edge of the stripe where white and black adjoin. Vertical strips are preferred to oblique or wavy-edged lines and taller figures are preferred to short ones. If one vertical stripes are presented the more complex figure is preferred. Such behaviour would be relevant to the locust in food finding. The ability to follow stripes is best developed in phytophagous insects. *Apis* responds to shapes and can be trained to come to any mark which contrasts with the background. Solid figures of different shapes are not differentiated from each other, nor are broken figures, but *Apis* readily differentiates between solid and broken patterns, showing a preference for broken figures which is not overcome by training.

The number of visits paid by bees to a particular pattern is proportional to the length of its contour, suggesting that the choice depends on the frequency of change of retinal stimulation as the

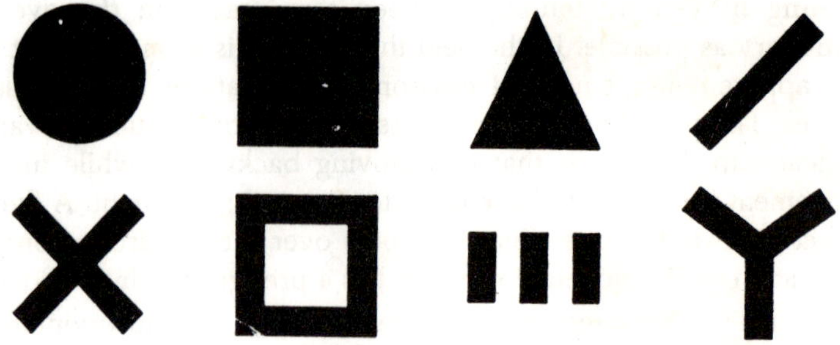

Fig. 9.20. Figures used to study form perception in honeybees.

bee moves, that is on the flicker effect which the pattern produces in the eye. related to this it is found that the setting of bees on flowers is improved if the flowers are moving slightly. At least some insects, particularly hunting insects, must have rather better vision than this suggests. For instance, the spider-hunting wasp *Scelophrom* must be able to recognise spiders from some distance, although olfaction is probably also important at close quarters. *Philanthus* (Hymenoptera) recognise landmarks, such as pine cones, in the vicinity of its nest and the removal of such landmarks makes it difficult for it to find the nest. Its vision must presumably be of high order for it to recognise such landmarks.

Movement Perception

The insect eye appeas to be better adapted for movement perception than for form perception. The system of small units, either ommatidia or rhabdomeres, which constitute the compound eye lends itself to the perception of changes in stimulation resulting from small movements of the object or the eye. Hence bees respond more readily to moving flowers than to stationary ones, dragonfly larvae respond to moving prey and most insect show a preference for more complex shapes causing more flicker. But if an object vibrates too quickly its movement may not be observed because the sensory units need time in which to recover from the previous sitmulus.

The stimulation of the eye by a succession ofstimuli is called a flicker effect and the highest number of separate stimuli which the eye can differentiate in unit time as the flicker threshold or the flicker fusion frequency. The flicker threshold varies with the type of eye. In slow eyes the threshold value lies between about 20/ second in *Tachycines* (Orthoptera) and 60/second in *Aeschna* larva, while in the fast eyes of *Apis* and *Calliphora* the threshold frequency approaches 300/second, the value varying with the light intensity. A high flicker threshold is well suited to a fast-flying insect since it facilitates the perception of features of the terrain passing rapidly beneath it.

Distance Perception

Most insects must be able to judge distance fairly accurately. This ability is obviously important in prey-catching insects and in grasshoppers jumping on to a perch, but it must also be important to most insects in making avoidance movements in the air and when they are landing. The ability to judge distances depends on binocular vision, and essentially on the simultaneous stimulation of ommatidia in the two eyes. If one eye is damaged the power is lost. Errors can arise in the estimate of distance due to the size of the ommatidial angle, since this is important in determining acuity. Fig shows the error of estimation which might arise if the ommatidial angle was 2° and bigger ommatidial angles will result in bigger errors. Possibly related to this is the fact that dragonfly larvae have smaller ommatidia on th inside of the eye.

These are the ommatidia used in judging the distance ot the prey. Error will also be larger if the distance of the insect from its prey is long relative to the distance between the eyes, and in may carnivorous insects which hunt visually, such as mantids and Zygoptera, the eyes are wide apart, possibly tending to reduce errors arising in this way. Insects, such as grasshoppers, which jump and need to judge distances accurately make peering movements while looking at their proposed perch. Peering movements are side to side sayings of the body with the feet still and the head vertical, but moving through an arc extending to or more or either side of the body axis. It is suggested that distance in this cae is estimated by the extent of movement over the retina; big movements indicate that the object is close to the insect, while small movements show that it is at a great distance.

Colour Vision

Colour is important in the lives of those insects which possess colour vision. Most flower-visiting insects, such as *Apis* and *Eristalis* (Diptera), exhibit preferences for blue or yellow and where red flowers are visited this may be because, as with the poppy, large amounts of ultraviolet light are relfected. It is also significant that the majority of flowers in the temperature zones, where the flowers are insect pollinated, are blue or yellow, with few pure reds, while in the trophics, where birds are common pollinators, red flowers are common. New Zealand with a sparse inset fanua is poor in indigenous coloured flowers. Colour is also important in the feeding of leaf-eating species, *Chrysomela* (Coleoptera) and various caterpillars being attracted to green. The reaction of an insect to colour may vary depending on its physiological state. Thus female *Pieris* at first show a preference for blue, purple and yellow, the colours of the flowers form which they feed, but when they are mature the preferred colours are green and green-blue, corresponding with the tendency to oviposit on leaves, *Macroglossa* (Lepidoptera) females show a similar change of preference. Colour vision also plays a part in the courtship behaviour of some insects and probably in the choice of background in cryptically coloured insects.

Reactions to Polarised Light

The light coming from a blue sky is polarised, and the degree of polarisation and the plane of maximum polarisation of light from different parts of the sky varies and is correlated with the position of the sun. Consequenlty it is possible to determine the position of the sun, even when it is obscured, form the composition of polarised light from a patch of blue sky. Certain insects are able to make use of this information in performing an astrotaxis. It is particularly important in the homing of social Hymenoptera, and is best known in *Apis* where the communication dances of workers may be orientated with respect to the sun even when the sun is obscured. In other insects where the ability to perceive polarised light exists it probably enables them to maintain a constant and steady orientation.

Dorsal Ocelli

Dorsal ocelli are found in adult insects and the larvae of hemimetabolous insects. Typically there are three, forming an inverted triangle antero-dorsally on the head although in Diptera and Hymenoptera they occupy a more dorsal position on the vertex. The median ocellus shows evidence of a paired origin since the root of the ocellar nerve is double and the ocellus itself is bilobed in Odonata and *Bombus* (Hymenoptera). Frequently one or all of the ocelli are lost and they are often in wingless forms. A typical ocellus has a single thickened cuticular lens, but in *Machilis* and *Periplaneta* there is no cuticular thickening, only a transparent area of cuticle.

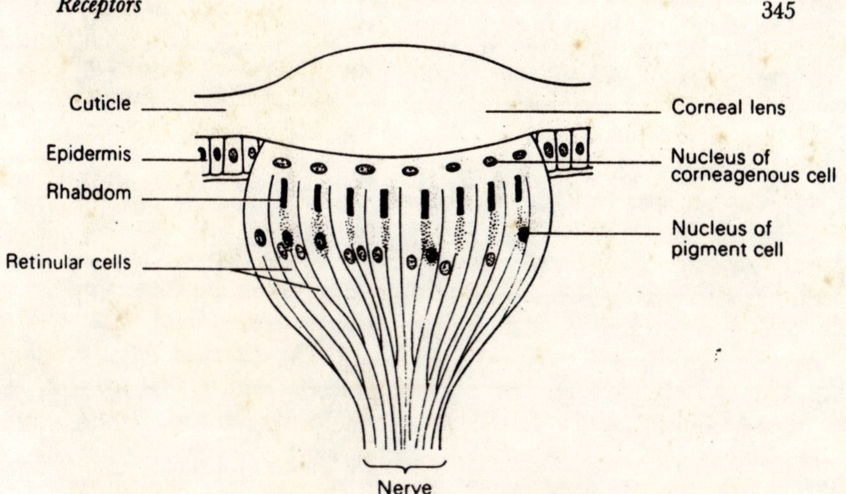

Fig. 9.21. Section of a dorsal ocellus (diagrammatic).

In Ephemeroptera the lens is formed from a numer of transparent cells, not from the cuticle. The epidermis beneath the lens is transparent and colourless, while beneath it are large numbers of neve cells arranged in groups of two or more. Distally, within each group, the cells form rhabdomeres similar to those of the compound eyes. Pigment may occur between the groups of sense cells or round the outside of the eye, but in some insects, as in the cockroach, pigment is lacking. Instead, the cockroach ocellus is backed by a reflecting tapetum which is probably formed of urate crystals. The fibres of the sense cells pass out at the back of the ocellus and extended at least halfway down the ocellar nerve, making repeated synaptic contacts with the axons of second order cells lying in the pars intercerebralis of the brain.